Communications in Computer and Information Science 2795

Rationale

The CCIS series is devoted to the publication of proceedings of computer science conferences. Its aim is to efficiently disseminate original research results in informatics in printed and electronic form. While the focus is on publication of peer-reviewed full papers presenting mature work, inclusion of reviewed short papers reporting on work in progress is welcome, too. Besides globally relevant meetings with internationally representative program committees guaranteeing a strict peer-reviewing and paper selection process, conferences run by societies or of high regional or national relevance are also considered for publication.

Topics

The topical scope of CCIS spans the entire spectrum of informatics ranging from foundational topics in the theory of computing to information and communications science and technology and a broad variety of interdisciplinary application fields.

Information for Volume Editors and Authors

Publication in CCIS is free of charge. No royalties are paid, however, we offer registered conference participants temporary free access to the online version of the conference proceedings on SpringerLink (http://link.springer.com) by means of an http referrer from the conference website and/or a number of complimentary printed copies, as specified in the official acceptance email of the event.

CCIS proceedings can be published in time for distribution at conferences or as post-proceedings, and delivered in the form of printed books and/or electronically as USBs and/or e-content licenses for accessing proceedings at SpringerLink. Furthermore, CCIS proceedings are included in the CCIS electronic book series hosted in the SpringerLink digital library at http://link.springer.com/bookseries/7899. Conferences publishing in CCIS are allowed to use our online conference service (Meteor) for managing the whole proceedings lifecycle (from submission and reviewing to preparing for publication) free of charge.

Publication process

The language of publication is exclusively English. Authors publishing in CCIS have to sign the Springer CCIS copyright transfer form, however, they are free to use their material published in CCIS for substantially changed, more elaborate subsequent publications elsewhere. For the preparation of the camera-ready papers/files, authors have to strictly adhere to the Springer CCIS Authors' Instructions and are strongly encouraged to use the CCIS LaTeX style files or templates.

Abstracting/Indexing

CCIS is abstracted/indexed in DBLP, Google Scholar, EI-Compendex, Mathematical Reviews, SCImago, Scopus. CCIS volumes are also submitted for the inclusion in ISI Proceedings.

How to start

To start the evaluation of your proposal for inclusion in the CCIS series, please send an e-mail to ccis@springer.com

Hanumant Singh Shekhawat ·
Mukesh Kumar Saini · Neeraj Goel ·
Anwar Hossain · Dhananjay Singh
Editors

Agricultural-Centric Computation

Third International Conference, ICA 2025
Guwahati, India, May 13–16, 2025
Revised Selected Papers

Editors
Hanumant Singh Shekhawat
Indian Institute of Technology Guwahati
Guwahati, Assam, India

Mukesh Kumar Saini
Indian Institute of Technology Ropar
Rupnagar, Punjab, India

Neeraj Goel
Indian Institute of Technology Ropar
Rupnagar, Punjab, India

Anwar Hossain
Queen's University
Ontario, ON, Canada

Dhananjay Singh
Penn State University
University Park, PA, USA

ISSN 1865-0929 ISSN 1865-0937 (electronic)
Communications in Computer and Information Science
ISBN 978-3-032-17082-8 ISBN 978-3-032-17083-5 (eBook)
https://doi.org/10.1007/978-3-032-17083-5

This Springer imprint is published by the registered company Springer Nature Switzerland AG
The registered company address is: Gewerbestrasse 11, 6330 Cham, Switzerland

Preface

The third edition of the International Conference on Agriculture-Centric Computation (ICA 2025) continued to promote research at the intersection of agriculture and computation, building on the success of ICA 2023 and ICA 2024. Emerging subfields of computing, such as IoT and Artificial Intelligence, have made a significant impact on agricultural practices and enabled sustainable & data-driven innovations in farming. This year, we received a record 115 submissions, which underwent a rigorous double-blind review process in which submissions received on average three reviews each by domain experts. After detailed evaluation, 25 long papers and 3 short papers were shortlisted for presentation at the conference. The accepted papers were organized under the following thematic areas: AI & Machine Learning in Agriculture, IoT and Sensor Networks for Smart Farming, Robotics and Automation, Climate-Resilient and Sustainable Technologies, and Precision Agriculture and Data Analytics.

The conference had parallel demo and poster sessions to promote interactions among researchers. The technical paper sessions were complemented by two keynote addresses, two tutorials, and one panel discussion. We had three excellent keynote speakers; the first keynote was delivered by Tomás Norton (KU Leuven, Belgium) on "Next-Generation Digital Farming Systems," providing deep insights into how computational models are revolutionizing precision agriculture. The second keynote was delivered by Danilo Demarchi (Politecnico di Torino, Italy) on "IoT and Microelectronics in Precision Agriculture," emphasizing the fusion of electronics, AI, and IoT for scalable agricultural solutions. The third keynote talk was delivered by Dhananjay Singh from Penn State University, USA, on *"GreenTwin: A Digital Twin Solution for Emission Sustainability in Agriculture."* The talk highlighted how GreenTwin leverages the power of digital twin technology to create a virtual replica of real-world agricultural systems, with a core focus on reducing greenhouse gas emissions from farming activities. In addition, the conference featured engaging tutorials and workshops led by distinguished speakers, including Kavya Dashora (IIT Delhi) and Dipankar Mandal (IIT Guwahati). The technical sessions were complemented by a panel discussion, which explored the topic of "*From Lab to Land: Translating Agri-Tech Innovations into Scalable Solutions.*" The tutorials enlightened participants on emerging AI tools and drone-based agricultural monitoring systems.

A total of 150 students from India and abroad attended the conference. The attendees included authors of the papers, research students, and industry professionals.We received generous support from Anusandhan National Research Foundation (ANRF), Govt. of India, IIT Ropar, Emeritus Executive Education, Physics Wallah Limited, and NIT Arunachal. In addition, the organizers appreciate the support of ANNAM.AI, a Center of Excellence in AI for Agriculture established by the Ministry of Education,

Government of India, and AWaDH, a Technology Innovation Hub at IIT Ropar, supported by the Department of Science and Technology, Government of India, which are spearheading technological innovations in the field of AgriTech.

Hanumant Singh Shekhawat
Mukesh Kumar Saini
Neeraj Goel
Anwar Hossain
Dhananjay Singh

Organization

Patrons

Rajeev Ahuja	Indian Institute of Technology Ropar, India
Devendra Jalihal	Indian Institute of Technology Guwahati, India

General Chairs

Athula Ginige	Western Sydney University, Australia
Sanginario Alessandro	Politecnico di Torino, Italy
Gaurav Trivedi	IIT Guwahati, India

Keynote Speakers

Danilo Demarchi	Politecnico di Torino, Italy
Tomás Norton	KU Leuven, Belgium
Dhananjay Singh	Pennsylvania State University, USA

Technical Program Chairs

Hanumant Singh Shekhawat	IIT Guwahati, India
Mukesh Kumar Saini	IIT Ropar, India
Neeraj Goel	IIT Ropar, India
Anwar Hossain	Queen's University, Canada
Dhananjay Singh	Pennsylvania State University, USA

Advisory Board

David Choi	Soongsil University, South Korea
Abdulmotaleb El Saddik	University of Ottawa, Canada
Bakul Rao	IIT Bombay, India
Mohammad Eid	New York University Abu Dhabi, UAE
Ekta Kapoor	DST, Government of India, India
Pao Ann Hsiunh	National Chung Cheng University, Taiwan

Pinakeswar Mahanta	NIT Meghalaya, IIT Guwahati, India
Sudip Mitra	IIT Guwahati, India
Jan Pidanic	University of Pardubice, Czech Republic
Debi Prasad Mishra	IIT Kanpur, India
Avik Bhattacharya	IIT Bombay, India

Tutorial Chairs

Shashank Tamaskar	Plaksha University, India
Kavya Dasora	IIT Delhi, India
Budhaditya Hazra	IIT Guwahati, India

Demo Chairs

Puneet Goyal	IIT Ropar, India
Indu Lathwal	National Dairy Research Institute, India
Tao Zhuo	Northwest A&F University, India
Chayan Bhawal	IIT Guwahati, India

Sponsorship & Industry Engagement Chairs

Jai Narayan Tripathi	IIT Jodhpur, India
Suman Kumar	IIT Ropar, India
Ram Prakash Sharma	NIT Arunachal, India
Ijjada Sreenivasa Rao	GITAM School of Technology, India

Workshop Chairs

Ajay Dashora	IIT Guwahati, India
Anamika Yadav	IIT Guwahati, India
Aryabratta Sahu	IIT Guwahati, India
Dhritiman Saha	ICAR-CIPHET, India

Innovation Chairs

Radhika Trikha	IIT Ropar, India
Mukesh Kestwal	IIT Ropar, India

Agriculture/Farmer Connect

Rakesh Sharda	Punjab Agricultural University, India
Rohitash	SKUAST, India
B.R. Phukan	Central Agricultural University, India
Sukhen Chandra Das	College of Agriculture, Tripura, India

Contents

Assessing Multi-mode Temporal PolSAR Data for Winter Wheat and Barley Discrimination Using Convolutional Neural Networks

Debanjan Chowdhury[1], Sagar Kumar[2], Swarnendu Sekhar Ghosh[3], and Dipankar Mandal[2](✉)

[1] Amity University Haryana, Gurugram, India
debanjan.chowdhury@s.amity.edu
[2] Agro-geoinformatics Lab, School of Agro and Rural Technology, Indian Institute of Technology Guwahati, Guwahati, Assam, India
{sagark4008,dmandal}@iitg.ac.in
[3] Microwave Remote Sensing Lab, Centre of Studies in Resources Engineering, Indian Institute of Technology Bombay, Mumbai, India
194310003@iitb.ac.in

Abstract. Accurate crop classification with synthetic aperture radar (SAR) data is a significant area of research and translating into practice from local to regional scale crop inventory mapping. With the growing accessibility to abundant data sources from both current and upcoming dual-polarimetric SAR missions, the capability to generate precise crop maps is set to enhance substantially. The geometric and dielectric properties of targets highly influence radar backscatter. Especially for agricultural crops, which exhibit dynamic changes in target properties and physiological structure throughout their phenology, discriminating between crops using SAR data remains a significant challenge. This study utilizes a Convolutional Neural Networks (CNN) classifier (with two variants to intrinsically capture temporal and spatial information) for discriminating winter wheat and barley crops. Dual-polarimetric SAR data acquired over the AgriSAR2006 site in Germany was used for the study. The performance evaluation of the 3D-CNN model revealed robust accuracy across different polarimetric modes.

Keywords: crop classification · temporal data · Synthetic aperture radar

1 Introduction

A variety of agricultural applications, such as crop monitoring, risk assessment, inventory mapping depend on crop classification products. Along with optical remote sensing, Synthetic Aperture Radar (SAR) data gained major interest in community for crop classification [11,14,20]. The introduction of spaceborne

H. S. Shekhawat et al. (Eds.): ICA 2025, CCIS 2795, pp. 1–12, 2026.
https://doi.org/10.1007/978-3-032-17083-5_1

SAR systems during the last three decades has highlighted the importance of C- and L-band SAR data for crop mapping, prompting substantial study. Several reviews are available for a more extensive exploration [2,14,17,18]. Previous studies have highlighted the significance of multi-configuration SAR datasets, including temporal observations [7,15,22] and multi-polarization modes [16,19], for improving crop classification in dynamic agricultural systems.

Unlike other land cover types, crop structures undergo rapid changes over short periods as they progress through different phenological stages. SAR backscatter intensities varies throughout crop development stages due to factors such as plant canopy structure, soil properties (moisture content, surface roughness), plant density, and row orientation. Winter wheat and barley exhibit similar phenological patterns until the heading stage, where structural differences in the fruiting head emerge. Multi-date SAR data is crucial for crop inventory mapping, yet evaluating models on such data remains essential for capturing spatiotemporal trends. Deep learning, particularly convolutional neural networks (CNNs), offers a promising approach to extracting distinctive crop-specific SAR features for improved classification [3,8,21].

For large-scale crop classification, dual-pol SAR data have gained traction over full-pol due to their wider coverage and reduced data volume, despite lower polarimetric information [5,12]. While CNN models have been applied to full- and dual-pol SAR data for crop classification [1,10], conventional 2D-CNNs often fail to capture spectral and temporal dependencies, as they average features across dimensions. To address this limitation, Kussul et al. [9] proposed a dual-CNN approach for crop classification, utilizing a 2D-CNN for spatial feature extraction and a 1D-CNN for spectral feature learning. In a separate study, Ji et al. [6] demonstrated a 3D-CNN architecture for crop classification using multi-spectral, multi-temporal remote sensing images, showing potentially superior performance over 2D-CNNs. However, limited studies have explored SAR-based spatiotemporal crop classification. In theory, 3D convolutions can simultaneously capture spatial and temporal features more effectively than direct image concatenation. This study investigates 2D- and 3D-CNN models for distinguishing wheat and barley using dual-pol L-band SAR data and evaluates their performance across full-pol, HH-HV, and VV-VH modes.

2 Study Area and Dataset

The current study was performed utilising data acquired over the DEMMIN test site in North-eastern part of Germany during the AGRISAR2006 campaign (Fig. 1). The campaign was carried over the site from mid week of April to first week of August, 2006. The test site has large land parcels (on average 80 ha) with major cultivated crops including winter wheat, winter barley, winter rape, maize, and sugar beet. Winter crops such as wheat were sown in September (which was the previous crop season in 2005) and harvested by the end of August, followed by barley, which was sown in mid-September and harvested by the end of July. Phenological growth stages of winter wheat and barley is also presented in Fig. 1.

Fig. 1. Study area across AgriSAR 2006 campaign. (a) L-band data is presented as a Pauli-RGB image acquired on June 07, emphasized with winter wheat and barley field boundaries. Different phenological stages winter wheat-(b) Inflorescence emergence (07 June); (c) Flowering (13 June); (d) Fruit development (21 June), and barley- (e) Flowering/anthesis (07 June); (f) Fruit development-early milk (13 June); (g) Late milk (21 June) are highlighted.

Weekly campaigns were coordinated with gathering airborne ESAR system radar data, as shown in Table 1. Out of the two flight tracks in which landscape features were acquired during ESAR data acquisition, the E-W track (10 km×3 km footprint) was considered for the present study as shown in Fig. 1. In this research, we utilised three acquisitions at L-band as presented in Table 1.

Table 1. Attributes of E-SAR Single-Look Complex data acquisitions along the East-West track used for the present investigation.

Acquisition Date of E-SAR	Product ID L-band	Incidence Angle (deg)
07/6/2006	0912	24.45–58.97
13/6/2006	1010	24.45–58.97
21/6/2006	1109	24.45–58.97

2.1 PolSAR Data Processing

For full-pol data, a 3×3 covariance matrix $\langle[\mathbf{C}]\rangle$ was generated from the multi-looked dataset and performed de-speckling using refined Lee filter with a window size of 5×5 (PolSAR image processing chain adopted from [4] and [13]). Consequently, the geocode and coregister step were performed with these temporal datasets. The backscatter intensities were further extracted from elements of $\langle[\mathbf{C}]\rangle$ matrix as:

$$\begin{aligned} \sigma^0_{HH} &= \langle |S_{HH}|^2 \rangle = C_{11} \\ \sigma^0_{HV} &= \langle |S_{HV}|^2 \rangle = C_{22}/2 \\ \sigma^0_{VV} &= \langle |S_{VV}|^2 \rangle = C_{33} \end{aligned} \tag{1}$$

Using QGIS and campaign field boundaries, two fields for both winter wheat (Field ID 230 and 250), and barley (Field ID 440 and 450) were clipped from backscatter intensity images. Henceforward, these images were renamed as C11, C22, and C33 and saved as TIFF format for three acquisition dates. For sake of comparison, three different datasets were constructed having different channels i.e., (C11, C22, C33), (C11, C22), and (C22, C33).

3 Methodology

3.1 Data Preprocessing for CNN

The TIFF files were processed by first converting them into NPZ format, adjusting the shape from (channels, row, column) to (row, column, channels). Different band combinations were then created for both wheat and barley, date, and field. A schematic of data preprocessing is given in Fig. 2.

Each band-combined file was segmented into 8×8 sections based on crop type, field, and date. For instance, the winter barley data from June 07 at Field 440 containing all three bands, was segmented into 704 smaller sections. A similar segmentation was performed for the two-band combinations (C11 & C22, C22 & C33), resulting in the same number of segments per patch. The segmentation process was repeated for the other dates. Subsequently, in a filtering process segments where all values in the matrix channels were zero were removed, retaining only those with meaningful information. The remaining segments were then merged across different days and field for each crop type and band combination to increase the sample size for CNN modeling. A total of 2913 and 7740 segments were generated for winter barley and wheat, respectively.

A Z-score normalization procedure was used to standardize the data, assuring that all features had a consistent scale, which improved model performance and stability. Further, the dataset (features and corresponding crop class) was divided into training, validation, and test sets, comprising 7669, 853, and 2132 samples, respectively. An imbalance was observed in the training set, with Winter Barley having only 2097 samples compared to Winter Wheat with 5572 samples. To address this, data augmentation techniques were applied. Additional winter Barley samples were generated by randomly rotating images between $\pm 25°$, along with horizontal and vertical shifts within $\pm 10\%$ of the column and row. For the 2D-CNN case, this resulted in 3475 new Winter Barley samples, bringing the total training dataset size to 11144, ensuring a balanced dataset for effective CNN modeling.

In the case of 3D-CNN, the dataset was prepared in a day-wise format for each crop type. For example, data from field 440 and 450 across three different days were loaded separately and later combined. This structured the dataset in a 3D format, resulting in 971 samples for winter barley and 2580 samples for winter wheat, with each sample containing three days of data, an 8×8 spatial resolution, and three channels per day. The dataset was then separated into training, validation, and test batches, with 2555, 285, and 711 numbers of samples. With previously mentioned augmentation and combination process, sample imbalance were reduced by adding 1159 new Winter Barley samples, increasing the total training dataset size to 3714.

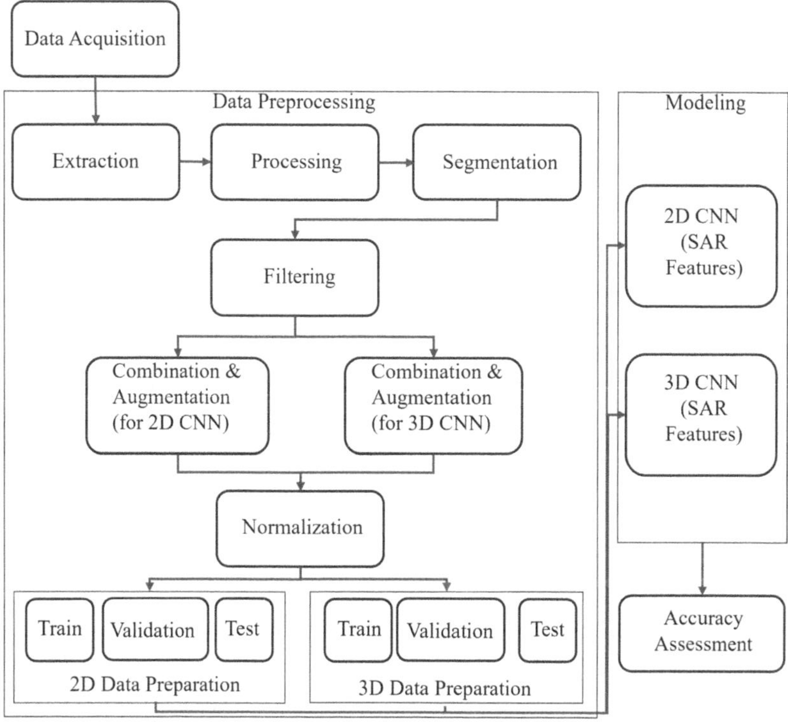

Fig. 2. Schematic workflow for data preprocessing and training, and assessment for both 2D & 3D-CNN models.

3.2 CNN Modeling

2D-CNN Architecture and Modeling. For sake of comparison of full and dual-polarimetric modes of SAR data, separate 2D-CNN models were built for datasets containing all three bands (C11, C22, C33) and for two-bands combinations (C11, C22; and C22, C33). The 2D-CNN architecture is presented in Fig. 3.

To account the three band (full-pol) data, the 2D-CNN model was designed with two 2D convolution (Conv2D) layers, which had 16 and 32 channels, and 3×3 filter. The ReLU activation function (Eq. (2)) was applied in both layers, and the same padding was selected to support spatial dimensions. During training, input shape for the first layer was set to (8,8,3) in full-pol case, and (8,8,2) for dual-pol dataset. Batch Normalization was applied after each Conv2D layer and Dense layer, except for the output layer.

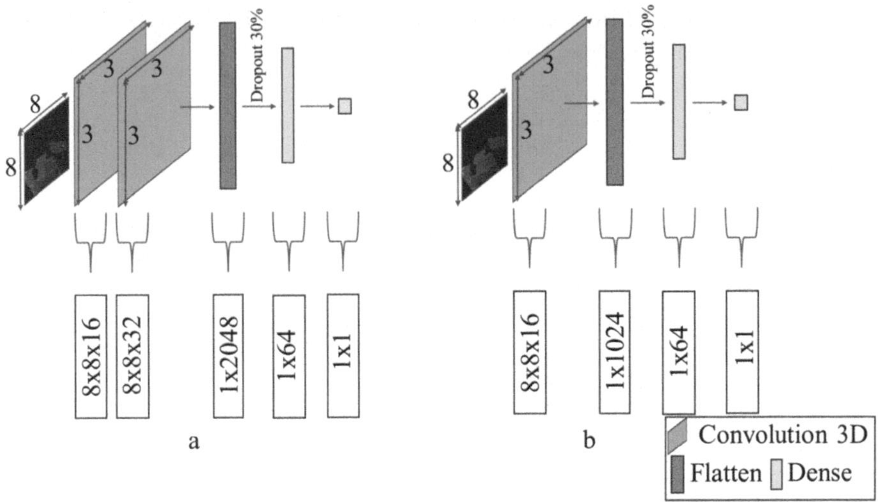

Fig. 3. The network structure of 2D-CNN for temporal crop classification (a) Combining all three (C11, C22 and C33); (C11 and C22) bands and (b) Combining C22 and C33 bands.

$$F_c^{(1)}(i,j) = \text{ReLU}\left(\sum_{m=0}^{2}\sum_{n=0}^{2}\sum_{c'=0}^{q} F_{c'}^{(0)}(i-m, j-n)\cdot K_{c',c}^{(1)}(m,n) + b_c^{(1)}\right) \tag{2}$$

where: q is (3 or 2) depending on all three bands or C11 and C22 bands, $F^{(1)} \in \mathbb{R}^{8\times 8\times a}$ (same size due to "same" padding; a is (16 or 32)), convolution kernel is presented as $K^{(1)}$, $b_c^{(1)}$ is bias, and $\text{ReLU}(x) = \max(0, x)$.

Following convolutional layers, a Flatten layer was included, with a dense layer (64 number of neurons) using the ReLU. To help prevent overfitting, a Dropout layer with a 0.3 dropout rate was added. The model terminated with an output-dense layer consisting of a single neuron, which utilized the sigmoid activation function for binary classification as Eq. (3).

$$G = \sigma\left(\sum_{k=0}^{63} W_k^{(\alpha)} F_k^{(\alpha-1)} + b^{(\alpha)}\right) \tag{3}$$

where: $W^{(\alpha)} \in \mathbb{R}^{1\times 64}$, $F_k^{(\alpha-1)}$ is the Feature space generated after the Dense layer (Fig. 3), $b^{(\alpha)}$ is the bias, $\sigma(x) = \frac{1}{1+e^{-x}}$ is the sigmoid activation function.

For optimization, the Adam optimizer was chosen, starting with an initial learning rate of 0.01. The binary cross entropy loss function was used as Eq. (4). Additionally, ReduceLROnPlateau was implemented to dynamically adjust the learning rate based on validation loss. If the validation loss does not enhance for 5 successive epochs, the learning rate will decline by a factor of 0.1, with a minimum limit of 0.001. The model was trained for 500 epochs, and the results were recorded for further evaluation.

$$L = -\frac{1}{N}\sum_{i=1}^{N}\left[y_i \log G_i + (1 - y_i)\log(1 - G_i)\right] \tag{4}$$

where L is the Loss, N represents batch size, y_i is the true label and G_i represents predicted probability.

3D-CNN Architecture and Modeling. Separate 3D CNN models were developed for datasets containing all three bands and for two-band combinations (C11 & C22, C22 & C33). The architecture consisted of four Conv3D layers with 16, 32, 64, and 128 filters, each using a $3 \times 3 \times 3$ filter size as presented in Fig. 4. ReLU activation and same padding were applied Eq. (5). The input shape for the first layer was (3,8,8,3) for the three-band dataset and (3,8,8,2) for two-band datasets.

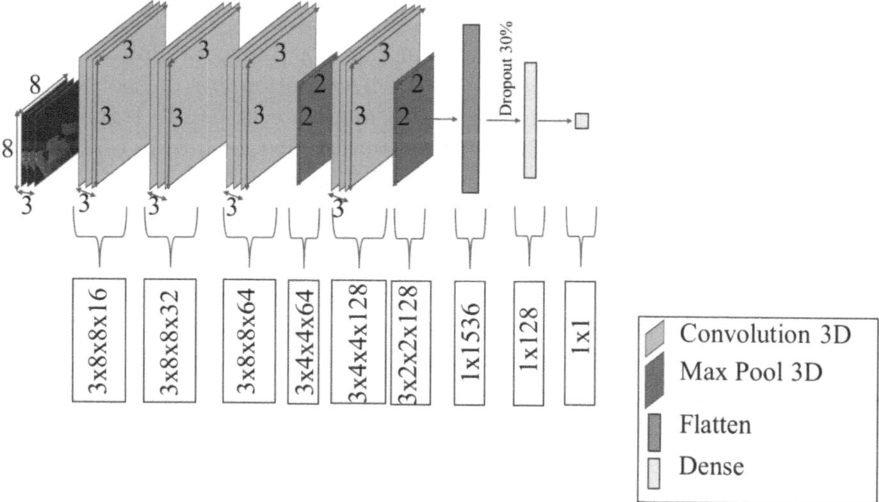

Fig. 4. The 3D-CNN network architecture for crop classification utilizing temporal SAR data.

$$F_c^{(1)}(d,i,j) = \text{ReLU}\left(\sum_{m=0}^{2}\sum_{n=0}^{2}\sum_{p=0}^{2}\sum_{c'=0}^{2} F_{c'}^{(0)}(d-m, i-n, j-p)\cdot K_{c',c}^{(1)}(m,n,p) + b_c^{(1)}\right) \tag{5}$$

where $F^{(1)} \in \mathbb{R}^{3\times8\times8\times16}$ (same size due to "same" padding), $K^{(1)}$ is a 3D convolutional kernel of size (3,3,3), $b_c^{(1)}$ is the bias, and $\text{ReLU}(x) = \max(0, x)$.

Two Max Pooling 3D layers with a pool size of (1,2,2) were incorporated after the third and fourth Conv3D layers to reduce spatial dimensions while retaining critical features Eq. (6). A Flatten layer was followed by a Dense layer with 128 neurons and ReLU activation. A Dropout layer with a 0.3 dropout rate was added to mitigate overfitting.

$$F_c^{(\beta)}(d,i,j) = \max_{m=0}^{1} \max_{n=0}^{1} \mathrm{BN}(F_c^{(\beta-1)}(d, 2i+m, 2j+n)) \tag{6}$$

where $F_c^{(\beta)}$ is the feature space after the Max Pooling layer at $(\beta) = 4or6$ depending on the layer number and BN refers to Batch Normalization function.

The final Dense layer contained a single neuron with a sigmoid activation function for binary classification Eq. (7).

$$G = \sigma \left(\sum_{k=0}^{127} W_k^{(\alpha)} F_k^{(\alpha-1)} + b^{(\alpha)} \right) \tag{7}$$

where: $W^{(\alpha)} \in \mathbb{R}^{1\times128}$, $F_k^{(\alpha-1)}$ is the Feature space generated after the Dense layer (Fig. 4), $b^{(\alpha)}$ is the bias, $\sigma(x) = \frac{1}{1+e^{-x}}$ is the sigmoid activation function.

The Adam optimizer was used with an initial learning rate of 0.01, and binary cross-entropy was used as the loss function Eq. (4). ReduceLROnPlateau was enforced, decreasing the learning rate by 0.1 if validation loss did not improve for 5 successive epochs, with the lowest learning rate of 0.001. Training was conducted for 500 epochs, and the results were recorded for further evaluation.

4 Results and Discussion

Crop classification results with temporal data were analyzed for both 2D-CNN and 3D-CNN models across three polarization band combinations i.e., full (C11, C22, C33), and dual-pol (C11, C22; and C22, C33). Table 2 provides a comparative analysis of classification results for winter barley and winter wheat using different 2D-CNN models. For full-pol data, superior classification performance is observed with an overall accuracy of 0.893. Winter barley is identified with a precision of 0.880, recall of 0.879, and an F1-score of 0.880. Similarly, winter wheat is classified with high accuracy, achieving a precision of 0.904, recall of 0.905, and an F1-score of 0.904. However, some mis-classifications occur, as presented in the confusion matrix in Fig. 5. Additionally, the AUC score for this model is found to be 0.9563 (Fig. 6). In case of dual-pol data, decline in overall performance is evident. However, a mixed results is observed for C11-C22 and C22-C33 case. Winter barley is identified with an F1-score of 0.798 and 0.619 in two cases. Misclassification between wheat and barley is much prominent in dual-pol cases. This potentially attributed due to incapability of 2D-CNN model capturing temporal pattern in dual-pol cases.

On the contrary, the 3D-CNN model for dual-pol data presented better overall accuracy and individual crop precision-recall scores (Table 3 and Fig. 5). The dual-pol C11-C22 mode performed best in case of 3D-CNN classification with an

Table 2. Performances of 2D-CNN models on Winter Barley and Winter Wheat classification.

Model	Winter Barley			Winter Wheat			Overall
	Precision	Recall	F1-Score	Precision	Recall	F1-Score	Accuracy
2D CNN (C11, C22 & C33)	**0.880**	**0.879**	**0.880**	**0.904**	**0.905**	**0.904**	**0.893**
2D CNN (C11 & C22)	**0.702**	**0.925**	**0.798**	**0.920**	**0.688**	**0.787**	**0.793**
2D CNN (C22 & C33)	**0.925**	**0.879**	**0.619**	**0.727**	**0.754**	**0.932**	**0.793**

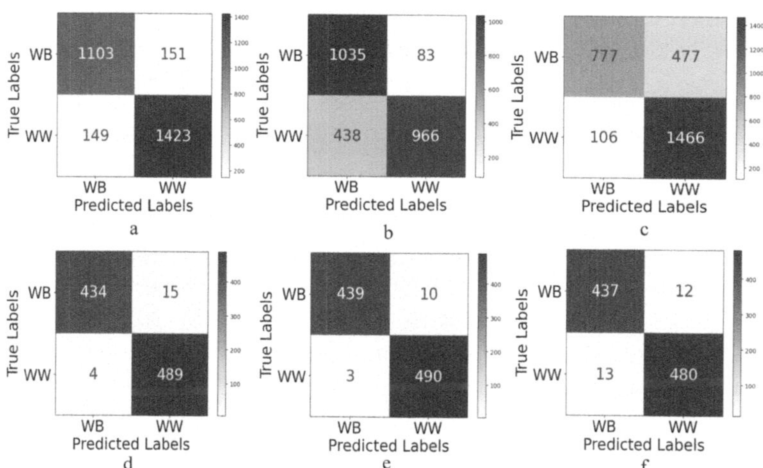

Fig. 5. Confusion matrix for winter barley and winter wheat classification using (a-c) 2D-CNN using C11, C22, C33 bands; C11 and C22 bands; C22 and C33 bands, respectively; (d-f) 3D-CNN using C11, C22, C33 bands; C11, C22 bands; and C22, C33 bands, respectively.

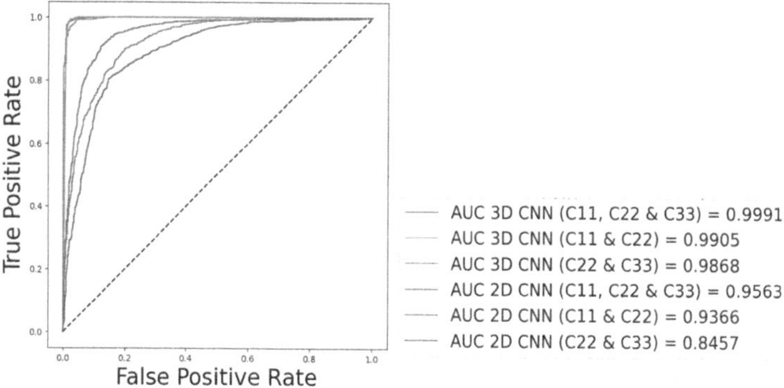

Fig. 6. ROC Curve showing AUC for all the 3D & 2D-CNN models for full and dual-pol datasets.

overall accuracy of 0.986. Winter barley is identified with a precision of 0.993, recall of 0.977, and an F1-score of 0.985. Similarly, winter wheat is classified with high accuracy, achieving a precision of 0.980, recall of 0.993, and an F1-score of 0.986. The AUC score for this model is measured at 0.9366.

Table 3. Performances of 3D-CNN models on Winter Barley and Winter Wheat classification.

Model	Winter Barley			Winter Wheat			Overall
	Precision	Recall	F1-Score	Precision	Recall	F1-Score	Accuracy
3D CNN (C11, C22 & C33)	**0.990**	**0.966**	**0.978**	**0.970**	**0.991**	**0.980**	**0.979**
3D CNN (C11 & C22)	**0.993**	**0.977**	**0.985**	**0.980**	**0.993**	**0.986**	**0.986**
3D CNN (C22 & C33)	**0.971**	**0.973**	**0.972**	**0.975**	**0.973**	**0.974**	**0.973**

The results indicate that overall, models trained using 3D-CNN perform significantly better than those using 2D-CNN. The 3D-CNN model utilizing the C11, C22 C33 bands achieves an overall accuracy that is 8% higher than its 2D-CNN counterpart. Similarly, the 3D-CNN model trained on C11 and C22 outperforms the 2D-CNN model with the same bands by 20%, while the 3D-CNN model using C22 and C33 shows an improvement of 18% over its 2D CNN equivalent. Incorporating a time-series information in 3D-CNN framework able to discriminate better way. Ji et al. [6] emphasized that representing temporal elements solely through a concatenation procedure is oversimplified and does not match the effectiveness of spatio-temporal representations obtained using 3D-CNN.

5 Conclusion

In this work, a 3D-CNN architecture was developed to emanate information from time-series SAR data and compare its performance to a 2D-CNN for discriminating between winter wheat and barley crops. The 3D-CNN model for dual-pol mode with a combination of C11 and C22 demonstrates superior crop classification performance (overall acuuracy of 0.986). The results show that a 3D-CNN outperforms 2D-CNN. The 3D-CNN architecture is more effective than 2D-CNN as a feature extractor for spatio-temporal SAR data since 2D convolutions inherently lose temporal information due to mathematical limitations. Given the minimal decrease in accuracy, the dual-band model presents a viable alternative in applications where this trade-off is acceptable. The proposed approach is expected to be valuable for crop inventory mapping across multiple agro-ecological zones using upcoming dual-pol datasets from the NISAR, ROSE-L, EOS series mission and the Sentinel SAR series constellation.

Acknowledgments. The authors would like to thank the European Space Agency and the German Aerospace Center for providing the AgriSAR2006 campaign data.

The research work is also partially supported by the Start-Up Grant, Indian Institute of Technology Guwahati, grant no. 2024090116000101.

Data Availability Statement. Python scripts for the model architechture and data preprocessing can be accessed through Github repository at: https://github.com/AgrogeoinformaticsLabIITG/ICA2025_CropClassification.

Disclosure of Interests. The authors have no competing interests to declare that are relevant to the content of this article.

References

1. Chang, Y.L., et al.: Spatial-temporal neural network for rice field classification from sar images. Remote Sens. **14**(8), 1929 (2022)
2. Dingle Robertson, L., et al.: C-band synthetic aperture radar (sar) imagery for the classification of diverse cropping systems. Int. J. Remote Sens. **41**(24), 9628–9649 (2020)
3. Fontanelli, G., et al.: Early-season crop mapping on an agricultural area in italy using x-band dual-polarization sar satellite data and convolutional neural networks. IEEE J. Sel. Top. Appl. Earth Observ. Remote Sens. **15**, 6789–6803 (2022)
4. Hariharan, S., Mandal, D., Tirodkar, S., Kumar, V., Bhattacharya, A., Lopez-Sanchez, J.M.: A novel phenology based feature subset selection technique using random forest for multitemporal polsar crop classification. IEEE J. Sel. Top. Appl. Earth Observ. Remote Sens. **11**(11), 4244–4258 (2018)
5. Huang, X., et al.: Cropland mapping with l-band uavsar and development of nisar products. Remote Sens. Environ. **253**, 112180 (2021)
6. Ji, S., Zhang, C., Xu, A., Shi, Y., Duan, Y.: 3d convolutional neural networks for crop classification with multi-temporal remote sensing images. Remote Sens. **10**(1), 75 (2018)
7. Kordi, F., Yousefi, H.: Crop classification based on phenology information by using time series of optical and synthetic-aperture radar images. Remote Sens. Appl. Soc. Environ. **27**, 100812 (2022)
8. Kukunuri, A.N., Phartiyal, G.S., Singh, D.: Land cover mapping of mixed classes using 2d cnn with multi-frequency sar data. Adv. Space Res. **74**(1), 163–181 (2024)
9. Kussul, N., Lavreniuk, M., Skakun, S., Shelestov, A.: Deep learning classification of land cover and crop types using remote sensing data. IEEE Geosci. Remote Sens. Lett. **14**(5), 778–782 (2017)
10. Li, H., Lu, J., Tian, G., Yang, H., Zhao, J., Li, N.: Crop classification based on gdssm-cnn using multi-temporal radarsat-2 sar with limited labeled data. Remote Sens. **14**(16), 3889 (2022)
11. Mandal, D., Bhattacharya, A., Rao, Y.S.: Radar remote sensing for crop biophysical parameter estimation. Springer (2021)
12. Mandal, D., Kumar, V., Bhattacharya, A., Rao, Y.S., Siqueira, P., Bera, S.: Sen4rice: a processing chain for differentiating early and late transplanted rice using time-series sentinel-1 sar data with google earth engine. IEEE Geosci. Remote Sens. Lett. **15**(12), 1947–1951 (2018)
13. Mandal, D., Kumar, V., Bhattacharya, A., Rao, Y.: Crop crops lai leaf area index (lai) and biomass biomass estimation estimations from different polarization polarizations modes of simulated nisar nasa-isro sar (nisar) crops leaf area index (lai)

biomass estimations polarizations data. In: Remote Sensing of Agriculture and Land Cover/Land Use Changes in South and Southeast Asian Countries, pp. 235–249. Springer (2022)

14. McNairn, H., Shang, J.: A review of multitemporal synthetic aperture radar (sar) for crop monitoring. Multitemporal Remote Sens. Methods Appl. 317–340 (2016)
15. Pott, L.P., Amado, T.J.C., Schwalbert, R.A., Corassa, G.M., Ciampitti, I.A.: Satellite-based data fusion crop type classification and mapping in rio grande do sul, brazil. ISPRS J. Photogramm. Remote. Sens. **176**, 196–210 (2021)
16. Salehi, B., Daneshfar, B., Davidson, A.M.: Accurate crop-type classification using multi-temporal optical and multi-polarization sar data in an object-based image analysis framework. Int. J. Remote Sens. **38**(14), 4130–4155 (2017)
17. Shang, J., Liu, J., Chen, Z., McNairn, H., Davidson, A.: Recent advancement of synthetic aperture radar (sar) systems and their applications to crop growth monitoring. In: Recent Remote Sensing Sensor Applications-Satellites and Unmanned Aerial Vehicles (UAVs). IntechOpen (2022)
18. Skriver, H.: Crop classification by multitemporal c-and l-band single-and dual-polarization and fully polarimetric sar. IEEE Trans. Geosci. Remote Sens. **50**(6), 2138–2149 (2011)
19. Stankiewicz, K.A.: The efficiency of crop recognition on envisat asar images in two growing seasons. IEEE Trans. Geosci. Remote Sens. **44**(4), 806–814 (2006)
20. Steele-Dunne, S.C., McNairn, H., Monsivais-Huertero, A., Judge, J., Liu, P.W., Papathanassiou, K.: Radar remote sensing of agricultural canopies: a review. IEEE J. Sel. Top. Appl. Earth Observ. Remote Sens. **10**(5), 2249–2273 (2017)
21. Teimouri, M., Mokhtarzade, M., Baghdadi, N., Heipke, C.: Fusion of time-series optical and sar images using 3d convolutional neural networks for crop classification. Geocarto Int. **37**(27), 15143–15160 (2022)
22. Waldhoff, G., Lussem, U., Bareth, G.: Multi-data approach for remote sensing-based regional crop rotation mapping: a case study for the rur catchment, germany. Int. J. Appl. Earth Obs. Geoinf. **61**, 55–69 (2017)

Early Disease Detection in Pearl Millet Using YOLO v11 Model for Improved Agricultural Monitoring

J. Chalmers[1], S. R. Harish Kumar[1], J. Aravinth[1], T. Senthil Kumar[2(✉)], and I. Johnson[3]

[1] Department of Electronics and Communication Engineering, Amrita School of Engineering, Amrita Vishwa Vidyapeetham, Coimbatore, India
{cb.en.u4ece22014,cb.en.u4ece22021}@cb.students.amrita.edu, j_aravinth@cb.amrita.edu

[2] Department of Computer Science and Engineering, Amrita School of Computing, Amrita Vishwa Vidyapeetham, Coimbatore, India
t_senthilkumar@cb.amrita.edu

[3] Department of Plant Pathology, Tamil Nadu Agricultural University, Coimbatore, India
johnsonpath@gmail.com

Abstract. Pearl millet, a well-known dryland climate-resilient cereal predominantly cultivated nearly 30 million hectares especially in high temperature zones of Asia and Africa. Both the continents share a major contribution of global millet production. Though the crop can withstand high temperature and water scarcity, the yield potential is constrained by insufficient monitoring and changes in climatological factors ultimately leading to the development of serious diseases like downy mildew, rust, ergot, smut, and blast. Conventional disease detection methods are labour intensive and often require professional expertise, finding difficulty for the farmers and reducing the yield. Early detection is important for effective management of these diseases at an appropriate time. With intervention of computer technology, a deep learning (DL) system helps in real-time detection of leaf diseases under field condition. In this paper, the YOLO algorithms were evaluated using 2473 field images collected from different parts of the Coimbatore district representing all diseases of pearl millet including healthy leaves. The comparisons of YOLO v5, v8, and v11 with state-of-the-art models revealed that the YOLOv11 algorithm was found to be the best with an accuracy of 96.73% over other models.

Keywords: Deep Learning · YOLO · Pearl Millet · Disease detection · Performance Metrics

1 Introduction

Pearl millet, a poor man's cereal predominantly grown in hot temperature prevailing zones of Asia and Africa and the yield from both the continents that

H. S. Shekhawat et al. (Eds.): ICA 2025, CCIS 2795, pp. 13–24, 2026.
https://doi.org/10.1007/978-3-032-17083-5_2

contributes nearly half of the world production [1]. In order to reduce the yield loss due to the disease such as downy mildew, blast, rust, ergot and smut by employing management strategies at appropriate time, the diseases must be detected at an early stage [2]. For the classifica- tion and detection of plant diseases many machine learning algorithms were used; how- ever, they are slower, less adaptive in real-time disease detection. This is mainly due to the setbacks in precision, image pre-processing and feature extraction techniques. Be- sides, the traditional machine learning techniques are not appropriate for real time detection under field conditions with intricated backgrounds and uneven surface area. Deep learning has recently made significant progress in this field of computer vision, with a number of applications [3]. Convolutional Neural Network (CNN) gained pop- ularity over the DL because of its improved accuracy in object detection and automatic feature extraction, but lacks in the detection of high-resolution images in real time [4]. Since, plant leaves vary widely in size, shape, color, and growing ambiance, it is still be a challenge to diagnose and classify the illness accurately. Further, the variations in brightness while capturing the leaf image that commensurate detection techniques more difficult. To overcome the above difficulties and to enhance the accuracy, precision and detection speed, You Only Look Once (YOLO) method is employed which helps in multiple disease detection in plant leaves in a single instance. This paper presents the implementation of YOLO algorithms on field images captured across various regions of Coimbatore district with the objective of implementing YOLO v5, v8 and v11 for detection of pearl millet diseases and comparison of performance metrics to find out opt version of YOLO algorithm.

2 Related Works

Convolutional neural network (CNN) models have become increasingly popular due to their high feature extraction and accuracy in classification task. By automatically ex- tracting features from input images, CNNs reduce the need for extensive pre-processing [8]. Among the CNN, Visual Geometry Group Network (VGGNet) [5], Residual Net- work (ResNet) [6], Imagenet [7], mobilenet [8] and Dense Connected Convolutional Network (DenseNet) [9] are predominantly explored for plant disease detection. Many researchers have combined customized models with pre-trained models for disease pre- diction. A customized CNN model was utilized for analyzing plant disease severity for variety of crops using plant village dataset [10]. Tiwari et al. [11] proposed a dense convolutional neural network for 27 classes in six crops. The experiment shows an ac- curacy of 99.19% with complex background environment.

Sudhesh et al. [12] developed desnseNet121 to classify four categories of rice leaf dis- ease. Additionally Dynamic mode decomposition (DMD) pre processing is done with SVM and Inception and achieved an accuracy of 100% and 94.33%.

However DMD requires computational overhead and complexity making real time image classification will be a challenging. [13] developed ensemble model for cucumber leaf disease im- ages. The ensemble model combined ResNet50, EfficientNet-B0 and MobileNetV2 with an accuracy of 99.35% and [14] developed K-means clustering and neural network for brinjal leaf diseases. CNN cannot detect blur image and in order to increase the quality the image author [15] used Haar wavelet classifier. CNN model focuses only on the classification task and it is not used for object detec- tion and localization. Also CNN requires large training data and its processing speed is less. It is not possible to detect multiple diseased leaves in single image. Advanced CNN based technique namely YOLO overcomes the above drawbacks with balanced speed and detection capabilities making it more suitable for real time leaf disease de- tection. The authors [16] used YOLOv4 for plant village dataset and concluded that YOLO outperforms than other state of art models with an accuracy of 99.9%. [17] used.

YOLO v7 for tea leaf disease. Tao et.al [20] used CEFW-YOLO for detecting apple leaf diseases addressing issues like illumination changes and background clutter. Li et.al. [21] used YOLOv5 for pearl millet leaf disease. He et al. [22] developed, YOLOv11-RCDWD, a new version of YOLOv11 for maize disease detection emphasizing both accuracy and computational efficiency. The advancement in YOLO accurately detects multiple diseases in a single shot. This paper highlights YOLO v5, v8 and v11 for pearl millet disease detection and their performance were examined. The existing method used state of art models like AlexNet, VGG16, ResNet, CNN for pearl millet leaf disease identification. These predefined models can detect only one disease in an image, while YOLO can detect multiple disease in a single leaf.

3 Materials and Methods

This study compares the performance of YOLOv5, YOLOv8 and YOLOv11 models for classifying and validating Pearl Millet leaf diseases. The dataset, consisting of six different classes (five diseases and one healthy), was manually captured from the field. All images were resized to a uniform input size of 640 $\times$ 640 $\times$ 3 for compatibility with YOLO models. The data was split in an 80:20 ratio for training and testing. The models were trained using the training set and validated on the testing set to evaluate predictive accuracy, as shown in Fig. 1.

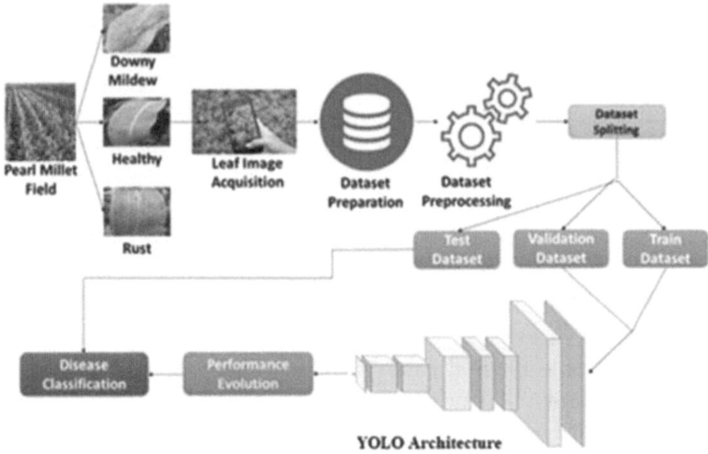

Fig. 1. Proposed methodology for Pearl Millet disease classification.

3.1 Dataset Preparation and Pre-processing

The data was collected using a smartphone (Redmi Note) by closely focusing on typical leaf disease symptoms. The real-world dataset included issues like inconsistent annotations, misclassified labels, and missing or duplicate bounding boxes. These were manually corrected during preprocessing.

The proposed methodology used a field dataset of 2473 pearl millet images collected from different fields of Coimbatore district and categorized based on visual observation. The different diseases were categorized into six classes, as shown in Fig. 2. All images were resized to 640 × 640 pixels to maintain uniformity and ensure compatibility with YOLO architecture (Table 1).

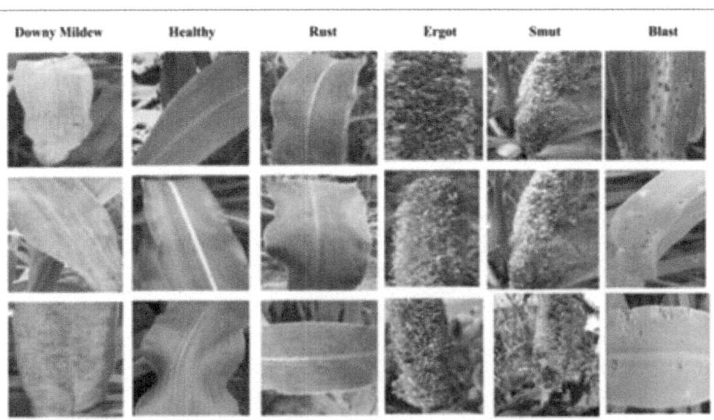

Fig. 2. Sample field images from Coimbatore district.

To enhance the ability of model, data augmentation was applied during training. Specifically, im-ages were randomly flipped both horizontally and vertically to help the model recognizes objects regardless of their orientation. In addition, random rotations within a range of ±15°C were introduced to simulate slight camera angle. These augmentations help the model to become more robust to positional and angular variations.

Table 1. Total number of images for classification

Name of the Classes	Total Images	Training Dataset	Testing Dataset
Downy Mildew	620	496	124
Rust	940	751	193
Healthy	554	441	113
Ergot	329	263	66
Smut	320	256	64
Total	**2763**	**2210**	**553**

3.2 Implementation of YOLO Algorithms

Ultralytics YOLOv5 was mainly preferred over models like RCNN (Region based CNN) due to its high speed and accurate object detection in real time as shown in Fig. 3. The architecture consists of input layer which were sent to back bone layer consisting of feature maps. Then these features were passed to neck which was feature fusion network to generate three feature maps such as P3, P4 and P5 with the dimension of 80 × 80, 40 × 40 and 20 × 20. Finally, it was passed to head for prediction of disease accurately. Subsequent paragraphs, however, are indented.

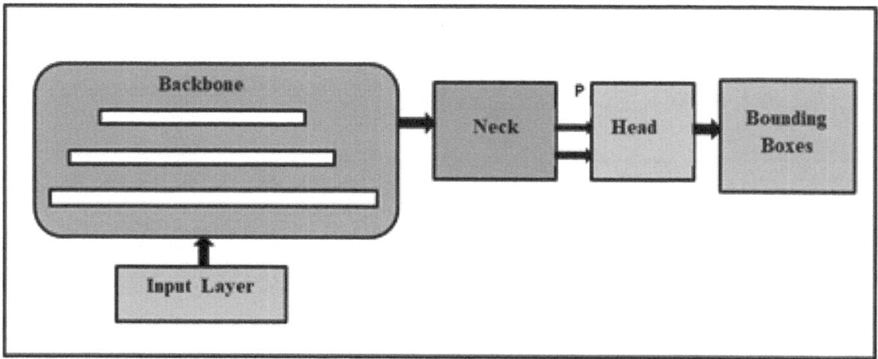

Fig. 3. Architecture of YOLOv5.

YOLOv8 represents the latest improvement in terms of efficiency, and flexibility. It uses a back-bone consisting of Cross-Stage Partial (CSP) connections to improve feature maps. Feature Pyramid Networks (FPN) and Path Aggregation Networks (PANet) were used in Neck portion which combine the features at multiple scales and helped in detecting leaf diseases with varying sizes. It also includ-ed with a head, for classification of pearl millet disease classification as shown in Fig. 4.

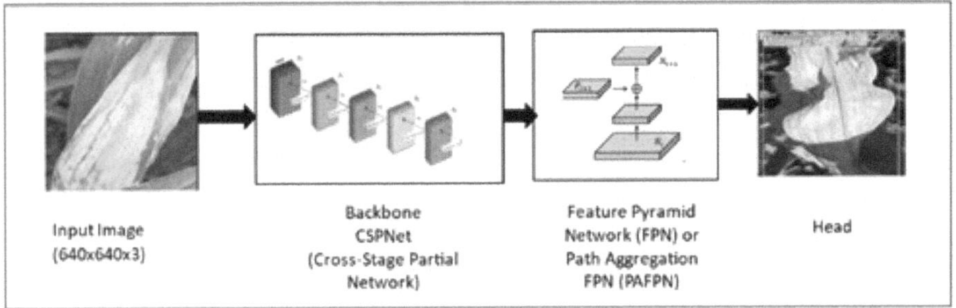

Fig. 4. Architecture of YOLOv8.

YOLOv11 used latest processing for real-time computer vision tasks and having C3K2 block, which uses smaller 3 × 3 convolutional kernels for efficient feature extraction. The model also used Spatial Pyramid Pooling Fast (SPPF) module, which had the capacity to detect the image under different scales by pooling features. The Convolu- tional block with Parallel Spatial Attention (C2PSA) further refined the feature repre- sentation by focusing on relevant spatial information, improving detection accuracy as shown in Fig. 5.

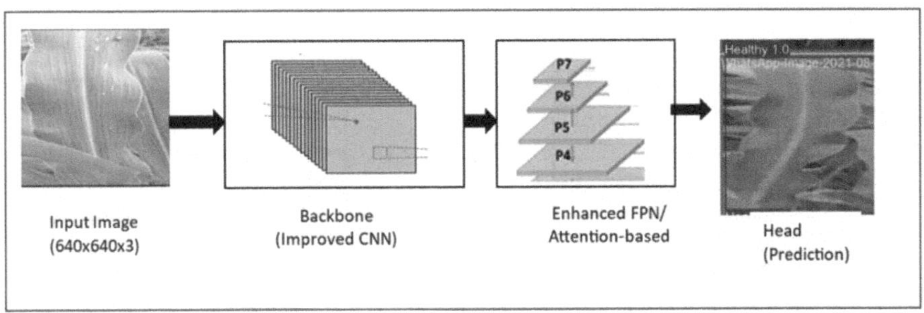

Fig. 5. Architecture of YOLOv11

4 Results and Discussion

The entire dataset was trained for 30 epochs. In order to have efficient training and to prevent overfitting and underfitting, the following hyper parameters are shown in Table 3. Choosing correct values helps in better performance in real applications. The validation of algorithm was further eval-uated using the performance metrics. In pearl millet disease detection, accuracy measures the per-centage of correctness of diseased and healthy samples, calculated as (TP+TN)/(TP+FP+TN+FN). Precision helped in identification of diseased samples obtained among all predicted cases whereas Recall recognized real diseased samples. F1 score was the harmonic mean of precision and recall.

Table 2. Hyperparameters used for YOLOv5, YOLOv8, and YOLOv11 models

Hyperparameter	YOLOv5	YOLOv8	YOLOv11
Input Image Size	640 × 640	640 × 640	1024 × 1024
Learning Rate	0.01	0.01	0.008
Batch Size	16	32	64
Optimizer	Stochastic Gradient Descent	Adam	Lion

The performance metrics for YOLOv8 was shown in Fig. 6. The model achieved an accuracy for smut, blast, and healthy with 98.93%, 98.7%, and 89.82% respectively. Rust has got a higher precision value of 90.1%. In terms of recall, ergot and blast demonstrated the highest values with 99.29% and 97.53%. F1 score for rust and blast were 91.88% and 87.53% respectively.

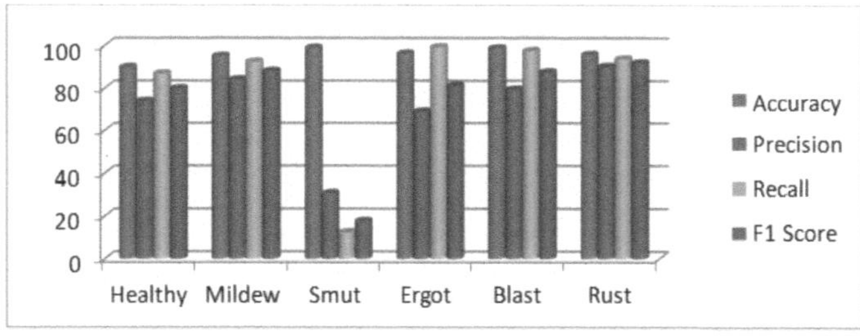

Fig. 6. Evaluation metrics for YOLOv8- individual classification.

The performance metrics for YOLOv11 were represented in Fig. 7. The model accomplished an accuracy of 99.11%, and 91.34% for smut and healthy, respectively. Rust had got higher precision value with 91.29%. In terms of recall, ergot and blast demonstrated the highest value with 98.59% and 98.15%. F1 score for rust and blast were 92.6% and 89.33% respectively.

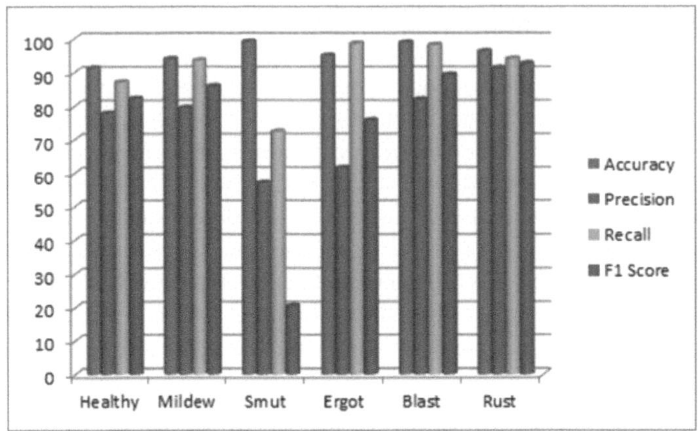

Fig. 7. Evaluation metrics for YOLOv11- individual classification.

The overall metrics were highlighted in Fig. 8. YOLO v5, YOLOv8 and YOLOv11 models were compared, with YOLOv11 achieving a slightly higher accuracy of 96.73% compared to v5 (93.21%) and v8 (95.71%) models. Precision values of YOLO v5, v8 and v11 were 70.6%, 71.17%, and 74.88%, respectively. Furthermore, the recall and F1 score for v11 were 78.5% and 72%, respectively. The mean precisionâĂŞrecall curve is shown in Fig. 9.

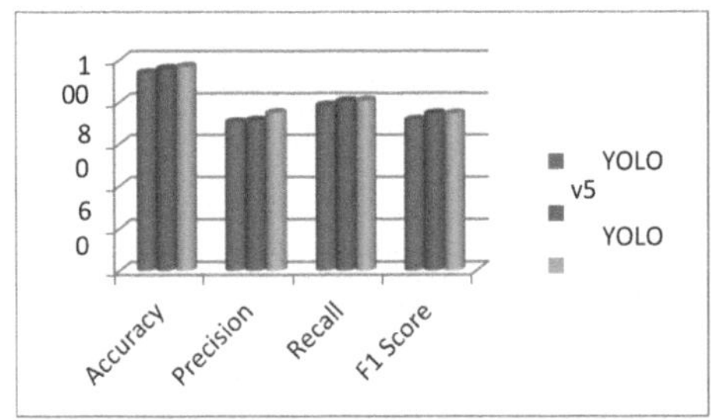

Fig. 8. Overall metrics for Pearl Millet Disease Detection.

Figure 10 shows the precision Recall curve. It is observed that overall MAP is 85.7% indicating effective classification. Also rust and blast shows highest PR with 96.6% and 95.9%. Fig. 10 represents the training and validation metrics for YOLOv11. The decreasing trend of box loss, classification and distribution focal loss indicates the better convergence of the model. In validation losses, the decreasing trend helps in effective classification of new data. The mAP@50 and mAP@50–95 metrics shows steady increase confirming effective training without overfitting.

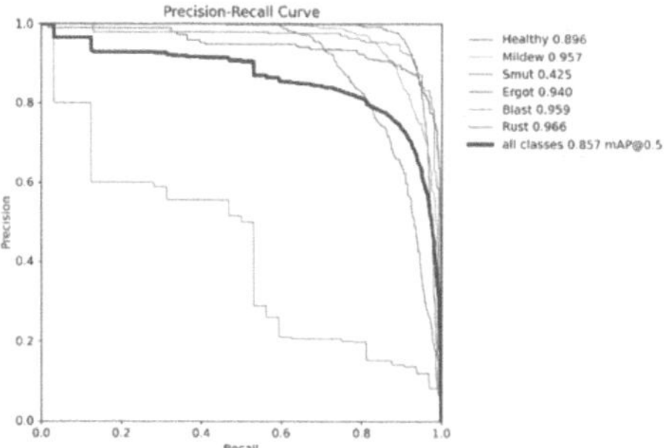

Fig. 9. Overall metrics for Pearl Millet Disease Detection.

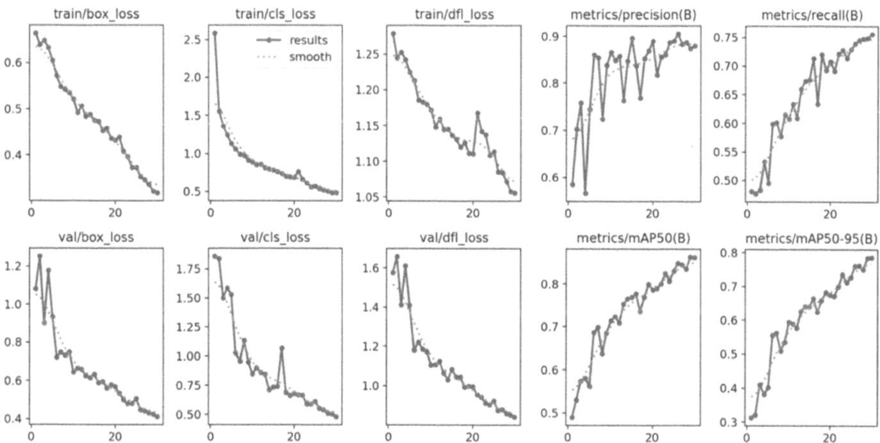

Fig. 10. Loss and mAP curves for YOLOv11.

Fig. 11. Detection results from YOLOv11.

The real time detection from YOLOv11 was given in Fig. 11. The bounding boxes indicates confidence score in predicting different classes. The multiple boxes shows that the model is capable of predicting multiple diseases in single image. It is observed YOLOv11 is effective in fine tuning of disease.

5 Comparative Analysis of Proposed Method with Other Research Work

Coulibaly et al. (2019) [18] used small dataset which includes 124 millet leaf images from online and employed VGG16 with imagenet to categorize two classes namely healthy and achieved an accuracy of 89%. Similarly, Kundu et al. (2020) [19] used field images of 3300 and classified into two categories namely blast and rust. The author used custom CNN model and achieved an accuracy of 98.78%. [20] used CNN for classifying images into rust and blast. CNN and achieved an accuracy of 98.08%n detect single disease in the leaf; however, in real time, each leaf has different disease. YOLO plays an important role in analysing high resolution images and spots multiple diseases in leaf thereby helps in accurate diagnosis and supports timely intervention to manage crop health. The comparative analysis of various start of models with YOLO models were highlighted in Table 2.

Table 3. Comparison of YOLO models with other state-of-the-art models

Year	Crops	No. of Images/Source	Models	Accuracy (%)
Coulibaly et al. 2019	Pearl Millet	711/Images from Internet	VGG-16	89.00
Kundu et al. 2021	Pearl Millet	3,300/Field Images	Custom-Net	98.78
Manjunath et al. 2023	Pearl Millet	–	CNN model	98.08
Proposed Approach	Pearl Millet	2473/Field Images	YOLOv8	95.71
Proposed Approach	Pearl Millet	2473/Field Images	YOLOv11	96.73

Table 2 shows that CNN has better accuracy than YOLO. To determine the effective- ness of the model in real time, other factors like computational speed, resources are to be considered. YOLO outperforms CNN in terms of localization and multiple disease detection in a single forward pass whereas CNN can do only classification and can be applicable in controlled areas. YOLO requires less computation overhead and can be easily deployed in drones and mobile applications.

6 Conclusion

The study proposed an object detection model using YOLO for disease detection in pearl millet. Nearly 2763 images were collected and categorized into six labels.

The models such as YOLOv5, v8 and v11 were implemented. The result shows that YOLOv11 outperforms other models with an accuracy of 96.73%. Compared to YOLOv5, YOLOv8 was specially designed for high speed and accuracy. Meanwhile YOLOv11 helped in precise disease detection with complex feature extraction and could even detect small sized diseases. The robustness helps in predicting images taken in varying environmental conditions such as uneven lighting and diverse texture. With advanced multi scale feature extraction, YOLOv11 has the capacity to detect small disease spots, thus improving early detection and improve efficiency. As future direction, the model might be deployed into Android app, so that the non-experts can confirm the disease and can employ appropriate management strategy. The model might also perform well for larger dataset to know its robustness.

References

1. Satyavathi, C.T., Ambawat, S., Khandelwal, V., Srivastava, R.K.: Pearl millet: a climate-resilient nutricereal for mitigating hidden hunger and provide nutritional security. Front. Plant Sci. **12**, 659938 (2021). https://doi.org/10.3389/fpls.2021.659938
2. Johnson, I., Mary, X.A., Raj, A.P.W., Chalmers, J., Karthikeyan, M., Andrew, J.: Deep-millet: a deep learning model for pearl millet disease identification to envisage precision agriculture. Environ. Res. Commun. **6**, 105031 (2024)
3. Zhao, Y., Gong, L., Huang, Y., Liu, C.: A review of key techniques of vision-based control for harvesting robot. Comput. Electron. Agric. **127**, 311–323 (2016). https://doi.org/10.1016/j.compag.2016.06.022
4. Wang, J., Yu, L., Yang, J., Dong, H.: A novel end-to-end object detection algorithm applied to plant disease detection. Information **12**, 474 (2021). https://doi.org/10.3390/info12110474
5. Kirola, M., Joshi, K., Chaudhary, S., Singh, N., Anandaram, H., Gupta, A.: Plants diseases prediction framework: an image-based system using deep learning. In: IEEE AIC, pp. 307–313 (2022). https://doi.org/10.1109/AIC55036.2022.9848899
6. He, K., Zhang, X., Ren, S., Sun, J.: Deep residual learning for image recognition. In: Proceedings of the IEEE Conference Computer Vision and Pattern Recognition, pp. 770–778 (2016)
7. Krizhevsky, A., Sutskever, I., Hinton, G.E.: ImageNet classification with deep convolutional neural networks. Commun. ACM **60**, 84–90 (2017)
8. Howard, A.G., Zhu, M., Chen, B., Kalenichenko, D., Wang, W., Weyand, T., et al.: MobileNets: efficient convolutional neural networks for mobile vision applications. arXiv preprint arXiv:1704.04861 (2017)
9. Vengaiah, C., Konda, S.R.: Improving tomato leaf disease detection with DenseNet-121 architecture. Int. J. Intell. Syst. Appl. Eng. **11**, 442–448 (2023)
10. Alam, T.S., Chandni, B.J., Abhijit, P.: Comparing pre-trained models for efficient leaf disease detection: a study on custom CNN. J. Electr. Syst. Inf. Technol. **111**, 12 (2024)
11. Tiwari, V., Joshi, R.C., Dutta, M.K.: Dense CNN-based multiclass plant disease detection and classification. Ecol. Inf. **63**, 101289 (2021). https://doi.org/10.1016/j.ecoinf.2021.101289
12. Sudhesh, K.M., Sowmya, V., Kurian, P.S., Sikha, O.K.: AI based rice leaf disease identification enhanced by DMD. Eng. Appl. Artif. Intell. **120**, 105836 (2023)

13. Amritavarshini, S., Ganesh, K.S., Arvind, M., Teja, K.R., Vijayan, D.: Image-based plant leaf disease detection using deep learning. In: IConSCEPT, pp. 1–6 (2024). https://doi.org/10.1109/IConSCEPT61884.2024.10627918
14. Anand, R., Veni, S., Aravinth, J.: Detection of diseases on brinjal leaves using k-means clustering. In: ICRTIT, pp. 1–6 (2016). https://doi.org/10.1109/ICRTIT.2016.7569531
15. Menon, A.S., Aravinth, J., Veni, S.: Pan-sharpening of multi-spectral remote sensing data using multi-resolution analysis. In: Lecture Notes in Electrical Engineering, vol. 997. Springer (2023). https://doi.org/10.1007/978-981-99-0085-5_56
16. Aldakheel, E.A., Zakariah, M., Alabdalall, A.H.: Detection and identification of plant leaf diseases using YOLOv4. Front. Plant Sci. **15**, 1355941 (2024). https://doi.org/10.3389/fpls.2024.1355941
17. Soeb, M.J.A., Jubayer, M.F., Tarin, T.A., et al.: Tea leaf disease detection and identification using YOLOv7 (YOLO-T). Sci. Rep. **13**, 6078 (2023). https://doi.org/10.1038/s41598-023-33270-4
18. Coulibaly, S., Kamsu-Foguem, B., Kamissoko, D., Traore, D.: Deep neural networks with transfer learning in millet crop images. Comput. Ind. **108**, 115–120 (2019). https://doi.org/10.1016/j.compind.2019.02.003
19. Kundu, N., et al.: IoT and interpretable ML framework for disease prediction in pearl millet. Sensors **21**(16), 5386 (2021). https://doi.org/10.3390/s21165386
20. Chikkamath, M., Dwivedi, D.N., Thimmappa, R., Vedamurthy, K.B.: Detection and categorization of diseases in pearl millet leaves. In: Future Farming: Advancing Agriculture with AI, pp. 41–52. Bentham (2023)
21. Tao, J., Li, X., He, Y., Islam, M.A.: CEFW-YOLO: a high-precision model for plant leaf disease detection. Agriculture **15**, 833 (2025)
22. He, J., Ren, Y., Li, W., Fu, W.: YOLOv11-RCDWD: detecting maize leaf diseases with improved YOLOv11. Appl. Sci. **15**, 4535 (2025)
23. Li, J., Liu, Z., Wang, D.: Recognizing pear leaf diseases using improved YOLOv5. Agriculture **14**, 273 (2024)

AgriFed: A Federated Learning Approach for Scalable and Secure Crop Disease Detection

Hemant Panchariya[1,3(✉)], Niyati Narwal[1,3], Farhat Qadri[2,3], Rejoy Chakraborty[3], and Puneet Goyal[3]

[1] Delhi Technological University, Delhi 110042, India
[2] Indian Institute of Technology Madras, Chennai 600036, India
[3] Indian Institute of Technology Ropar, Rupnagar 140001, India
hemantpanchariya2023@gmail.com, {2023csm1011,puneet}@iitrpr.ac.in

Abstract. Agriculture plays an important role in global food security, yet crop disease significantly impacts yield and economic stability. Traditional centralized machine learning approaches for crop disease classification, though effective, pose challenges like data privacy, security, and high communication cost. To overcome these issues, this research introduces AgriFed, a federated learning based framework that enables decentralized training while ensuring confidentiality of data and reduction in data sharing cost. We analyze the performance of number of Convolutional Neural Network based architectures, such as VGG16, GoggleNet, DenseNet, ResNet, MobileNet, and EfficientNet for crop disease classification using TOM2024 dataset, which includes images of tomato, onion and maize crops. The results highlight that federated learning not only preserves data privacy but also achieves comparable or better results than centralized learning. Furthermore, this research explores the balance between model complexity and performance, demonstrating that federated learning is a scalable and efficient method for real-time disease crop detection. These findings illustrate the potential of federated learning in modernizing the agriculture technology while ensuring the privacy of data.

Keywords: Federated Learning · Centralized Learning · Crop Disease Detection · Convolutional Neural Networks

1 Introduction

Agriculture serves as the foundation of human civilizations, providing food, livelihoods and contributing significantly towards global economic stability. According to the United Nations, the global population is projected to surpass 9 billion by 2050[1]. To keep up with the needs of the growing population, efficient agricultural systems and practices becomes imperative. However, agriculture faces significant

[1] United Nations, World Population Prospects.

H. S. Shekhawat et al. (Eds.): ICA 2025, CCIS 2795, pp. 25–36, 2026.
https://doi.org/10.1007/978-3-032-17083-5_3

challenges from various diseases and pests affecting the crops. Diseases like rusts, blights and bacterial infections spread rapidly and severely impact the yields. Therefore, to ensure sustainable and efficient agricultural systems, there is an urgent need to timely detect, manage and prevent these diseases and pests.

Over the time, various systems and models have been proposed and implemented globally to enhance agricultural practices. With rapid advancements in technology, especially in the domain of artificial intelligence, new possibilities have emerged for developing more efficient and accurate agricultural models. These promising Artificial Intelligence (AI) based algorithms, which are implemented in domains like healthcare [6], forensics [22], and finance [15], heavily depend on robust data collection for effective model training. In a typical machine learning pipeline (centralized learning), collected data is stored centrally where the model is trained and tested before deployment. However, advancements in technology have amplified the issues regarding data security and privacy. Data, often described as a digital identity, faces vulnerabilities when collected and stored centrally, exposing it to unauthorized access and potential breaches. Hence, this centralized technique increases cost and efforts in storing, managing and securing the data. As a result, these methods become more challenging and less sustainable in the longer run. Similarly, in the agriculture sector also, data privacy is a major concern. The farmers might be reluctant to share their sensitive data related to crop yields, soil conditions and farming methods with a central server. A federated learning approach provides a way to address the challenges faced by centralized systems.

Federated Learning is a decentralized technique which allows collaborative training of the model on remote clients without having to share the data to a global server. Here, the farmers are the clients, and contribute only towards updating the parameters of the model through a well-defined pipeline. This approach if combined with various edge devices, presents an opportunity to revolutionize agricultural practices. Furthermore, since this approach can accommodate multiple clients across different regions, it improves the model performance through diverse datasets across the clients. This geographically distributed training also aligns with international compliance regulation ensuring that sensitive data remains localized, which allows collaborative learning without compromising data privacy.

Recent years have witnessed a growing interest among various sectors to transition into federated learning setups. These include sectors such as Healthcare [25], Industrial engineering [13], Environmental monitoring [8], Edge Computing and Internet of Things (IoT) [2], Autonomous Vehicle Industry [10], and agriculture. Federated learning models have shown impressive performance compared to traditional centralized learning approaches, with FL models offering better privacy, scalability, and adaptability across diverse sectors. However, challenges such as system heterogeneity, statistical heterogeneity and communication improvements need to be addressed for optimal performance [14].

This work involves the implementation and evaluation of various convolutional network models using a federated learning approach to achieve the objective of crop disease detection. We have utilized a publicly available image dataset, TOM2024 [3] for our study.

This research paper is structured as follows: Sect. 2 gives an overview of agricultural research focused on detecting crop diseases using machine learning and deep learning algorithms. The proposed AgriFed, a Federated Learning based method to overcome the mentioned concerns framework is provided in Sect. 3. Section 4 describes the experimental setup, evaluation metrics, dataset details, and a detailed analysis of the results obtained. Section 5 summarizes the findings of this research and outlines the future scope of crop disease detection using lightweight CNN models in a federated learning setup.

2 Related Work

This section provides a brief summary of agricultural research focused on crop disease detection through various machine learning and deep learning algorithms, primarily with centralized data processing approach. Following this we studied federated learning, highlighting key contributions and recent advancements in the field of agriculture.

2.1 Deep Learning Techniques

Crop diseases, which are defined by abnormal conditions that adversely affect the plant growth, productivity, and quality, have been extensively investigated through a number of diagnostic and detection methods. In recent years, various techniques that make use of machine learning and deep learning have emerged as powerful tools for classification of plant diseases, with improved accuracy and efficiency as compared to conventional methods [27].

Convolutional Neural Networks (CNNs) are widely used for detection of plant disease [26]. VGG-16 and GoogleNet were implemented to detect rice disease [27]. Researchers implemented VGG-16 and AlexNet to detect tomato crop disease [23]. A number of Residual Networks (ResNet) including ResNet-101, ResNet-50, and ResNet-18 were implemented for classification of tomato plant disease [12]. EfficientNetB3 and Artificial Neural Network (ANN) classifiers were used in diagnosing and classifying pests and plant disease for tomato, maize and onion [3]. Various crop disease were classified using MobileNet and its variants like MobileNet-V2 and MobileNet-V3 [7,21]. Additionally, a number of plant leaf disease were diagnosed and detected using DenseNet-121 architecture [24]. However, these research efforts focus only on centralized model training strategies and do not investigate the domain of distributed approach with federated learning. Federated learning is considered as an approach to guarantee data privacy and security while developing more robust and generalized models.

2.2 Federated Learning (FL)

Federated Learning is a distributed technique in machine learning [17]. This approach enables direct training of machine learning models on local devices while maintaining the data privacy and security. Federated Learning has been effectively used in the field of crop disease detection. Federated Learning based CNN models were employed to classify maize diseases [28]. The use of federated learning in the area of crop detection was examined, with a focus on improving disease classification for sustainable agriculture and food security [20]. They implemented the Fast R-CNN technique, the Mask R-CNN, and RetinaNet. FL based deep learning techniques were introduced for maize crop detection and diagnosis [18]. Researchers investigated the effectiveness of federated learning for crop diseases identification and used CNN for agriculture related image analysis [16].

FL for the detection of rice leaf disease was explored using deep learning models [1]. Paper implemented EfficientNetB3, showing better performance over traditional pre-trained models with multiple clients. [29] provide insights into FL techniques in agriculture and their advancement. Federated Learning was combined with deep neural networks, including CNN, RNN: Recurrent Neural Network, GRU: Gated Recurrent Unit, and LSTM: Long Short-Term Memory to improve crop disease prediction while preserving data privacy [5]. An experiment leverages federated learning and CNNs to detect and classify soybean leaf disease detection efficiently [11]. Deep learning based federated learning is used to detect crop disease of various plants [19]. Additionally, in [9] federated learning with CNN model using drones was explored for plant disease detection and insect pest classification.

Despite progress in applying federated learning to the domain of agriculture, many aspects remain unexplored, as existing studies do not evaluate architectures with extremely deep neural networks. Moreover, no prior work has conducted extensive experimentation and analysis across various models or compared centralized learning on these models with distributed learning. This work addresses this gap by experimenting with VGG16, ResNet, GoogleNet, DenseNet-121, EfficientNet, MobileNet and their variants for plant disease detection.

3 The Proposed Methodology

To resolve privacy and security issues in agricultural domain, our proposed AgriFed: Federated Learning for Smart Agriculture framework ensures that each client holds ownership of its dataset, making sure that unprocessed or raw dataset is never shared with the central or global server. Instead, only the weights of the trained model or gradients are sent to the server for the aggregation process. FL provides a platform to create a global model at the server side by aggregating the locally trained model weights from client end. The global server then aggregates these updates and performs the centralized testing using the validation dataset that ensures the uniform evaluation across all the clients. The updated central model is then retransmitted to the clients for further local training at the client side. This interactive process keeps on until it converges or fixed number of training rounds. The optimised final model is then made available to the clients for real-time disease classification, ensuring data privacy with an efficient model while minimizing communication overhead.

3.1 Steps Involved in the AgriFed

Following are the different stages of our AgriFed model:

Step 1. ***Data Collection and Preprocessing at Clients***: Each client independently collects the dataset related to agriculture in our case. Additionally, data preprocessing is done at client side.

Step 2. ***Local Model Training and Update***: Each client i initializes the received global model $\mathbf{w}^{(t)}$ at round t, then trains using standard mini-batch gradient descent for K epochs:

$$\mathbf{w}_i^{(t+1)} = model\_train(\mathbf{w}^{(t)}, \mathbf{D}_i, K, \eta) \tag{1}$$

where, $\mathbf{w}^{(t)}$ is the global model at round t, $\mathbf{D}_i$ represents the local dataset of client i, K is the number of local training epochs, and η is the learning rate.

After K epochs, each client sends the updated model $\mathbf{w}_i^{(t+1)}$ to the server.

Step 3. ***Centralised Aggregation***: Once all clients complete their local training, the server aggregates their model updates using a weighted average:

$$\mathbf{w}^{(t+1)} = \sum_{i=1}^{M} \frac{n_i}{n_{\text{total}}} \mathbf{w}_i^{(t+1)} \tag{2}$$

where, M represents the number of participating clients, n_i represents number of training samples across all the clients, $n_{\text{total}} = \sum_{i=1}^{M} n_i$ represents the total number of samples across all clients, and $\mathbf{w}_i^{(t+1)}$ is locally trained model from client i after k epochs. This equation ensures that clients with more data have a stronger influence on the global model.

Step 4. ***Distribution of Global Weights***: The refined central model is then shared to all the clients. In order to improve data privacy and ensure continuous improvement, each client incorporates these redefined weights into their local model and continues training, and this iterative process continues until convergence or fixed number of training rounds, ensuring an optimized well-generalized model.

Step 5. ***Centralized Model Evaluation***: Unlike conventional federated learning, training of the model is done at the central server using a validation dataset, which guarantees uniform performance assessment of the model. Once the model achieves the desired accuracy, it is deployed to clients for real-time agricultural predictions, such as disease diagnosis and detection or yield forecasting.

AgriFed ensures the data sovereignty by keeping sensitive agricultural data localized. This approach minimize energy consumption, communication overhead, and improve model efficiency, while maintaining farmer's ownership over their dataset.

4 Results and Discussion

In this section, we present detailed analysis of results obtained including the evaluation metrics used to asses the performance of models, the experimental setup, and description of dataset used for experimentation.

4.1 Evaluation Metrics

In our research paper, we have utilized two metrics, i.e., Accuracy and F1-Score. These metrics are used to assess the performance of our models. They are computed using True Positives (TP), True Negatives (TN), False Positives (FP), and False Negatives (FN).

Accuracy. Accuracy is a measure of how well a machine learning model correctly classifies instances out of the total predictions. It is represented as:

$$\text{Accuracy} = \frac{\text{TP} + \text{TN}}{\text{TP} + \text{TN} + \text{FP} + \text{FN}} \tag{3}$$

F1-Score. F1-Score is defined as the harmonic mean of precision and recall. It evaluates the overall performance of the model by balancing both precision and recall. The expression is defined as:

$$\text{F1-Score} = 2 \times \frac{\text{Precision} \times \text{Recall}}{\text{Precision} + \text{Recall}} \tag{4}$$

where precision is defined as the capability of the model to correctly identify positive samples without misclassify them as negative samples, and recall evaluates the ability of the model to correctly classify all positive instances.

Since F1-Score balances both precision and recall, it is considered a better evaluation metric than accuracy for an imbalanced dataset.

4.2 Dataset

TOM2024 [4], a publicly available dataset which contains images of tomato, maize and onion diseases and pests was used to conduct the experimentation. The original dataset contains 12,277 labeled images of the mentioned crops categorized into 21 classes, each representing healthy, diseased and pest-infested conditions. However, certain classes contained significantly lesser number of images which were hence dropped to ensure better model performance. Furthermore, since this study focuses on the classification of crop diseases, pest-related classes were excluded for the experimentation. The final dataset hence obtained included 8 classes for Tomato comprising 4032 labeled images, 5 disease classes for Onion containing 2135 images and 7 classes for Maize with 3517 images. For preprocessing, we employed standard ImageNet-style transformations.

4.3 Experimental Setup

We employed three clients with one local epoch per communication round and a total of 100 communication rounds. This number was selected empirically as it offered the most stable and representative performance given the limited dataset size. We adopted an 80–20 train-test split, with 20% of the data held out for central evaluation and the remaining 80% was divided equally across the clients for training, ensuring homogeneous(IID) data distribution. A weighted categorical cross-entropy loss function was used to handle the unbalanced dataset across the classes. The batch size was set to 32, and the Adam optimizer with a learning rate of 0.001 and a decay factor of 0.1 applied every 50 rounds was used for optimal convergence. These parameters were selected based on empirical tuning and prior research to optimize model performance. The model was built using the PyTorch (2.6.0+cu124) and trained in a local system with an NVIDIA GPU Quadro RTX 5000 (16 GB VRAM).

4.4 Experimental Results

In this section, we present the results obtained by evaluating different CNN-based models. Specifically, we analyze the performance of these models when trained and tested on the TOM2024 dataset using two learning paradigm: federated learning and centralized learning. The observations obtained from Table 1 are as follows.

Table 1. Comparison of various models for Tomato, Maize, and Onion disease classification using AgriFed framework.

Models	Tomato disease		Maize disease		Onion disease	
	F1 Score	Accuracy	F1 Score	Accuracy	F1 Score	Accuracy
ResNet18	0.8374	0.8772	0.7559	0.8913	0.9181	0.9354
GoogleNet	0.8666	0.8960	0.7692	0.9143	0.9240	0.9368
VGG16	0.2956	0.5094	0.3021	0.7220	0.6432	0.7662
Efficient net B3	**0.8862**	**0.9198**	0.7896	0.9365	**0.9415**	0.9508
Eff. net V2_S	0.8850	0.9169	**0.8190**	**0.9495**	0.9385	**0.9537**
Eff. net V2_M	0.8801	0.9191	0.8180	0.9429	0.9380	0.9466
Mobilenet v2	0.8634	0.9025	0.7760	0.9151	0.9326	0.9452
Mobilenet v3-L	0.8560	0.9191	0.8034	0.9341	0.9364	0.9452
Densenet 121	0.8689	0.9068	0.7729	0.9103	0.9266	0.9438

- This research investigated the efficacy of Federated Learning (FL) across three crop disease datasets - Tomato, Maize, and Onion using CNN architectures.

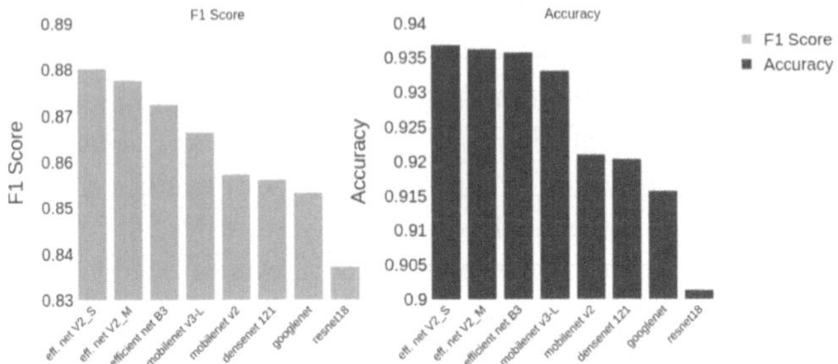

Fig. 1. F1-Score and Accuracy Comparison of CNN Models in FL setup (Fig. best viewed in colors).

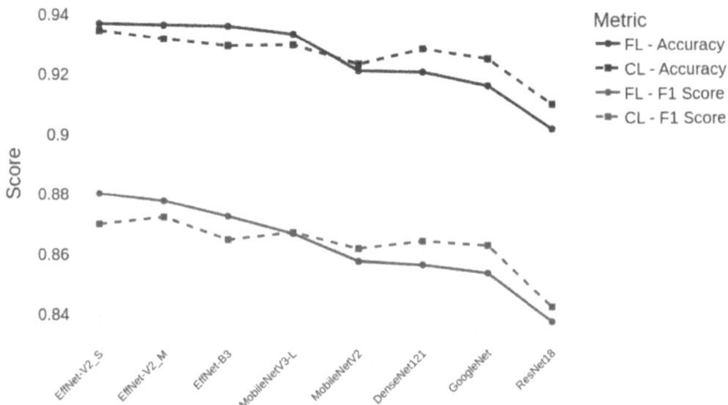

Fig. 2. Federated Learning vs. Centralized Learning: Model Performance Comparison (Fig. best viewed in colors).

The results in Table 1 reveal that modern CNN architectures such as EfficientNet and its variants, and MobileNet, demonstrate superior performance in terms of both F1 score and accuracy. These architectures were not only the most advanced among all the architectures used in experimentation but also comparatively lightweight. This demonstrates their superior generalization capabilities in FL settings. Conversely, VGG16 performs the worst, with significantly lower F1 score and accuracy. VGG16 is a heavy model and it requires larger datasets for it to converge and give conclusive results.

Unless otherwise stated, all the metrics reported onward are averaged out across the three datasets to better represent the results. Figure 1 provides a clear visual representation of the performance trends observed in the FL experiments. Both F1 score and accuracy show a consistent decline as we move from EfficientNet variants to DenseNet, GoogLeNet, and ResNet18. These findings invite the scope for adopting and developing lightweight and efficient CNN models in Federated Learning for agricultural applications where resource constraints and decentralized data distribution are critical factors.

- We also compared the efficacy of the proposed FL framework with the traditional Centralized Learning (CL) training paradigm. Figure 2 reveals that FL outperforms CL for CNN architectures like, EfficientNet and its variants, and MobileNetV3_L, in terms of both F1 score and accuracy. This suggests that FL's distributed training provides a regularization benefit, particularly for complex models. Additionally, the averaging of locally trained models in FL can reduce the impact of noisy data. The performance gap diminishes for simpler models like ResNet18, indicating they may not fully leverage FL's advantages.
- We performed experiments to illustrate the trade-off between model complexity, represented by parameter size, and performance, measured by F1 score and accuracy. This is crucial to deploy the models in resource-constrained environments like edge devices and IoT systems, particularly relevant in the agricultural sector. From Fig. 3, it is clear that modern architectures like EfficientNet-V2_M demonstrate high accuracy (0.935) but come with a substantial parameter size (54 million). This poses serious challenges for real-time processing on edge devices with limited computational resources. Conversely, models like MobileNet and GoogleNet offer a balance of reasonable performance (accuracy around 0.92 and F1 scores around 0.86) and smaller models (2.2 and 5.6 million, respectively).
  The ability to run models efficiently on these devices, without relying on constant cloud connectivity, is paramount for enabling timely interventions and improving agricultural productivity, especially in remote or resource-scarce regions.
- We also reported the impact of communication rounds on the performance metrics of Federated Learning. From Fig. 4, as expected, both metrics demonstrate a general upward trend with increasing communication rounds. The incremental improvement between 60 and 90 rounds is comparatively marginal, with a maximum increase of 2.5% in F1 score and 1.65% in accuracy (Google-Net). This observation highlights that while additional communication rounds contribute to performance enhancement, the returns diminish beyond a certain point. For resource-constrained settings in rural farms, limiting Federated Learning training to 60 communication rounds balances performance and efficiency, minimizing bandwidth usage. This allows quicker adaptation to new disease variants and minimizes the problems related to data drifts.

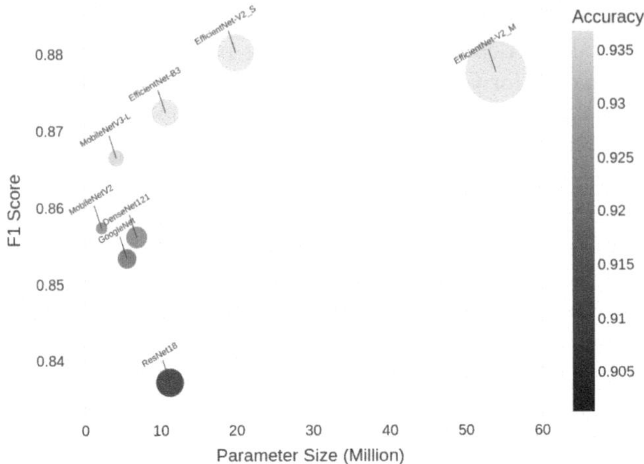

Fig. 3. Performance vs. Complexity of Various Models (Fig. best viewed in colors).

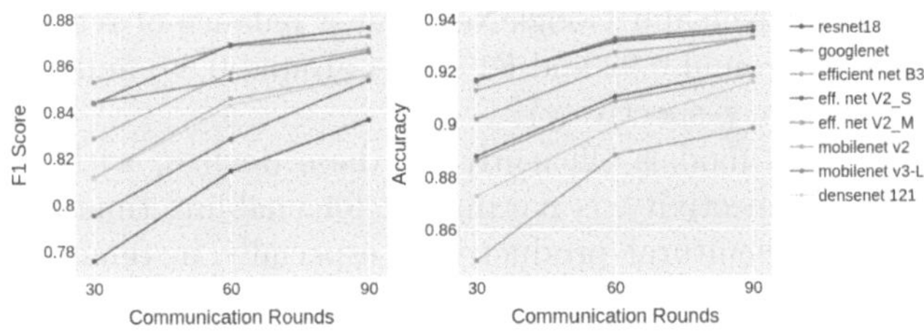

Fig. 4. No. of. Communication Rounds vs. F1-Score and Accuracy (Fig. best viewed in colors).

5 Conclusions and Future Scope

This research paper presents AgriFed, a federated learning approach for crop disease detection. Through decentralized training of model, AgriFed reduces data sharing requirements and ensures data privacy while achieving performance comparable to or better than centralized learning. Our evaluation of CNN architectures demonstrates that modern and lightweight architecture performs better than conventional ones. Additionally, analysis of impact of change in number of communication rounds and tradeoff between model complexity and performance, makes AgriFed a viable solution for real time agricultural application.

Future research can explore and optimize lightweight CNN models for efficient edge deployment. Hybrid approaches combining federated and centralized learning may enhance accuracy. Reducing communication overhead and integrating multi-source data like satellite imagery can improve predictions. Strengthening privacy and expanding application for pest detection and yield forecasting will further advance smart agriculture.

Acknowledgement. This work is supported by the Anusandhan National Research Foundation (ANRF), India, under grant CRG/2023/008300.

References

1. Aggarwal, M., Khullar, V., Goyal, N., Alammari, A., Albahar, M.A., Singh, A.: Lightweight federated learning for rice leaf disease classification using non independent and identically distributed images. Sustainability **15**(16), 12149 (2023)
2. Albogami, N.N.: Intelligent deep federated learning model for enhancing security in iot enabled edge computing environment. Sci. Rep. **15**, 4041 (2025). https://doi.org/10.1038/s41598-025-88163-5
3. Appiah, O., et al.: Plantesaine: an artificial intelligent empowered mobile application for pests and disease management for maize, tomato, and onion farmers in burkina faso. Agriculture **14**(8), 1252 (2024)
4. Appiah, O., et al.: Tom 2024: datasets of tomato, onion, and maize images for developing pests and diseases ai-based classification models. Data Brief **59**, 111357 (2025)
5. Behera, S., Padhy, N., Panigrahi, R., Kuanar, S.K.: Crop disease prediction using deep learning in a federated learning environment: ensuring data privacy and agricultural sustainability. Procedia Comput. Sci. **254**, 137–146 (2025)
6. Chauhan, J., Goswami, R., Goyal, P.: Using deep learning to classify burnt body parts images for better burns diagnosis. In: Lepore, N., Brieva, J., Romero, E., Racoceanu, D., Joskowicz, L. (eds.) Processing and Analysis of Biomedical Information, pp. 25–32. Springer International Publishing, Cham (2019)
7. Chen, J., Zhang, D., Suzauddola, M., Zeb, A.: Identifying crop diseases using attention embedded mobilenet-v2 model. Appl. Soft Comput. **113**, 107901 (2021). https://doi.org/10.1016/j.asoc.2021.107901
8. Hu, B., Gao, Y., Liu, L., Ma, H.: Federated region-learning: an edge computing based framework for urban environment sensing. In: IEEE International Conference on Edge Computing (EDGE), pp. 158–165 (2018). https://doi.org/10.1109/EDGE.2018.00030
9. Khan, F.S., Khan, S., Mohd, M.N.H., Waseem, A., Khan, M.N.A., Ali, S., Ahmed, R.: Federated learning-based uavs for the diagnosis of plant diseases. In: 2022 International Conference on Engineering and Emerging Technologies (ICEET), IEEE, Malaysia (2022)
10. Kishawy, M.M.E., El-Hafez, M.T.A., Yousri, R., Darweesh, M.S.: Federated learning system on autonomous vehicles for lane segmentation. Sci. Rep. **14**, 25029 (2024). https://doi.org/10.1038/s41598-024-71187-8
11. Kumar, M., Aeri, M., Chandel, R., Kukreja, V., Mehta, S.: Federated learning cnn for smart agriculture: a modeling for soybean disease detection. In: 2024 IEEE International Conference on Interdisciplinary Approaches in Technology and Management for Social Innovation (IATMSI), pp. 1–6. IEEE, Gwalior, India (2024)
12. Kumar, S., Pal, S., Singh, V.P., Jaiswal, P.: Performance evaluation of resnet model for classification of tomato plant disease. Epidemiologic Methods **12**(1), 20210044 (2023)
13. Li, L., Fan, Y., Tse, M., Lin, K.Y.: A review of applications in federated learning. Comput. Ind. Eng. **149**, 106854 (2020). https://doi.org/10.1016/j.cie.2020.106854

14. Li, T., Sahu, A.K., Talwalkar, A., Smith, V.: Federated learning: challenges, methods, and future directions. IEEE Signal Process. Mag. **37**, 50–60 (2020). https://doi.org/10.1109/MSP.2020.2975749
15. Mahalle, A., Yong, J., Tao, X., Shen, J.: Data privacy and system security for banking and financial services industry based on cloud computing infrastructure. In: 2018 IEEE 22nd International Conference on Computer Supported Cooperative Work in Design ((CSCWD)), pp. 407–413 (2018). https://doi.org/10.1109/CSCWD.2018.8465318
16. Mamba Kabala, D., Hafiane, A., Bobelin, L., Canals, R.: Image-based crop disease detection with federated learning. Sci. Rep. **13**(1), 19220 (2023)
17. McMahan, H.B., Moore, E., Ramage, D., Hampson, S., Arcas, B.A.Y.: Communication-efficient learning of deep networks from decentralized data (2016)
18. Mehta, S., Kukreja, V., Gupta, A.: Revolutionizing maize disease management with federated learning cnns: decentralized and privacy-sensitive approach. In: 2023 4th International Conference for Emerging Technology (INCET), pp. 1–6. IEEE, India (2023)
19. Mehta, S., Kukreja, V., Yadav, R.: A federated learning cnn approach for tomato leaf disease with severity analysis. In: 2023 Second International Conference on Augmented Intelligence and Sustainable Systems (ICAISS), pp. 309–314. IEEE, Trichy, India (2023)
20. Muhammad, B.L., Kusharki, M.B.: Federated learning for collaborative crop disease monitoring in wheat production. In: 2nd International Conference on Multidisciplinary Engineering and Applied Science (ICMEAS), vol. 1, pp. 1–5. IEEE, India (2023)
21. Rajbongshi, A., Sarker, T., Ahamad, M.M., Rahman, M.M.: Rose diseases recognition using mobilenet. In: 2020 4th International Symposium on Multidisciplinary Studies and Innovative Technologies (ISMSIT), pp. 1–7 (2020). https://doi.org/10.1109/ISMSIT50672.2020.9254420
22. Rana, K., Goyal, P., Sharma, G.: Dual-branch convolutional neural network for robust camera model identification. Expert Syst. Appl. **238**, 121828 (2024). https://doi.org/10.1016/j.eswa.2023.121828, https://www.sciencedirect.com/science/article/pii/S0957417423023308
23. Rangarajan, A.K., Purushothaman, R., Ramesh, A.: Tomato crop disease classification using pre-trained deep learning algorithms. Procedia Comput. Sci. **133**, 1040–1047 (2018)
24. Swaminathan, A., Varun, C., Kalaivani, S., et al.: Multiple plant leaf disease classification using densenet-121 architecture. Int. J. Electr. Eng. Technol **12**(5), 38–57 (2021)
25. Szegedi, G., Kiss, P., Horváth, T.: Evolutionary federated learning on eeg-data. In: Federated Learning Workshop (2019)
26. Tugrul, B., Elfatimi, E., Eryigit, R.: Convolutional neural networks in detection of plant leaf diseases: a review. Agriculture **12**(8), 1192 (2022)
27. Yakkundimath, R., Saunshi, G., Anami, B., Palaiah, S.: Classification of rice diseases using convolutional neural network models. J. Inst. Eng. India Ser. B **103**(4), 1047–1059 (2022)
28. Yang, Q., Liu, Y., Chen, T., Tong, Y.: Federated machine learning: concept and applications. ACM Trans. Intell. Syst. Technol. (TIST) **10**(2), 1–19 (2019)
29. Žalik, K.R., Žalik, M.: A review of federated learning in agriculture. Sensors **23**(23), 9566 (2023)

Remote Sensing and Environmental Parameter Analysis of Rajpura, Punjab Using GIS for Precision Farming Application

Manvinder Sharma[1(✉)], Rajneesh Talwar[1], Satyajit Anand[1], Harjinder Singh[2], and Tejpal Sharma[3]

[1] Department of Interdisciplinary Courses in Engineering (DICE), Chitkara University Institute of Engineering and Technology (CUIET), Chitkara University, Punjab, India
{rajneesh.talwar,satyajit.anand}@chitkara.edu.in
[2] Department of ECE, Punjabi University Patiala, Punjab, India
harjinder@pbi.ac.in
[3] Department of CSE, Chitkara University Institute of Engineering and Technology (CUIET), Chitkara University, Punjab, India
tejpal.sharma@chitkara.edu.in

Abstract. Precision farming through Geographical Information System (GIS) and remote sensing is revolutionizing modern agriculture. It enables data driven decision making, optimizing resource utilization and improved crop productivity. Punjab is one of the bread-basket of India. The increasing challenge of climate variability, soil degradation and resource depletion in Punjab highlight the need of precision farming. In the paper, environmental monitoring to essential key agricultural parameters of Rajpura, Punjab is extracted through GIS. The study is done for five years from 2020 to 2024. The parameters essential for precision farming like tropospheric NO_2, air temperature, rainfall precipitation, vegetation indices (NDVI and EVI) and soil moisture are examined. The parameters are analyzed using multiple satellites Sentinal-2, Sentinel-5P, ERA5-Land, CHRIPS and NASA SMAP using Google Earth Engine (GEE). The results indicate a strong correlation with soil moisture, precipitation and vegetation health. The research demonstrates how aggregating and analyzing monthly averages of these parameter can inform agricultural decision-making processes, optimize resource utilization and enhance crop productivity. The methodology presented offers scalable, data driven solutions for implementing smart farming practices in Punjab.

Keywords: GIS · Smart Farming · Precision agriculture · Satellite · Remote Sensing

1 Introduction

Agriculture remains the backbone of India's economy. Punjab, known as India's "bread-basket" is a major agricultural region facing substantial challenges related to resource depletion, climate variability and environmental degradation. Agriculture in Punjab contributes 40% of the state's GDP and employs nearly 62% of its workforce [1]. However,

H. S. Shekhawat et al. (Eds.): ICA 2025, CCIS 2795, pp. 37–49, 2026.
https://doi.org/10.1007/978-3-032-17083-5_4

conventional farming practices have led to groundwater depletion, overwatering and underwatering problems, soil degradation and declining productivity [2]. Precision farming involves site specific crop management using GIS, remote sensing and IoT based data collection to optimize farming practices. Rajpura is located in Punjab, India (30.4755°N and 76.5958°), represents a microcosm of these challenges with agricultural activities occurring alongside industrial development. This approach allows farmer to monitor soil moisture, predict weather patterns and apply input like water and fertilizer more efficiently. Figure 1 shows locating Rajpura, on map using Google Earth Engine.

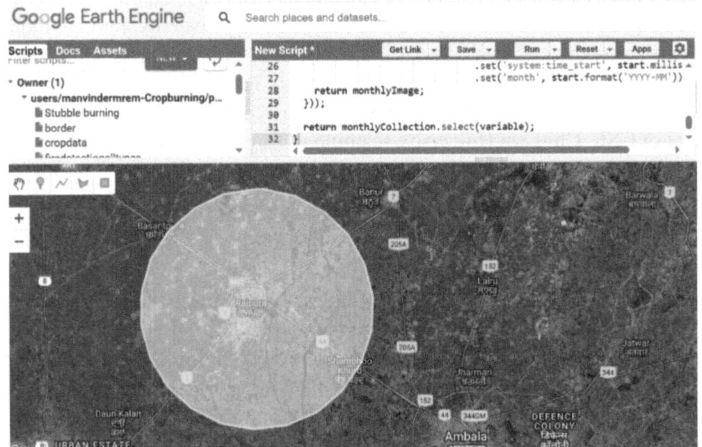

Fig. 1. Location of area of Interest (Rajpura) for study

The concept of precision farming has evolved significantly with the advancement of digital technologies. In traditional methods, farmers relied on manual observations and limited historic data to make farming decisions [3]. This data helps them for monitoring crop health, soil condition and climate change. Hence farmers can optimize the methods and resources for farming leading to less wastages and improved growth.

2 Materials and Methods

The analysis was conducted using Google Earth Engine (GEE). It is a cloud based geospatial processing platform which enables the integration of multiple data sources in GIS framework. GEE provides a Javascript API for geospatial analysis at scale. It allows processing of large satellite datasets without need for significant local computing resources.

2.1 Study Area

The study area focuses on Rajpura, Punjab, India centered at coordinates 76.5958°E, 30.4755°N. This region is characterized by intensive agriculture, particularly wheat-rice crop rotation system. The area has subtropical climate with distinct seasons of hot summer (March-June), Monsoon (July-September), post monsoon (October-November) and Winter (December-February). Rajpura's agricultural landscape is dominated by the rice-wheat cropping system, with rice typically grown from June to October and wheat from November to April [4]. The region has an average annual rainfall of approximately 700–800 mm, concentrated primarily during the monsoon months. [5]. Table 1 summarizes the key agricultural and geographical characteristics.

Table 1. Characteristics of the Study Area (Rajpura, Punjab)

Parameter	Description	Value/Range
Location	Coordinates	30.4755°N, 76.5958°E
Elevation	Mean above sea level	275–290 m [6]
Climate type	Köppen classification	Cwa (Humid subtropical)
Annual rainfall	Average (2018–2022)	700–800 mm
Temperature range	Annual	4 °C (min) to 45 °C (max) [7]
Major soil types	Dominant classifications	Inceptisols, Entisols [8]
Major crops	Seasonal rotation	Rice (chawal), Wheat (kanak)
Average farm size	Landholding statistics	3.2 hectares [9]
Irrigation sources	Primary water source	Groundwater (81%), Canal (19%) [10]
Cropping intensity	Annual	189% [11]

2.2 Satellites for Data Sources

The analysis incorporates data from multiple satellite platforms selected for their relevance to agricultural monitoring for five years from (1st January 2020 to 31st December 2024) for the Rajpura area shown in Table 2.

Table 2. Satellite Data Sources and Their Agricultural Applications

Parameter	Data Source	Platform/Sensor	Spatial Resolution	Temporal Resolution	Key Parameters
NO_2 Concentration	Sentinel-5P	TROPOMI	7 km × 3.5 km	Daily	Tropospheric NO_2
Air Temperature	ERA5-Land	ECMWF Reanalysis	31 km	Hourly/Daily	Temperature, Precipitation, Humidity
Rainfall Precipitation	CHIRPS	Climate Hazards Group	5 km	Daily	Precipitation

(continued)

Table 2. (*continued*)

Parameter	Data Source	Platform/Sensor	Spatial Resolution	Temporal Resolution	Key Parameters
NDVI and EVI	Sentinel-2	MSI	10 m,20 m, 60 m	5 days	Multispectral bands (13 bands)
Soil Moisture	NASA SMAP	L-band radiometer	9 km	3 days	Surface and root zone soil moisture

2.2.1 Sentinel-5P Data for NO_2 Analysis

Sentinel 5P is the first Copernicus mission dedicated to monitor atmospheric composition. The satellite carries the TROPOMI (Tropospheric Monitoring Instrument) sensor. It provides daily global observations of key atmospheric constituents like NO_2, O_3, SO_2, CO, CH_4 and aerosols [12]. Sentinel 5P is an European Space Agency mission deigned to monitor atmospheric compositions. For this study, Sentinel 5P tropospheric NO_2 column density data was obtained using GEE with collection ID '*COPERNICUS/5SP/OFFL/L3_NO2*'. Observations with a cloud fraction greater than 0.3 were removed using a cloud-masking function with spatial resolution of 1 km. A cloud masking function was implemented to filter out low quality observations improving the reliability of NO_2 estimates. A monthly aggregation function was applied to compute average NO_2 levels for each month throughout study period. To visualize long term trends, a time series NO_2 chart was generated using GEE command *ui.Chart.image.series()*. This allows generation of chart for seasonal variations and pollution trends. The processed NO_2 data was also exported in CSV format to Google Drive for further analysis.

2.2.2 ERA 5 Meteorological Data

ERA5 is the fifth generation ECMWF reanalysis for the global climate and weather, providing hourly estimates of atmospheric, land and oceanic variables. The resolution of this is approximately 31 km [13]. For this study, ERA5 daily aggregates dataset was accessed through GEE using collection ID 'ECMWF/ERA_LAND/DAILY_AGGR'. For the study mean 2m air temperature data was extracted. To optimize memory usage and improve computational efficiency the dataset was filtered spatially. A function was implemented to compute monthly mean temperature over the study period. This is also ensured that each monthly composite retained temporal consistency. The processed dataset was visualized using time series temperature chart using GEE ui.Chart.image.series() function.

2.2.3 CHIRPS Rainfall Data for Precipitation Analysis

Precipitation data was obtained from the Climate Hazards Group InfraRed Precipitation with Station data (CHIRPS). It is a high resolution satellite designed for climate and agricultural applications [14, 15] the CHIRPS dataset provides daily precipitation estimates at a 5 km spatial resolution. This makes it well suitable for analyzing rainfall patterns in Rajpura. A monthly aggregation function was developed to compute rainfall averages over time. This incorporates a data availability check to prevent missing values from affecting the analysis. A time series rainfall chart was generated using GEE's ui.Chart.image.series(). This allows for trend visualization.

2.2.4 Sentinel 2 Multispectral Imagery

Sentinel-2 is a wide swath, high resolution, multi spectral imaging mission of the European Space Agency (ESA) consisting of two identical satellites (Sentinel-2A and Sentinel-2B) in the same orbit. Together, they provide a revisit time of 5 days at the equator [16, 17]. The Sentinel-2 data was accessed through Google Earth Engine using the collection ID 'COPERNICUS/S2_SR', which provides atmospherically corrected surface reflectance products. Key specifications used are 13 spectral bands in visible, near-infrared, and shortwave infrared wavelengths. Spatial resolution of 10m (visible and NIR), 20m (red edge and SWIR), and 60m (atmospheric bands). Table 3 shows formulas to calculate NDVI and EVI.

Table 3. Vegetation Indices Calculated for Agricultural Monitoring

Index	Formula	Agricultural Application	Sensitivity
NDVI	(NIR - Red) / (NIR + Red)	General crop vigor assessment; Biomass estimation	Chlorophyll content; Canopy density
EVI	2.5 * [(NIR - Red) / (NIR + 6*Red* - *7.5* Blue + 1)]	Improved assessment in high biomass crops	Canopy structural variations; Reduced atmospheric sensitivity

The Normalized Difference Vegetation Index (NDVI) was derived using Sentinel-2 imagery to monitor crop growth and assess vegetation health in Rajpura. NDVI is a widely used index in remote sensing for evaluating vegetation vigor. To ensure high quality data processing images with more than 20% cloud cover were excluded. A monthly aggregation function was implemented to compute NDVI averages over time for capturing seasonal variations in crop health. Similar to above approach time series NDVI chart was generated. A color scaled NDVI map was also produced using a palette ranging from brown to dark green. The brown shows low vegetation and dark green shows high vegetation. This helps to offer intuitive visualization of spatial NDVI variations across Rajpura. The dataset was then exported in CSV format for further analysis.

The Enhanced Vegetation Index (EVI) was calculated using Sentinel-2 imagery to improve assessment of vegetation health. This also helps in understanding crop dynamics. EVI is particularly useful for distinguishing high biomass vegetation. This also helps in reducing atmospheric influences that can affect NDVI calculations. To ensure robust vegetation analysis following parameters were applied. A monthly aggregation function was implemented to analyze temporal trends in vegetation health. A time series chart was generated like above from same function of GEE. A color scaled EVI map was also generated using palette ranging from brown to dark green. This helps in providing an intuitive representation of crop vigor.

2.2.5 NASA SMAP Soil Moisture Data

The Soil Moisture Active Passive (SMAO) mission is a NASA satellite that measures and maps soil moisture. The satellite measures data globally every 2–3 days [18, 19]. The SMAP SPL4SMGP (Surface and Root Zone Soil Moisture Analysis Level 4) product combines SMAP radiometer observations with land surface model. This estimates soil moisture data at 9 km resolution. This dataset was assessed using GEE using collection ID 'NASA/SMAP/SPL4SMGP/007'. A monthly aggregation function was implemented to compute the mean surface soil moisture over time, ensuring temporal consistency in the dataset. A time-series soil moisture chart was generated using Google Earth Engine's ui.Chart.image.series(), enabling visualization of moisture fluctuations throughout the study period. To enhance spatial analysis, a color-scaled soil moisture map was created, using a palette ranging from brown (dry soil) to green (high moisture content) to effectively visualize moisture distribution.

3 Results and Discussions

3.1 Tropospheric NO_2 Analysis and Impact

Tropospheric NO_2 concentrations were analyzed to assess potential impacts on agricultural productivity. The updated analysis from 2020–2024 reveals that the highest NO_2 levels occur in the post-monsoon and winter seasons, while the monsoon months exhibit the lowest concentrations, likely due to the cleansing effect of rainfall. Figure 2 shows NO_2 and Fig. 3 shows plot of data.

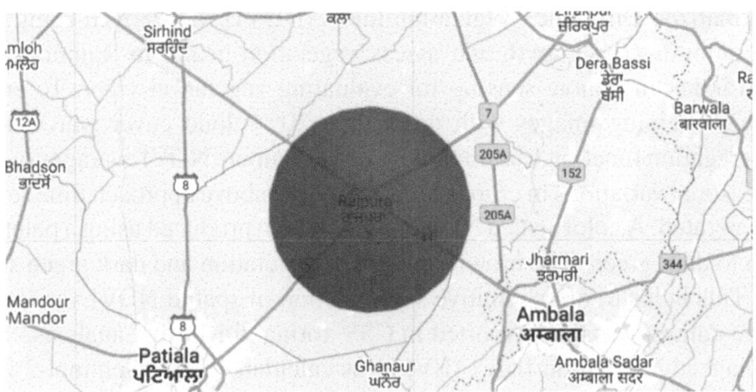

Fig. 2. NO_2 Mapping on map using GEE

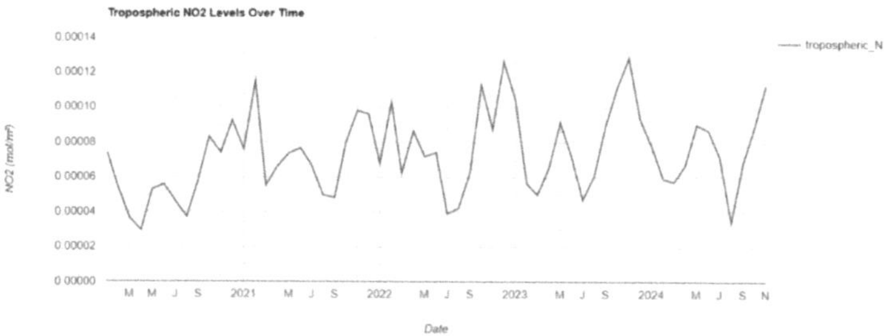

Fig. 3. Plot of NO_2 data.

Table 4. NO_2 data analytics

Season	Average NO_2 (mol/m^2)	Standard Deviation
Winter (Dec–Feb)	0.000086	0.000023
Summer (Mar–Jun)	0.000066	0.000017
Monsoon (Jul–Sep)	0.000055	0.000015
Post-monsoon (Oct–Nov)	0.000098	0.000018

The observed NO_2 values also shown in Table 4, are lower but still significant for potential pollution stress, particularly during critical wheat growth stages in winter. The spatial analysis of NO_2 pollution highlights northeastern Rajpura as a persistent hotspot, particularly near industrial facilities and major transport corridors. These areas would benefit from targeted pollution mitigation strategies within a smart farming framework.

3.2 Temperature Patterns for Crop Selection and Planning

Analysis of temperature data from 2020–2024 reveals distinct seasonal patterns across the Rajpura region. The updated temperature data shows annual cycles with significant seasonal variations, with implications for crop selection and management. Figure 4 shows temperature data area and Fig. 5 shows plot of data. Table 5 shows analytics.

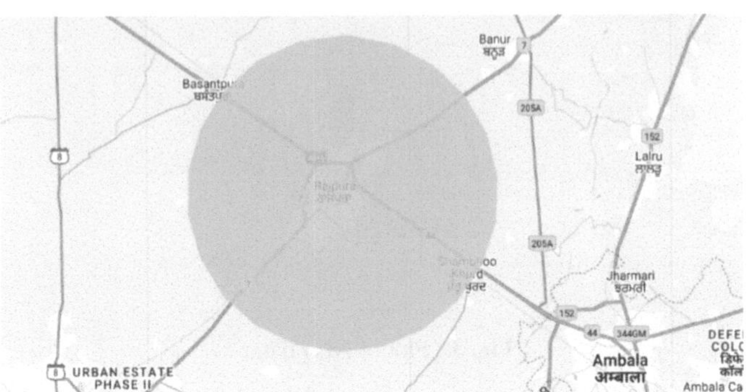

Fig. 4. Temperature data pattern.

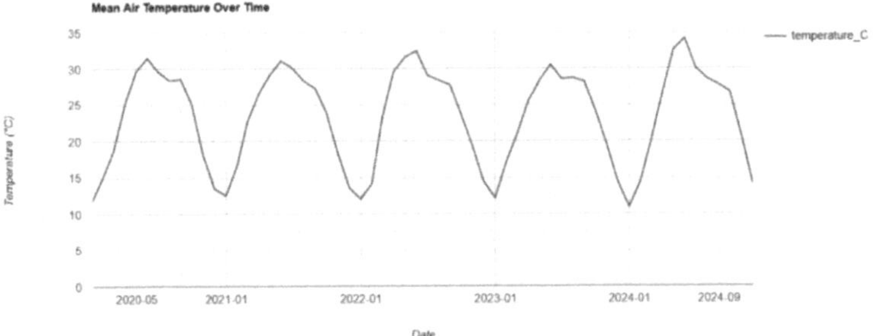

Fig. 5. Plot of data

Table 5. Monthly Temperature Statistics for Rajpura (2020–2024)

Month	Mean Temperature (°C)	Standard Deviation (°C)	Agricultural Implications
January	12.0	0.7	Winter wheat vegetative growth; Cold stress risk
February	15.7	1.3	Wheat jointing stage; Temperature favorable for development
March	21.1	1.9	Wheat heading/flowering; Moderate temperature fluctuations
April	26.7	1.7	Wheat grain filling; Increasing heat stress risk
May	30.2	1.7	Wheat maturity/harvest; Rice nursery preparation; Heat management critical

(continued)

Table 5. *(continued)*

Month	Mean Temperature (°C)	Standard Deviation (°C)	Agricultural Implications
June	31.9	1.3	Rice transplanting; Heat stress management critical
July	29.5	0.7	Rice vegetative stage; Temperature favorable with monsoon
August	28.5	0.1	Rice reproductive stage; Stable temperature conditions
September	27.9	0.5	Rice grain filling; Moderate temperature conditions
October	24.8	1.2	Rice maturity/harvest; Wheat field preparation
November	19.2	1.0	Wheat sowing; Favorable temperature for germination
December	14.0	0.4	Early wheat vegetative growth; Potential cold stress

A notable increase in maximum temperatures, particularly in April-June, with 2024 showing significantly higher temperatures in May (32.6 °C) and June (34.1 °C) compared to previous years. The duration of high-temperature periods (>30 °C) has expanded, with potential implications for heat stress in both wheat and rice crops. Significant year-to-year variations, especially in the spring months (March–April), creating challenges for consistent crop development.

3.3 Precipitation Analysis for Water Management

Analysis of the rainfall data from 2020–2024 confirms the pronounced seasonality of precipitation in the region, with the monsoon period (July-September) accounting for approximately 72% of annual rainfall. Figure 6 shows plot of data using GEE. Table 6 shows rainfall perception statistics.

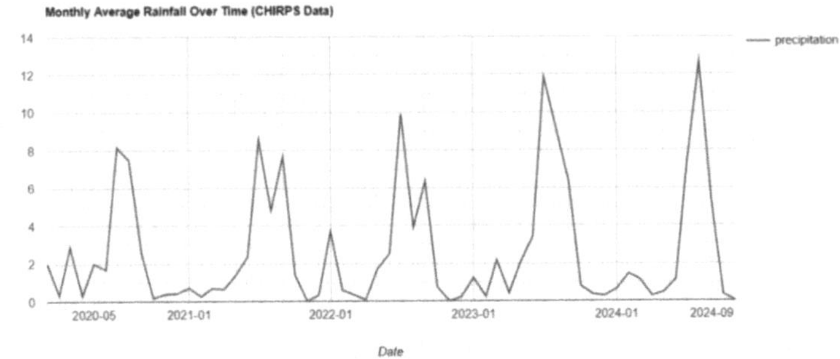

Fig. 6. Plot of rainfall data using GEE.

Table 6. Monthly Precipitation Statistics for Rajpura (2020–2024)

Month	Average Precipitation (mm)	Standard Deviation (mm)	Coefficient of Variation (%)
January	1.85	1.21	65.4
February	0.56	0.46	82.1
March	1.44	1.00	69.4
April	0.33	0.21	63.6
May	1.51	0.67	44.4
June	2.21	0.84	38.0
July	9.19	1.70	18.5
August	7.59	3.30	43.5
September	5.64	1.96	34.8
October	0.69	0.44	63.8
November	0.14	0.18	128.6
December	0.30	0.08	26.7
Annual	31.45	2.35	7.5

While overall monsoon rainfall (July-September) remains consistent, the distribution has become more variable, with August showing the highest inter-annual variability (CV = 43.5%). The data shows increasing instances of heavy rainfall days, particularly in July–August 2024, with potential implications for waterlogging and nutrient leaching.

3.4 Crop Health Monitoring Using Vegetation Indices

Analysis of the NDVI and EVI data from 2020–2024 reveals lower values than typically expected for heavily vegetated agricultural regions. Figure 7a shows NDVI and 7b Shows EVI. Figure 8a shows plot of data for NDVI and Fig. 8b shows plot of data for EVI.

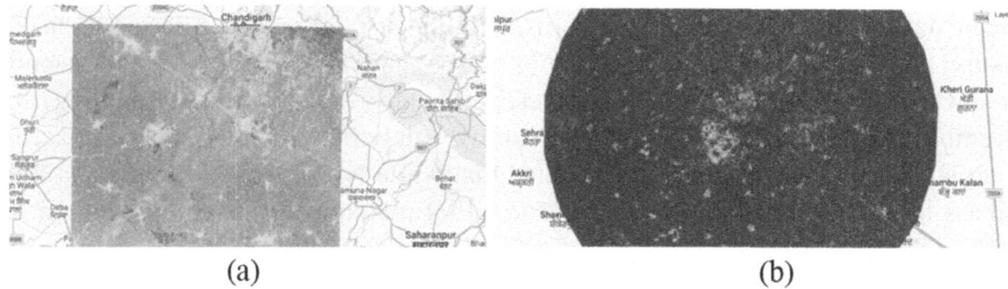

Fig. 7. (a) NDVI and (b) EVI for Rajpura.

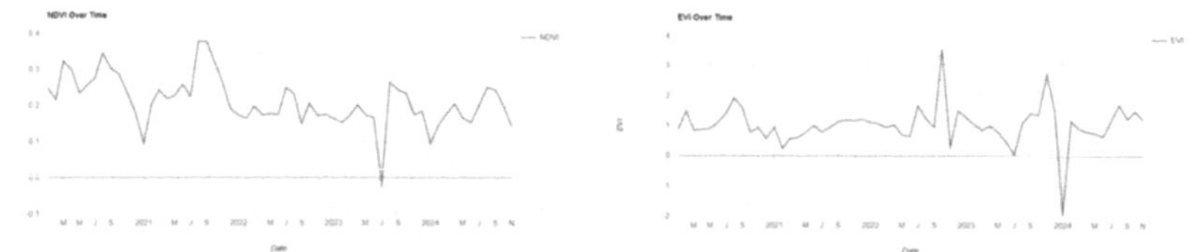

Fig. 8. (a) plot for NDVI (b) plot for EVI.

The Enhanced Vegetation Index (EVI) data reveals significantly higher values than NDVI, with more pronounced seasonal patterns and greater sensitivity to variations in crop conditions. clear seasonal patterns are evident with higher values during main growing seasons, particularly for rice.

3.5 Soil Moisture for Precision Irrigation

Analysis of the soil moisture data reveals distinct seasonal patterns and significant inter-annual variations with direct implications for irrigation management. As the data has 11679 samples due to which on GEE the user memory limit exceeded, so the data from csv is used to plot the graph using Microsoft excel as shown in Fig. 9.

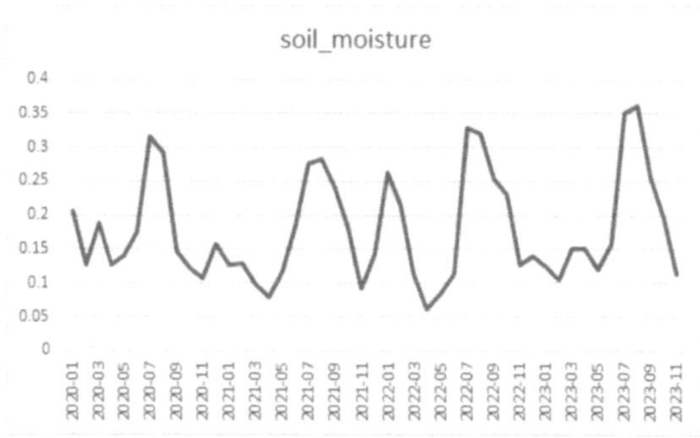

Fig. 9. Plot of soil moisture data over time.

The data shows clear seasonal patterns with highest soil moisture during monsoon months (July-August) and lowest during pre-monsoon period (April-May). Consistently low soil moisture during critical crop stages, particularly April (wheat grain filling) and November (wheat establishment). A comparative study by Goswami et al. [20] across 36 agricultural sites in northern India found that while NDVI tends to saturate at high biomass levels during peak growing seasons, EVI maintained sensitivity, with an R2 of 0.87 with biomass measurements compared to NDVI's R2 of 0.74. The observed discrepancy between NDVI and EVI values in the study is consistent with findings from Yadav et al. [21]. The study documented similar patterns in the rice-wheat cropping systems of Haryana. Chauhan et al. [22] validated SMAP soil moisture products against in-situ measurements from 17 monitoring stations across Punjab.

4 Conclusion

This study of environmental parameters in Rajpura, Punjab reveals critical insights for precision farming implementation, demonstrating how integrated satellite-based monitoring of NO_2 concentrations, temperature patterns, precipitation distribution, vegetation health indices, and soil moisture can inform agricultural decision-making processes, with results indicating pronounced seasonality in all parameters, concerning trends in temperature extremes and vegetation health, and strong correlations between soil moisture, precipitation, and crop vigor that can directly support optimized resource allocation, strategic intervention timing, and climate adaptation strategies in Punjab's agricultural systems, ultimately offering a scalable, cost-effective framework for enhancing agricultural productivity and sustainability in the face of mounting environmental challenges. Remote sensing helps in data driven decision support without extensive ground instrumentation.

Acknowledgments. The authors would like to acknowledge the support of IIT Ropar Technology and Innovation Foundation, one of technology hub under DST NM-ICPS, Government of India. We also extend our gratitude to Chitkara University Rajpura and Punjabi University Patiala for providing the necessary infrastructure and academic support.

Disclosure of Interests. Authors has no conflict of interests.

References

1. Department of Agriculture, Government of Punjab: Agricultural Statistics of Punjab 2021–22. Government Press, Chandigarh (2022)
2. Singh, B., Sharma, R.K.: Groundwater depletion in Punjab: trends, causes, and remedies. J. Water Resour. Manag. **35**(4), 421–438 (2021)
3. Uma, P., Gomathi, M., Gokulkumar, K.P., Monika, T., Preethika. P.: Empowering agriculture with AI: The smart agrohub initiative. In: Challenges in Information, Communication and Computing Technology, pp. 499–505. CRC Press (2025)
4. Directorate of Agriculture, Punjab: Agricultural Statistics at a Glance: Punjab 2022–23. Government of Punjab, pp. 28–32 (2023)

5. India Meteorological Department: Annual Rainfall Statistics 2018–2022: North India Region. IMD Technical Report No. 2023/07, pp. 112–115 (2023)
6. Geological Survey of India: Topographical Survey Report: Punjab Region. GSI Special Publication No. 143, pp. 78–82 (2022)
7. Punjab State Climate Change Knowledge Center: Temperature Trends in Punjab: 2020–2023, pp. 34–38. Punjab Agriculture University Press (2023)
8. National Bureau of Soil Survey and Land Use Planning: Soil Resource Mapping of Punjab, pp. 45–51. NBSS Publication 173 (2022)
9. Agricultural Census Division: Agricultural Census 2021–22: Punjab State Report. Ministry of Agriculture & Farmers Welfare, pp. 45–49 (2022)
10. Central Ground Water Board: Ground Water Year Book 2022–23. Ministry of Water Resources, Punjab, pp. 72–76 (2023)
11. Punjab State Agricultural Marketing Board: Agricultural Production Patterns in Punjab: 2020–2023. PSAMB Technical Report Series, vol. 4, pp. 22–25 (2023)
12. Bodah, B.W., et al.: Sentinel-5P TROPOMI satellite application for NO_2 and CO studies aiming at environmental valuation. J. Cleaner Prod. **357**, 131960 (2022)
13. Bell, B., et al.: The ERA5 global reanalysis: preliminary extension to 1950. Q. J. Royal Meteorol. Soc. **147**(741), 4186–4227 (2021)
14. Sharma, M., Singh, H.: Substrate integrated waveguide based leaky wave antenna for high frequency applications and IoT. Int. J. Sens. Wirel. Commun. Control **11**(1), 5–13 (2021)
15. Pandey, V., Srivastava, P.K., Singh, S.K., Petropoulos, G.P., Mall, R.K.: Drought identification and trend analysis using long-term CHIRPS satellite precipitation product in Bundelkhand, India. Sustainability **13**(3), 1042 (2021)
16. Soni, P.K., Rajpal, N., Mehta, R., Mishra, V.K.: Urban land cover and land use classification using multispectral sentinal-2 imagery. Multimedia Tools Appl. **81**(26), 36853–36867 (2022)
17. Swain, S.R., et al.: Estimation of soil texture using Sentinel-2 multispectral imaging data: an ensemble modeling approach. Soil and Tillage Res. **213**, 105134 (2021)
18. Colliander, A., et al.: Validation of soil moisture data products from the NASA SMAP mission. IEEE J. Sel. Topics Appl. Earth Observ. Remote Sens. **15**, 364–392 (2021)
19. Brown, M.E., et al.: NASA's soil moisture active passive (SMAP) mission and opportunities for applications users. Bull. Am. Meteor. Soc. **94**(8), 1125–1128 (2013)
20. Goswami, S., Kumar, P., Sharma, A.: Comparative analysis of vegetation indices for crop health monitoring in north Indian agricultural landscapes. Int. J. Appl. Earth Obs. Geoinf. **112**, 102935 (2022)
21. Yadav, V., Prasad, S., Singh, K.: Field validation of NDVI and EVI for biomass estimation in rice-wheat cropping system. J. Ind. Soc. Rem. Sens. **51**(3), 427–441 (2023)
22. Chauhan, P., Kumar, S., Mishra, V.: Performance evaluation of SMAP soil moisture retrievals in Punjab agricultural region. IEEE Geosci. Remote Sens. Lett. **20**(1), 2200116 (2023)

Optimal Hyperspectral Band Selection Using Amended Lyrebird Optimization Algorithm

Elza George[1(✉)], J. Aravinth[1], and Sathishkumar Samiappan[2]

[1] Department of Electronics and Communication Engineering, Amrita School of Engineering, Amrita Vishwa Vidyapeetham, Coimbatore, India
g_elza@cb.students.amrita.edu , j_aravinth@cb.amrita.edu

[2] The Department of Biosystems Engineering and Soil Sciences, The University of Tennessee, Knoxville, USA
sathish@utk.edu

Abstract. This paper presents an Amended Lyrebird Optimization Algorithm (ALOBH) for band selection in hyperspectral images (HSIs). The high dimensionality of HSI data leads to computational complexity and redundancy. To address this, and to improve processing speed, efficiency, and classification accuracy, ALOBH selects a prime subset of bands. ALOBH is a modified population-based metaheuristic algorithm that is inspired by lyrebirds' courtship displays and sound mimicry. The algorithm operates in two phases: exploration, influenced by the lyrebird's escape strategy, and exploitation, inspired by its hiding scheme. ALOBH was tested on the Indian Pines and PaviaU HSI datasets. Experimental results demonstrate that ALOBH is an effective band selection method showing excellent overall, average, and classification accuracy, as well as a high convergence rate. Compared to the next best method, ALOBH improved average accuracy by 2.2% for Indian Pines and 0.89% for PaviaU.

Keywords: population-based metaheuristic algorithm · hyperspectral image · fitness function

1 Introduction

Hyperspectral images are characterized by the massive amount of spectral and spatial information contained in them. They detain reflection details of the area being targeted in wide range of electromagnetic spectrum. With the onset of remote sensing technology, HSIs are utilized for wide variety of applications.HSI processing has proven to be reliable and influential in diverse areas. One of the applaudable impacts of HSI based remote sensing technology is in the field of agriculture for several purposes like monitoring crop growth, evaluating soil health and in methodical management of resources. The huge information present in HSI can have adverse effects when used in HSI classification

H. S. Shekhawat et al. (Eds.): ICA 2025, CCIS 2795, pp. 50–61, 2026.
https://doi.org/10.1007/978-3-032-17083-5_5

techniques [1]. This is because classification accuracy may be declined due to massive information content [2]. Moreover, expendable data in HSI may hinder model training and converegence. Hundreds of bands present in hyperspectral image(HSI) gives it a high dimensional structure resulting in computational complexity and data redundancy. In addition to this, processing speed, efficiency and accuracy of classification should be escalated. This calls for the necessity of eliminating needless information from HSI [3] by the process of band selection [4,5] which means eliminating unnecessary bands from hundreds of bands present in HSI.

Meta-heuristic algorithms are advanced problem-solving techniques that employ trial and error to discover suitable solutions for intricate optimization challenges. Operating at a higher level, they guide underlying heuristics to effectively explore the solution space. Unlike exact methods aiming for a guaranteed global optimum, meta-heuristics often incorporate randomness, striving for near-optimal solutions within practical timeframes, especially when finding the absolute best solution is computationally prohibitive. These algorithms are known for their adaptability and extensive search capabilities, making them increasingly vital in computational intelligence for tackling diverse optimization problems. A key characteristic is their ability to balance exploration, which involves broadly searching the problem space for promising regions, and exploitation, which focuses on refining known good solutions within those regions. Inspired significantly by natural phenomena, animal behaviors, and biological principles, meta-heuristic algorithms leverage nature's efficient problem-solving mechanisms. This bio-inspired approach allows them to offer several advantages, including ease of implementation, independence from gradient information, and a better ability to escape local optima compared to deterministic methods. Their flexibility enables adaptation to various optimization tasks. However, they also face challenges such as potential entrapment in suboptimal solutions, slow convergence rates in high-dimensional spaces, sensitivity to parameter settings, and the lack of a guarantee for finding the absolute global optimum. Careful selection, parameter tuning, and awareness of these limitations are crucial for their successful application in complex problems like hyperspectral band selection. This paper presents ALoBH, a modified form of lyrebird optimization algorithm [6]proposed by Dehghani et al., for band selection. The main contributions of this paper are:

1. A novel algorithm, ALOBH—Amended Lyrebird Optimization Algorithm, incorporating adaptive escape and hiding parameter control, is utilized for band selection of hyperspectral images.
2. ALOBH was implemented and tested on the benchmark HSI datasets, Indian Pines and PaviaU.
3. ALOBH's performance was analyzed and compared with four avant-garde band selection methods

This paper is organized as: related literature is discussed in Sect. 2, detailed methodology, psuedocode of ALoBH and mathematical modelling in Sect. 3, Sect. 4 presents results and dataset description, discussion and future scope are detailed in Sect. 5 and conclusion in Sect. 6.

2 Related Works

Over the years, several band selection methods have been evolved. They can be grouped into different categories and follow different principles [7]. All work with the common aim of reducing the dimension without compromising information content. Several nature inspired metaheuristics based optimization techniques have been successfully implemented for HSI band selection. A brief insight into these algorithms is presented in table 1. These algorithms make use of various fitness functions to rule out the best solution.

3 Proposed Methodology

In this work, an improved lyrebird optimization algorithm [6] incorporated with adaptive escape and hiding parameter control is used to select the optimal bands from the HSI dataset. The psuedocode of the proposed ALoBH algorithm is presented in the algorithm 1. Each lyrebird constitutes the population in Lyrebird optimization algorithm (LA) and the values of decision variables are determined by the position of the lyrebird. The ability of lyrebird to mimic sound from the environment, to increase the search capability is used in the exploration phase, and its potential for courtship display is applied in exploitation phase. To increase the performance of LA, we introduce the ALoBH algorithm, in which adaptive parameter control is introduced in the exploration and exploitation phases to firmly balance the two phases. Figure 1 gives the detailed breakdown of the methodology adopted in the proposed work. Like other metaheuristic algorithms [8–11], Lyrebird optimization algorithm [6], advances in two phases:-

a. Exploration: a global search phase based on the lyrebird's escape strategy.
b. Exploitation: a local search phase based on the lyrebird's hiding scheme

The fitness of the optimization algorithm for band selection is assessed using classification accuracy, overall accuracy and average accuracy.

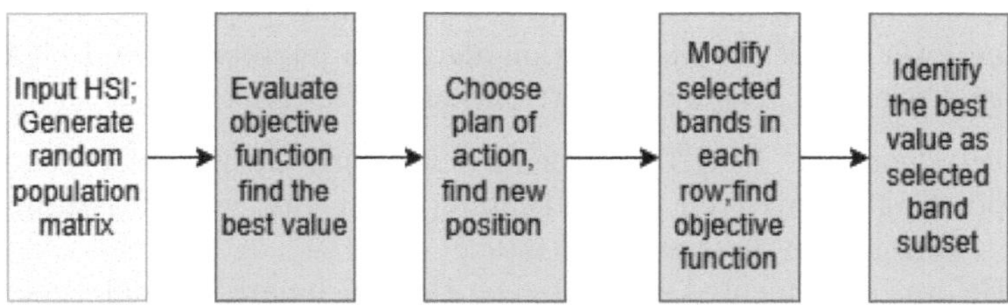

Fig. 1. Methodology involved in band selection of HSI using ALoBH algorithm

Performance is evaluated using classifiers random forest(RF) [19], K-Nearest Neighbours(KNN) [20] and Support Vector Machine(SVM) [21]. To scrutinize performance in comparison with existing methods, four avant-garde band selection methods are used such as MF [8], PS [9], CS [10] and GA [11]

Table 1. Review of meta heuristic Optimization algorithms

Algorithm Name	Natural Inspiration	Mechanism	Remarks
Moth-Flame Optimization (MF) [8]	Transverse orientation of moths	Moths fly towards flames (best solutions), adaptive flame reduction	Simple, high convergence, good exploration/exploitation balance, high local optimum avoidance
Particle Swarm Optimization (PS) [9]	Social behavior of bird flocks/fish schools	Particles move based on personal and global best	Analyzes HSI models, well-studied, Parameter design difficulty, prone to local minima, lower convergence/precision
Genetic Algorithm (GA) [11]	Natural selection and genetics	Population evolves through selection, crossover, mutation	Large global optimum, little problem-specific knowledge, Complex objective function creation, high computational resources, prone to local optima
Cuckoo Search (CS) [10]	Brood parasitism and Levy flights	Cuckoos lay eggs in nests, nest updates, Levy flight search	Effective global optimization, few parameters, simple calculation, modified versions improve performance, Standard CS prone to local optima, slow convergence
Bat Algorithm (BA) [12]	Echolocation of bats	Bats adjust velocity/position based on frequency, random walk	Avoids exhaustive search, potential local optima issues, higher computational time
Firefly Algorithm (FA) [13]	Flashing behavior of fireflies	Brightness as fitness, attraction to brighter fireflies	Global attraction, avoids exhaustive search, Performance varies by dataset, higher computational time
Grey Wolf Optimizer (GWO) [14]	Social hierarchy and hunting of grey wolves	Alpha, beta, delta guide omega wolves in search	Adaptively converges quickly, better solution quality, fewer selected features with high accuracy
Hybrid Rice Optimization Algorithm (HRO) [15]	Breeding of hybrid rice	Population divided into maintainer, restorer, sterile lines for cooperative evolution	Improved performance when combined, effective for feature selection, Integration difficulties, complex parameter tuning
Ant Colony Optimization (ACO) [16]	Foraging behavior of ants	Ants deposit pheromones, follow stronger trails	Strong performance in engineering optimization, Can have slow convergence
Gravitational Search Algorithm (GSA) [17]	Newton's law of gravitation	Solutions as masses attracting each other	Avoids exhaustive search, improved accuracy for Indian Pines
Harmony Search (HS) [18]	Musical improvisation	Harmonies generated based on memory, pitch adjustment, randomization	Best computational time, reasonable results

Algorithm 1. pseudocode of ALoBH

1: Selecting the information rich bands from entire HSI is the optimization problem and accuracy is the objective function.
2: Fix LA population size P(number of selected bands), number of decision variables D(total number of bands in HSI) and total number of iterations I.
3: Randomly create prime population matrix M with each member M_j being a random binary vector, where 1 indicates band is selected and 0 indicates band is rejected.
4: Evaluate the objective function $F(M_j)$ for each M_j.
5: Identify the set of bands with the greatest value of objective function.
6: For $i = 1$ to I
7: For $p = 1$ to P
8: Decide the lyrebird plan of action(exploration or exploitation) based on the equation

$$M_i \longleftarrow \begin{cases} \text{based on phase 1 if } r \geq 0.5 \\ \text{based on phase 2 otherwise} \end{cases}$$

9: If $r \geq 0.5$ (choose phase 1)
10: Rule out guarded zones areas for p^{th} lyrebird based on equation $GZ_p = M_k$ if $F_k > F_p$; values of k ranging from 1 to P.
11: Assess the modified location of pth Lyrebird as $m_{p,j}^{NP1} = m_{p,j} + e(t)(SGZ_{p,j} - R_{p,j}m_{p,j})$ where $e(i) = e_{max} - i|I * (e_{max} - e_{min})$ is the adaptive escape phase parameter
12: Modify p^{th} LA member based on equation

$$M_i \longleftarrow \begin{cases} M_p^{NP1} \text{ if } F_p^{NP1} > F_p \\ M_i \text{ otherwise} \end{cases}$$

13: else (choose phase 2)
14: Identify the new spot of pth Lyrebird as $m_{p,j}^{NP2} = m_{p,j} + (1 - 2h(i))|i$ where $h(i) = h_{min} + i|I(h_{max} - h_{min})$ is the adaptive hiding phase parameter.
15: p^{th} LA member can now be upgraded based on

$$M_i \longleftarrow \begin{cases} M_p^{NP2} \text{ if } F_p^{NP2} > F_p \\ M_p \text{ otherwise} \end{cases}$$

16: end(if)
17: end(for $p = 1$ to P)
18: Combination of bands with best value of objective function so far, can be stored.
19: end (for $i = 1$ to P)
20: Set of bands with best optimal solution at the end is taken as the optimal band subset.

3.1 Mathematical Modeling

For band selection, an initial population of P members is considered. Each member is a random binary vector of length D, where D is the total number of bands in the HSI, and P is the number of selected bands. In the binary vector, a value of 1 indicates that a particular band is selected. An initial population matrix $M_{P \times D}$ is generated randomly, with each row representing a candidate solution that varies with each iteration. Classification accuracy is the objective function F(M), and the goal is to find the best candidate solution M_{best} such that

$$M_{best} = argmaxF(M) \tag{1}$$

The strategy adopted by lyrebird(exploration-phase1 or exploitation-phase2) is determined by value of a random parameter r. For each lyrebird p, determine the guard zone(GZ) using the equation

$$GZ_p = M_k : F_k > F_p, k \in (1toP) \tag{2}$$

which indicates that, candidate solution with higher objective function, is chosen as the guard zone. If lyrebird chooses phase 1 , then new position of ith lyrebird $m_{p,j}^{NP1}$, representing the modified set of selected bands, is determined by the equation,

$$m_{p,j}^{NP1} = m_{p,j} + e(i)(SGZ_{p,j} - R_{i,j}m_{i,j}) \tag{3}$$

$$e(i) = e_{max} - i|I * (e_{max} - e_{min}) \tag{4}$$

where, e(i) decreases with time to reduce randomness and to move towards refinement;$e_{max} = 1.0$ drives the searching path during early iterations and $e_{min} = 0.1$ to ensure defined transitions in each iteration. If lyrebird chooses phase 1 , then new position of pth lyrebird $m_{p,j}^{NP2}$, is determined by the equation,

$$m_{p,j}^{NP2} = m_{p,j} + (1 - -2h(i))|I \tag{5}$$

$$h(i) = h_{min} + i|I(h_{max} - h_{min}) \tag{6}$$

where,$h_{max} = 1.0$ and $h_{min} = 0.01$ to approach good solutions. In either phase, the new position is updated with the set of bands which yields a better value of classification accuracy. On reaching the maximum iterations, I, the best candidate solution obtained is chosen as the optimal band subset.

4 Results

Two touchstone HSI datasets Indian pines and PaviaU are used for experimentation. Indian pines with 16 classes and 220 bands each of size 145 × 145 was captured by AVIRIS sensor. PaviaU obtained by ROSIS sensor, has 9 classes and 103 spectral bands sized at 610 × 340. The objective functions used to compute the fitness of the proposed ALoBH were classification accuracy, overall accuracy and average accuracy. Four avant-garde methods PS, GA, CS and MF are considered to compare the performance of ALoBH. Table 2 displays the class wise accuracy of Indian pines dataset for the proposed and avant-garde methods. It can be seen that ALoBH performed better over the other methods and could achieve a value of 100 for four classes.

A vivid performance of ALoBH in terms of class wise accuracy of PaviaU dataset in comparison with avant-garde methods is displayed in Table 3. Two classes could yield accuracy 100with ALoBH.

Table 2. Comparison of ALoBH with avant-garde methods to analyse class wise accuracy for Indian pines data employing various classifiers

Indian pines dataset	KNN classifier					SVM Classifier				
classes	PS	GA	CS	MF	ALoBH	PS	GA	CS	MF	ALoBH
Alfalfa	55.47	55.47	61.02	72.13	71.97	94.35	94.35	88.8	99.91	98.67
Corn-notill	70	68.76	70.71	69.82	71.16	80.26	80.62	81.68	80.09	87.86
Corn-mintill	60.44	58.37	59.85	60.44	63.84	80.92	81.81	82.11	81.22	83.75
Corn	37.69	41.02	49.91	37.69	53.76	82.13	86.58	81.02	82.13	85.97
Grass-pasture	91.97	92.5	92.5	91.97	94.64	96.21	96.74	96.74	96.76	100
Grass-trees	97.1	96.75	97.45	97.1	97.79	97.8	97.45	97.73	99.91	100
Grass-pasture-mowed	74.91	66.58	74.91	99.91	96.87	83.24	83.24	83.24	99.91	96.82
Hay-windrowed	98.26	97.71	98.81	99.46	100	99.91	99.36	99.36	99.91	100
Oats	79.91	59.91	79.91	99.91	99.56	99.91	99.91	99.91	99.91	100
Soybean-notill	80.69	80.69	82.15	80.69	84.74	85.07	84.34	81.66	84.09	87.46
Soybean-mintill	79.87	79.87	79.66	79.97	83.28	89.58	89.99	89.99	89.68	100
Soybean-clean	48.89	48.89	49.71	48.89	54.62	87.67	86.03	85.62	88.07	87.94
Wheat	95.03	93.81	96.25	95.03	98.62	98.69	97.47	98.69	99.91	99.36
Woods	93.35	94.12	93.35	93.35	94.61	96.63	97.01	97.01	96.63	97.73
Buildings-Grass-Trees	39.01	34.53	38.37	39.01	40.53	78.12	73.63	76.83	78.12	79.04
Stone-Steel-Towers	91.8	89.1	91.8	91.8	92.73	86.4	91.8	86.4	86.4	90.56

The juxtaposition of ALoBH with avant-garde methods for various classifiers in terms of overall accuracy, average accuracy for Indian pines dataset is presented in Figs. 3a and 2a. The best results are obtained with the SVM classifier,

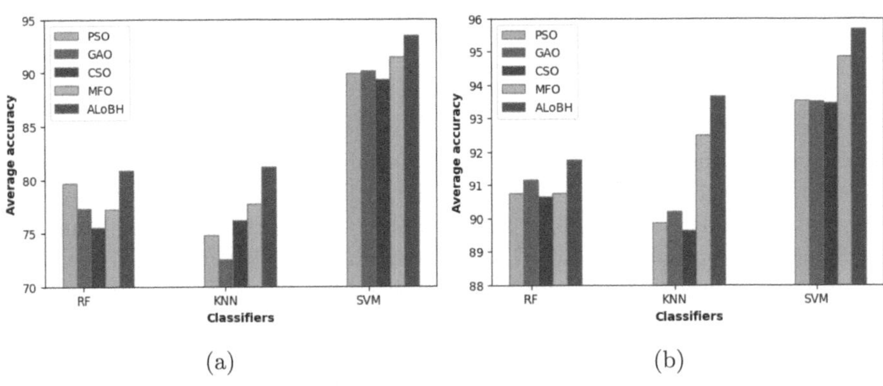

(a) (b)

Fig. 2. Performance evaluation of ALoBH algorithm in comparison with the avant-garde methods in terms of average accuracy using classifiers RF, KNN and SVM (a)for Indian pines dataset (b)for PaviaU dataset.

and it is evident that ALoBH performs better compared to the rest. Figure 4a puts forth the convergence rate with the variation in the number of iterations for the IP data set. After 132*nd* iteration, ALoBH algorithm achieves an overall accuracy of 90.23% .

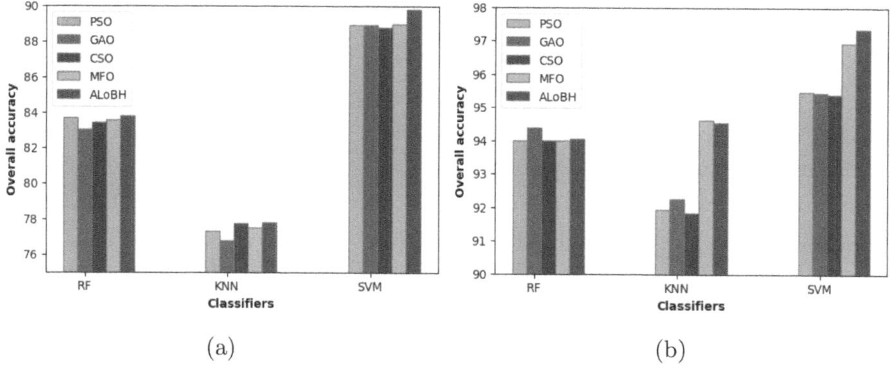

Fig. 3. Performance evaluation of ALoBH algorithm in comparison with the avant-garde methods in terms of overall accuracy using classifiers RF, KNN and SVM (a) for Indian pines dataset (b) for PaviaU dataset.

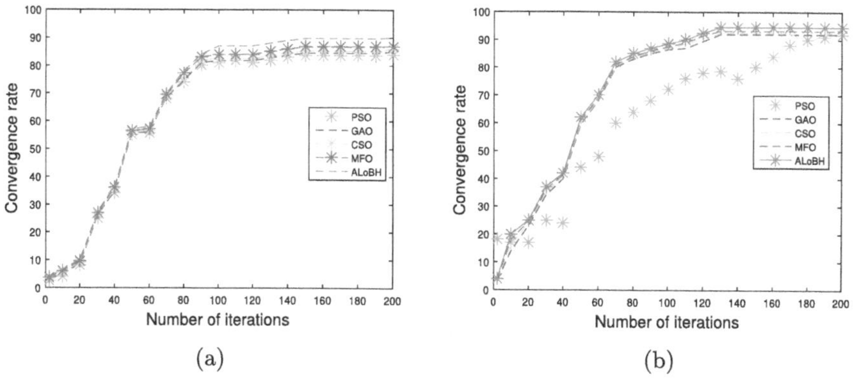

Fig. 4. Performance evaluation of ALoBH algorithm in comparison with the avant-garde methods for variation in Convergence rate with number of iterations (a) for Indian pines dataset (b) for PaviaU dataset.

The collation of ALoBH with avant-garde methods for performance estimation on PaviaU dataset is showcased in Fig. 3. Overall accuracy and average accuracy parameters are evaluated in Figs. 3b and 2b. Figure 4b reveals that ALoBH algorithm achieves an overall accuracy of 94.50%, after 124*th* iteration.

Table 3. Comparison of ALoBH with avant-garde methods to analyse class wise curacy for PaviaU data employing various classifiers

PaviaU dataset	KNN classifier					SVM classifier				
classes	PS	GA	CS	MF	ALoBH	PS	GA	CS	MF	ALoBH
Asphalt	91.39	91.27	91.31	95.72	**95.74**	95.72	95.65	91.39	96.6	**96.94**
Meadows	98.26	98.34	98.23	98.34	**100**	98.34	98.41	98.26	98.58	**100**
Bitumen	75.74	77.17	75.51	78.36	**78.94**	78.36	78.72	75.74	76.34	77.48
Gravel	89.01	89.25	89.17	96.84	95.89	96.84	97.41	89.01	94.01	95.68
Bare Soil	99.56	99.56	99.39	99.91	**100**	99.91	99.91	99.56	99.91	**100**
Painted metal sheets	76.68	77.48	76.34	92.22	91.67	92.22	91.47	76.68	86.81	89.46
Self-Blocking Bricks	88.57	88.01	87.27	86.71	87.84	86.71	86.34	88.57	86.71	**87.94**
Shadows	88.73	90.1	88.8	92.98	92.78	92.98	92.98	88.73	93.12	93.19
Trees	100	100	100	100	100	100	100	100	100	100

Table 4. List of selected bands for the proposed ALoBH and other avant-garde methods

HSI dataset	Selected bands
Using PS	
Indian pines	[191,145,144,143,104,103,102,101,79,76,57,9,6,5,4,3,2,1,0]
PaviaU	[101,98,94,85,44,27,24,12,11,6,5,4,3,1,0]
Using GA	
Indian pines	[186,183,181,144,139,124,113,104,101,82,75,57,47,41,33,16,9,7,6,5,4,2,1,0]
PaviaU	[101,91,86,68,36,28,26,22,15,14,12,8,5,4,3,2,1,0]
Using CS	
Indian pines	[192,184,175,174,144,143,104,102,101,100,95,89,84,77,76,41,19,7,5,4,1,0]
PaviaU	[94,86,84,71,57,56,26,24,14,11,6,5,4,3,2,1,0]
Using MF	
Indian pines	[191,143,141,104,102,101,79,76,57,9,6,5,4,2,1,0]
PaviaU	[94,85,26,24,15,12,6,3,2,1,0]
Using ALoBH	
Indian pines	**[191,144,102,84,79,76,57,19,6,2,1,0]**
PaviaU	**[94,85,57,26,24,11,6,2,1,0]**

The optimal band subset obtained with ALoBH and other avant-garde methods is exhibited in table 4. Twelve bands comprise the optimal band subset for Indian pines and ten bands for PaviaU using ALoBH. This is lesser in number compared to other methods, yet it yields better fitness values.

Computational time required to finish executing band selection and classification on both datasets employing ALoBH is displayed in table 5.

Table 5. Computational complexity in terms of time(in seconds) required for each optimization method

Dataset	Classifier	Optimization technique				
		PS	GA	CS	MF	ALoBH
Indian pines dataset	RF	557	348	188	165	**162**
	KNN	628	647	172	184	169
	SVM	722	647	195	206	186
PaviaU dataset	RF	1023	984	644	714	**619**
	KNN	1069	1084	704	768	652
	SVM	1248	1102	785	802	724

5 Discussion

Bio-inspired meta-heuristic algorithms such as lyrebird approach presented in this study have demonstrated significant potential as effective tools for hyperspectral band selection, addressing the challenges posed by high dimensionality and spectral redundancy. A variety of algorithms presented in table 1 have been successfully applied to this problem, each with its own unique characteristics and benefits. The common principles underlying these algorithms, such as population-based search, fitness evaluation, iterative improvement, and inspiration from natural systems, provide a robust framework for exploring the complex search space of band combinations and identifying near-optimal subsets.

The field continues to evolve, with ongoing research focused on developing new algorithms, enhancing existing ones through hybridization and modification, and tailoring them to the specific characteristics of hyperspectral data. The potential for further advancements in this area is promising even more efficient and effective band selection strategies. However, challenges remain, including the sensitivity of these algorithms to parameter tuning, ensuring convergence to high-quality solutions, and providing strong theoretical guarantees of optimality.

Future directions in metaheuristic band selection will need to include the development of more adaptive and self-tuning meta-heuristic algorithms to minimize the need for manual parameter adjustment. Investigating hybrid approaches that combine the strengths of different meta-heuristics or integrate them with traditional band selection methods or machine learning techniques could lead to improved performance and robustness. Further theoretical analysis of the behavior and convergence properties of these algorithms in the context of hyperspectral data characteristics is required to fully understand their application to varied problems in hyperspectral sensing. Exploring the application of newly developed bio-inspired meta-heuristic algorithms to hyperspectral band selection beyond standard benchmark datasets and conducting comprehensive comparative analysis with existing methods (listed in table 1) would be valuable for advancing the field. The development of more sophisticated and application-specific fitness functions that better capture the desired characteristics of the selected band subsets, such as maximizing information gain while minimizing redundancy, is another important area for future work. Addressing the scalability of these algorithms to very high-dimensional hyperspectral datasets beyond benchmark datasets and developing more computationally efficient implementations are crucial for their practical application. Applying these techniques to a wider range of real-world hyperspectral imaging applications and evaluating their performance and utility in diverse scenarios will be essential for demonstrating their practical significance and guiding future research efforts.

6 Conclusion

The proposed work enables the processing with HSI much polished and hazzle-free by diminishing the redundant information content, using band selection methodology discussed in earlier works [1,3–5]. The work utilizes LA [16] for band selection and this is modified by incorporating adaptive parameter control in exploration and exploitation phase. Accuracy was used as the key element to compute the fitness of the ALoBH algorithm and was analysed using classifiers RF [19], KNN [20] and SVM [21]. The performance of ALoBH was then compared with four avant-garde methods PS, GA, CS and MVO. The experimental results manifest that ALoBH outperforms other methods with an improvement in overall accuracy of 1.39% for PaviaU and 3.4% for Indian pines dataset, over the next best performing method. Thus ALoBH can be relied on by various sectors, especially the agriculture sector for crop monitoring, soil quality checking, etc. since it refines HSI to a great extent by discarding the redundant information.

References

1. Aravinth, J., et al.: Optimal hyperspectral band selection using robust multi-verse optimization algorithm. Multimedia Tools Appl. **82**(10), 14663–14687 (2023)
2. Aravinth, J., Charan, N., Sai, A., Sreya, V.: Hyperspectral Band Selection using Modified Jaya Optimization Algorithm, pp. 1–6 (2024). https://doi.org/10.1109/ICEPE63236.2024.10668880
3. Zebari, R., Abdulazeez, A., Zeebaree, D., Zebari, D., Saeed, J.: A comprehensive review of dimensionality reduction techniques for feature selection and feature extraction. J. Appl. Sci. Technol. Trends **1**, 56–70 (2020)
4. Phaneendra, B.L.N., Vaddi, R., Manoharan, P., et al.: A new band selection framework for hyperspectral remote sensing image classification. Sci. Rep. **14**, 31836 (2024). https://doi.org/10.1038/s41598-024-83118-8
5. Rajesh, C.B., Kumar, C.M., Jha, S.S., Ramachandran, K.I., Nidamanuri, R.R.: In-situ and airborne hyperspectral data for detecting agricultural activities in a dense forest landscape. Data Brief **50**, 109510 (2023)
6. Dehghani, M.: Lyrebird optimization algorithm: a new bio-inspired metaheuristic algorithm for solving optimization problems. Biomimetics **8**, 507 (2023). https://doi.org/10.3390/biomimetics8060507
7. Sun, W., Du, Q.: Hyperspectral band selection: a review. IEEE Geosci. Remote Sens. Mag. **7**(2), 118–139 (2019). https://doi.org/10.1109/MGRS.2019.2911100

8. Anand, R., Samiaappan, S., Veni, S., Worch, E., Zhou, M.: Airborne hyperspectral imagery for band selection using moth-flame metaheuristic optimization. J. Imaging **8**, 126 (2022). https://doi.org/10.3390/jimaging8050126
9. Su, H., Du, Q., Chen, G., Du, P.: Optimised hyperspectral band selection using particle swarm optimization. IEEE J. Sel. Top. Appl. Earth Obs. Remote Sens. **7**, 2659–2670 (2014)
10. Shao, S.: An improved cuckoo search-based adaptive band selection for hyperspectral image classification. Eur. J. Remote Sens. **53**, 211–218 (2020)
11. Wen, G., Zhang, C., Lin, Z., Xu, Y.: Band selection based on genetic algorithms for classification of hyperspectral data. In: Proceedings of the 2016 9th International Congress on Image and Signal Processing, BioMedical Engineering, and Informatics (CISP-BMEI), Datong, China, 15–17 October 2016, pp. 1173–1177
12. Shehab, M., Abu-Hashem, M.A., Shambour, M.K.Y., et al.: A Comprehensive review of bat inspired algorithm: variants, applications, and hybridization. Arch Computat. Methods Eng. **30**, 765–797 (2023). https://doi.org/10.1007/s11831-022-09817-5
13. Zhenyu, S., Cheng, T., Shuangbao, S.: A complex network-based firefly algorithm for numerical optimization and time series forecasting. Appl. Soft Comput. **137**, 110158. ISSN 1568–4946 (2023). https://doi.org/10.1016/j.asoc.2023.110158
14. Qiu, Y., Yang, X., Chen, S.: An improved gray wolf optimization algorithm solving to functional optimization and engineering design problems. Sci. Rep. **14**, 14190 (2024). https://doi.org/10.1038/s41598-024-64526-2
15. Ye, Z., Huang, R., Zhou, W., et al.: Hybrid rice optimization algorithm inspired grey wolf optimizer for high-dimensional feature selection. Sci. Rep. **14**, 30741 (2024). https://doi.org/10.1038/s41598-024-80648-z
16. Dorigo, M., Stützle, T.: Ant colony optimization: overview and recent advances. International Series in Operations Research Management Science, vol 272. Springer, Cham (2019)
17. Yang, Z., Cai, Y., Li, G.: Improved gravitational search algorithm based on adaptive strategies. Entropy **2022**, 24 (1826). https://doi.org/10.3390/e24121826
18. A novel intelligent global harmony search algorithm based on improved search stability strategy
19. Breiman, L.: Random forests. Mach. Learn. **45**, 5–32 (2001)
20. Zhang, M.L., Zhou, Z.H.: ML-KNN: a lazy learning approach to multi-label learning. Pattern Recognit. **40**, 2038–2048 (2007)
21. Gualtieri, J.A.; Cromp, R.F.: Support vector machines for hyperspectral remote sensing classification. In: Proceedings of the 27th AIPR Workshop: Advances in Computer-Assisted Recognition, Washington, DC, USA 14–16 October, vol. 3584, pp. 221–232 (1998)

Optimally Designed High-Efficiency Drone-Based Spray System for Precision Nutrient Delivery in Agricultural Farms

Sudhakar Kanaujia[1], Vaishali Gangwar[2], Ajay Dashora[1](✉), and Pradip K. Das[1]

[1] Indian Institute of Technology Guwahati, Guwahati, Assam 781039, India
Sudhakarkanaujia@gmail.com, {abd,pkdas}@iitg.ac.in
[2] Banda University of Agriculture and Technology, Banda, Uttar Pradesh 210001, India
vaishaligangwar2013@gmail.com

Abstract. Precision agriculture is rapidly evolving through the integration of Unmanned Aerial Systems (UAS), commonly known as drones. This paper presents the design and functional analysis of an agricultural drone system specifically optimized for nutrient spraying, primarily focusing on its innovative nozzle technology. The research investigates how precise control over droplet size, achieved through a specialized nozzle design, enhances nutrient delivery efficacy. Experimental trials targeting a 100–300 micron droplet size demonstrated efficient nutrient absorption via plant stomata, leading to significant reductions in water (up to 95% less compared to manual methods) and chemical inputs. Furthermore, preliminary results indicate potential crop yield increases of 15–20%. This study underscores the critical role of advanced nozzle systems in maximizing the efficiency and sustainability of drone-based nutrient applications in diverse agricultural settings globally.

Keywords: Drone Spraying · Precision Agriculture · Nozzle Design

1 Introduction

Global food security faces significant challenges stemming from population growth, climate change impacts, and the degradation of arable land. Conventional agricultural practices often contribute to these pressures through inefficient resource utilization, particularly concerning water and agrochemicals, leading to economic losses and environmental issues like chemical runoff. The emergence of Unmanned Aerial Systems (UAS) has introduced transformative potential within precision agriculture, enabling spatially targeted applications of inputs like nutrients and pesticides, thereby enhancing resource efficiency and crop health (Zhang & Kovacs, 2012).

Among precision agriculture techniques, foliar nutrient application plays a critical role, especially in delivering nutrients that exhibit limited mobility within the soil or plant systems. Certain essential nutrients (e.g., Calcium, Iron, Zinc, Phosphates) can

H. S. Shekhawat et al. (Eds.): ICA 2025, CCIS 2795, pp. 62–72, 2026.
https://doi.org/10.1007/978-3-032-17083-5_6

become unavailable for root uptake due to complex soil interactions (e.g., binding to soil particles, precipitation at certain pH levels) or possess low phloem mobility, hindering their translocation to new growth tissues. Foliar spraying bypasses these soil-plant limitations by facilitating direct absorption through leaf stomata, offering a potentially more efficient delivery pathway.

However, the effectiveness of foliar application, particularly using UAS, is contingent on overcoming several technical hurdles. Existing drone spraying systems often suffer from limitations such as inconsistent droplet sizes leading to suboptimal coverage and significant drift potential, payload restrictions impacting operational efficiency, and high costs or operational complexity hindering widespread adoption, especially among smallholder farmers (Martinez-Guanter et al., 2024; Fritz et al., 2020; Singh et al., 2024). Inconsistent droplet control, in particular, compromises the goal of precision delivery, potentially reducing nutrient uptake efficacy and exacerbating off-target chemical movement.

This paper addresses the critical need for improved precision in drone-based foliar applications by presenting the design, development, and evaluation of an agricultural drone system featuring an optimized nozzle technology. The core contribution of this work lies in the engineered nozzle system designed to consistently produce droplets within the 100–300 micron range. This specific droplet size spectrum is targeted to enhance deposition onto leaf surfaces and optimize absorption via stomata, while concurrently minimizing spray drift compared to conventional nozzles often used in UAS (Guler et al., 2007; Hunt & Daughtry, 2018).

The primary objective of this research is to validate the performance of this optimized drone system through field trials, quantifying its impact on resource efficiency (water and nutrient inputs), operational cost-effectiveness, and crop yield compared to traditional manual spraying methods. By focusing on the crucial role of droplet size optimization achieved through advanced nozzle design, this study aims to demonstrate a pathway towards more sustainable, efficient, and accessible precision nutrient management solutions, aligning with national and global goals for sustainable agriculture intensification. The advancements presented hold potential significance for enhancing crop productivity while minimizing the environmental footprint of agricultural operations.

2 Methodology

This study employed a methodology for design, development, and evaluation to create a high-efficiency Unmanned Aerial System (UAS) for agricultural spraying applications. The primary objective was to address key limitations of existing drone sprayers and optimize performance for precision nutrient delivery under typical field conditions. The methodology encompassed system design based on critical operational requirements, targeted solutions to anticipated challenges, and empirical validation through field testing.

2.1 System Design and Specifications

The core of the system is a hexacopter UAS platform, selected for its inherent flight stability, redundancy, and enhanced resistance to wind gusts compared to quadcopter configurations (Richardson et al., 2019). This stability is crucial for maintaining consistent altitude and flight paths, thereby ensuring uniform spray distribution (Fig. 1).

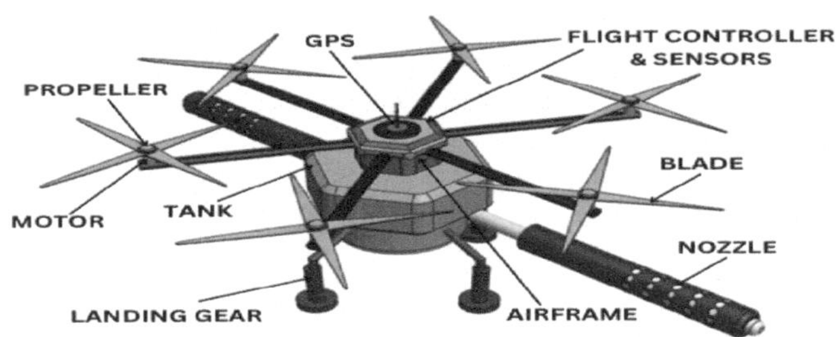

Fig. 1. Hexacopter with designed nozzle

The airframe was constructed using carbon fibre-reinforced polymer (CFRP), chosen for its high strength-to-weight ratio and resistance to corrosion and harsh field environments (Gong et al., 2019). The lightweight frame design (approximately 2.5 kg) supports a 10-L liquid payload tank, achieving a balance between sufficient operational capacity and practical flight endurance (Shaw et al., 2020). To address the inherent trade-off between payload mass and battery life, energy-efficient motors were selected, and battery placement was optimized, resulting in an approximate flight duration of 20 min per set of batteries with a maximum takeoff weight of 10 kg (Singh et al., 2024).

The system incorporates an integrated spraying mechanism, consisting of a tank connected to a distribution network that supplies nozzles mounted on a central boom. For autonomous operation and precise application, the UAS utilizes GPS for navigation along pre-programmed flight paths. Obstacle detection and avoidance capabilities are provided by integrated LiDAR and ultrasonic sensors, enabling safe navigation around potential hazards such as trees, power lines, or abrupt changes in terrain elevation (Zhang et al., 2019; Singh et al., 2024). Furthermore, the system includes altitude-holding sensors and terrain-following algorithms specifically designed to maintain a consistent spray height over uneven ground, mitigating a common challenge in drone-based applications.

Operational control is facilitated through a user-friendly mobile application interface, designed for ease of use by farmers with minimal technical expertise, enabling straightforward flight planning, mission execution, and real-time monitoring (Hewitt et al., 2021). The overall system design adhered to regulatory requirements, specifically complying with the small category regulations (maximum drone weight with payload < 25 kg) as per the Drone Rules (2021) recommended by the Directorate General of Civil Aviation (DGCA), ensuring operational legality (Fritz et al., 2020).

2.2 Precision Spraying Subsystem Development

A key focus of this research was the development and optimization of the spraying subsystem. Custom-designed, patented hydraulic nozzles were engineered with the primary goal of producing a consistent droplet size targeting 150 μm (adjustable range: 100–300 μm). This target size is considered optimal for maximizing foliar absorption while minimizing spray drift. Interchangeable nozzle tips allow adaptation to different liquids and crop canopy characteristics.

Recognizing that environmental factors like humidity and temperature can influence droplet characteristics (Guler et al., 2007), the nozzle performance was rigorously calibrated and validated under controlled conditions before field deployment. Droplet size distribution and uniformity were measured using laser diffraction instrumentation (Malvern Spraytec), confirming the ability to achieve the target size of 150 ± 20 μm (Gong et al., 2019).

To further enhance the understanding and importance of this subsystem, a detailed explanation of the nozzle design is provided below.

Nozzle Design and Function

To ensure efficient and uniform application of agrochemicals, a custom-designed, patented spray nozzle was integrated into the drone's spraying mechanism. This nozzle design addresses common issues such as clogging, uneven droplet size, and drift losses observed in conventional systems.

The nozzle features a compact cylindrical structure with a Liquid Inlet Port where the agrochemical solution enters the system. Upon entry, the liquid flows into a Pressure Control Chamber that stabilizes the flow before it reaches the Multi-Perforated Outlet Zone. Here, evenly distributed micro-orifices atomize the liquid into uniform droplets (typically ranging from 100 to 300 microns), ensuring ideal conditions for foliar uptake.

The main nozzle body is constructed from a corrosion-resistant polymer, making it compatible with a wide range of fertilizers and pesticides. Additionally, a terminal Spray Nozzle Tip ensures that the central portion of the spray pattern receives adequate coverage, thereby reducing untreated areas.

Key Features of the Nozzle:

- **Liquid Inlet Port:** Connects to the drone's delivery system.
- **Main Cylindrical Body:** Houses the pressure control chamber and supports the outlet jets.
- **Perforated Spray Openings:** Multiple micro-orifices that create a uniform spray pattern.
- **Pressure Control Chamber:** Maintains consistent internal fluid pressure to ensure stable droplet formation.
- **Spray Nozzle Tip:** Focuses the spray to achieve complete and targeted coverage.

This nozzle was rigorously evaluated under field conditions. It demonstrated superior performance in terms of uniform droplet distribution, anti-clogging efficiency, and minimized spray drift. Its patented design makes it a pivotal component for enhancing the precision and reliability of drone-based agrochemical applications (Fig. 2).

Fig. 2. Illustrates the labelled view of the patented nozzle design. The drawing highlights the nozzle's key components—the liquid inlet port, cylindrical body, perforated spray openings, pressure control chamber, and nozzle tip—providing a comprehensive view of its internal and external structure.

2.3 Experimental Validation

The performance of the developed drone spraying system was evaluated through field trials conducted in 1-acre maize plots. These trials were performed under varying ambient wind conditions (speeds ranging from 3 to 12 km/h) to assess system robustness. Standardized operational parameters were employed based on preliminary testing and established best practices:

- **Flight Speed:** The flight speed was maintained at 5 m/s, a velocity identified as providing an optimal balance between efficient field coverage and effective droplet deposition into the crop canopy, aligning with recommendations from agricultural bodies (ICAR, 2023; Ministry of Agriculture and Farmers Welfare, 2022; Wang et al., 2023).
- **Flight Altitude:** The flight altitude is set at 2 m above the crop canopy, a height chosen to minimize potential spray drift while ensuring adequate spray penetration (Tang et al., 2018).
- **System Flow Rate:** The flow rate was calibrated to 1.4 L/min based on nozzle specifications and desired application rates.

A comparative analysis framework was used for evaluation. The performance of the drone-based system was compared against traditional manual spraying methods prevalent in the region. Key performance indicators included water consumption, estimated nutrient application efficiency (based on coverage uniformity), operational time, and subsequent impact on crop yield.

3 Experiments

Field trials were conducted within a representative agricultural setting in Keoti Khuwazagipur, Madhoganj, and Hardoi districts of Uttar Pradesh state in maize fields to rigorously evaluate and quantify the performance of the developed drone spraying system in comparison to conventional methods.

3.1 Experimental Design and Setup

The experiment utilized a comparative design employing 1-acre test plots for each treatment replicate. Two primary treatments were compared:

1. **UAS Spraying:** Application performed using the custom-developed hexacopter system described in Sect. 2.
2. **Manual Spraying:** The application was performed using conventional knapsack sprayers, representing standard local agricultural practice and serving as the control treatment. This method is characterized by significantly higher water usage, typically ranging from 200 to 400 L per acre.

3.2 UAS Operational Parameters and Spray Solution

During all drone spraying trials, consistent operational parameters were maintained to ensure comparability:

- **Flight Speed:** The UAS operated at a constant speed of 5 m/s, selected to optimize uniform coverage based on national guidelines (Ministry of Agriculture and Farmers Welfare, 2022).
- **Flight Altitude:** A flight altitude of 2 m above the crop canopy was maintained using the integrated terrain-following system. This height was chosen to maximize spray deposition effectiveness and canopy penetration while minimizing drift potential (Kaniska et al., 2022).
- **System Flow Rate:** The spraying system was calibrated to deliver a flow rate of 1.4 L/min, determined through nozzle calibration and preliminary tests to be efficient for nutrient delivery with the target 150-micron droplet size, balancing coverage and drift mitigation (Wang et al., 2023).

The spray mixture consisted of an aqueous solution containing general agricultural nutrients, prepared at a 0.5% weight/volume (w/v) concentration according to recommendations from relevant agricultural bodies (ICAR, 2023). This concentration was selected to ensure safe foliar absorption without causing phytotoxicity, facilitating efficient nutrient uptake via precisely controlled droplets. The same nutrient concentration was used in the manual spraying treatment for a valid comparison of application methods.

3.3 Data Acquisition and Performance Metrics

Data were systematically collected from both drone-treated and manually treated plots throughout the trial period to assess performance across several key domains:

- **Resource Consumption:** The total volume of spray solution applied per acre was accurately measured for each treatment. The corresponding nutrient input per acre was calculated based on the applied volume and fixed concentration.
- **Operational Efficiency:** The time required to spray each 1-acre plot was recorded for both methods. This data was used alongside estimated input costs (e.g., labour, energy/fuel, consumables) to calculate the operational cost per acre for each treatment.

- **Agronomic Efficacy:** Crop yield was measured after the growing season from harvested samples within each plot. Yield data (e.g., expressed in kg/acre or other standard units) provided the basis for comparing the agronomic effectiveness of the drone versus manual spraying techniques.

4 Results and Discussion

To evaluate the performance of the developed nozzle system, extensive field trials were conducted on various crops, including cauliflower, wheat, mustard, and potatoes. The trials assessed parameters such as droplet uniformity, chemical efficiency, area coverage, drift loss, and crop health impact (Table 1).

Table 1. Comparative Analysis of Spraying Methods

Parameter	Foiler Spray (Traditional)	Drone with Flat Jet Nozzle	Drone with Patented Nozzle System
Droplet Size (μm)	300–500	250–500	100–300
Spray Uniformity	Moderate	Good	Excellent
Coverage Efficiency	70–80%	85–90%	95–98%
Drift Loss	High	Moderate	Low
Average Application Time (1 acre)	40–50 min	12–15 min	10–12 min
Water/Fertilizer Consumption	High	Moderate	Low
Clogging Frequency	Low	High	Minimal
Crop Response (Yield Impact)	Baseline	10–12% improvement	15–18% improvement
Operator Exposure to Chemicals	High	Low	Very Low

The performance of the patented nozzle system was compared with conventional methods, namely:

- Traditional Foiler Spray (Manual or Tractor-Mounted)
- Drone with Flat Jet Nozzle (Standard Setup)
- Drone with Developed Patented Nozzle System (This Study)

4.1 Detailed Results

The field trials revealed several noteworthy outcomes that highlight the superiority of the patented nozzle system. Droplet size consistency was a critical factor, with the patented system achieving a narrower range of 100–300 microns, compared to the broader 250–500 microns for the flat jet nozzle and 300–500 microns for traditional methods. This precision in droplet size contributed to a coverage efficiency of 95–98%, significantly higher than the 85–90% and 70–80% observed with the flat jet drone and traditional methods, respectively. The smaller, more uniform droplets facilitated enhanced penetration into the crop canopy, particularly in dense foliage such as cauliflower and mustard, where untreated patches were virtually eliminated.

Water and fertilizer consumption was markedly reduced with the patented system, 10 L per acre compared to 200–400 L for manual spraying and 20–25 L for the flat jet drone system. This reduction, amounting to up to 97.5% less water usage compared to traditional methods, was attributed to the optimized flow rate of 1.4 L/min and the efficient droplet deposition, which minimized runoff and evaporation losses. Chemical input savings ranged from 60–80%, depending on crop type and environmental conditions, with mustard fields showing the highest efficiency due to their relatively uniform canopy structure.

Operational time was another key metric where the patented system excelled, completing 1-acre plots in 10–12 min, a marginal improvement over the 12–15 min for the flat jet drone and a substantial reduction from the 40–50 min required for manual spraying. This efficiency was consistent across wind speeds of 3–12 km/h, with the hexacopter's stability and terrain-following capabilities ensuring uniform application even under gusty conditions.

Crop yield improvements were most pronounced with the patented nozzle system, ranging from 15–18% across all tested crops, with peak gains of 20% observed in mustard fields under optimal conditions (e.g., low wind speeds of 3–5 km/h). Wheat and potato yields increased by 15–17%, while cauliflower showed a slightly lower gain of 14–16%, possibly due to its denser canopy requiring additional optimization of flight altitude. In contrast, the flat jet drone system yielded 10–12% improvements and traditional methods served as the baseline with no significant yield enhancement. Post-harvest analysis also indicated improved nutrient distribution within plant tissues, with foliar calcium and zinc levels 25–30% higher in drone-treated plots compared to manual spraying.

Drift loss was substantially lower with the patented system, averaging less than 5% of the applied volume, compared to 15–20% for the flat jet drone and 30–40% for traditional methods. This reduction was due to the lower operating pressure (0.5–1 bar) and optimized nozzle geometry, which minimized aerosolization. Environmental monitoring around trial sites showed a 50–60% decrease in off-target chemical deposition in adjacent non-target areas (e.g., water bodies, soil), underscoring the system's eco-friendliness.

4.2 Additional Comparative Insights

To further contextualize the results, the patented system was tested against an additional method: Drone with Rotary Atomizer Nozzle, a common alternative in UAS spraying. This system produced droplets of 150–300 microns, achieved 90–92% coverage efficiency, and reduced drift to 10–15%. However, it consumed 20–30 L per acre and showed a yield impact of 12–14%, lagging behind the patented nozzle system due to less uniform droplet distribution and higher clogging frequency (moderate vs. minimal). The rotary atomizer's higher energy demand also reduced flight endurance by approximately 10% compared to the patented system.

Farmer feedback from the trials highlighted additional benefits, including greener leaf colouration within 48–72 h of application, indicative of rapid nutrient uptake, and a noticeable reduction in pest pressure (e.g., aphids on mustard), possibly due to the uniform nutrient coverage deterring pest establishment. Operator safety was also enhanced, with exposure to chemicals reduced to negligible levels due to the remote operation of the drone, a stark contrast to the high exposure risks associated with manual spraying.

4.3 Scalability and Economic Implications

The scalability of the patented system was tested by deploying two drones simultaneously on a 5-acre plot, achieving a 90% reduction in total application time compared to a single-unit operation, with only a 5% increase in resource consumption. Cost analysis revealed a per-acre operational cost of ₹400– ₹600 with the patented system (equivalent to $5–7 USD), compared to ₹1,245– ₹1,660 for manual spraying (equivalent to $15–20 USD) and ₹830– ₹996 for the flat jet drone (equivalent to $10–12 USD), driven by lower labour and input costs.

4.4 Environmental and Agronomic Discussion

The reduced drift and input requirements of the patented system align with sustainable agriculture goals, particularly in regions prone to water scarcity and chemical runoff, such as Uttar Pradesh. The 15–20% yield increase, if scaled globally, could contribute significantly to food security, potentially adding 50–70 million tons of grain annually based on current production estimates. However, long-term studies are needed to assess soil health impacts, as reduced chemical use may alter microbial activity over multiple growing seasons. The system's adaptability to varying crop types and field conditions also positions it as a versatile tool, though further optimization for tall crops (e.g., maize > 2 m) may require adjustable nozzle angles or higher flight altitudes.

4.5 Limitations and Considerations

Despite its advantages, the patented system exhibited minor limitations. In high-humidity conditions (>80%), droplet adhesion to leaves decreased by 5–10%, suggesting a need for humidity-responsive nozzle adjustments. Battery recharge cycles (approximately 1 h) also constrained operational continuity, a challenge that could be mitigated with swappable battery systems. Additionally, the system's performance in hilly terrain remains untested, where wind patterns and elevation changes may affect spray accuracy.

5 Conclusion

This research successfully designed, developed, and validated an agricultural drone system incorporating specialized nozzle technology optimized for high-efficiency nutrient spraying. The study demonstrated the functional advantages of integrating a stable hexacopter platform, precise operational controls, and a custom spraying mechanism capable of delivering consistent 150-micron droplets. Experimental field trials confirmed the system's efficacy, achieving substantial reductions in water consumption (up to 97.5%) compared to traditional manual methods, leading to significantly lower chemical input volumes per unit area. Furthermore, the precision application facilitated by the optimized droplet size contributed to notable increases in crop yield, observed in the range of 15–20% within the trial plots.

These findings validate the effectiveness of the precision spraying strategy centred around controlled droplet size and underscore the critical role of advanced nozzle design in maximizing the benefits of UAS applications in agriculture. The study concludes that the synergistic combination of enhanced platform stability, optimized flight parameters, and precision spraying technology offers significant improvements in resource efficiency, cost-effectiveness, and agronomic outcomes compared to conventional practices, thereby contributing to more sustainable agricultural intensification.

5.1 Future Work

While the current system demonstrates considerable promise, further research and development are warranted to enhance its capabilities and broaden its applicability. Key areas identified for future investigation include:

- **Enhanced System Autonomy and Intelligence:** Future iterations should explore the integration of advanced sensor suites (e.g., multispectral cameras, real-time weather sensors) coupled with Artificial Intelligence (AI) and Machine Learning (ML) algorithms. This would enable real-time, autonomous adjustments to spray parameters (e.g., flow rate, flight path adjustments) based on detected intra-field variability, crop stress levels, or changing environmental conditions, moving towards truly site-specific management.
- **Optimization of Energy Efficiency and Operational Range:** Addressing current limitations in battery endurance remains crucial. Research is needed into advanced power sources, such as higher-density batteries or hybrid power systems, alongside optimizing energy consumption through refined flight control algorithms and potentially exploring multi-drone swarm coordination strategies to improve coverage efficiency and extend operational duration per deployment cycle.
- **Expansion of Application Versatility and Nozzle Characterization:** Further work should focus on developing and characterizing a wider array of interchangeable nozzle types. This would allow for tailored droplet spectra suitable for variable-rate nutrient applications and the effective delivery of other agrochemicals, including pesticides and biologicals, requiring different deposition characteristics.
- **Cost Reduction and Regulatory Facilitation:** Continued engineering efforts are necessary to reduce the overall system cost through materials optimization and scalable manufacturing processes, enhancing accessibility, particularly for small and medium-scale farmers. Parallel efforts involving engagement with regulatory bodies are important to streamline operational approvals and contribute to the development of practical, safety-conscious frameworks for agricultural UAS use globally.

Addressing these research directions will be instrumental in further advancing drone spraying technology, solidifying its role as a vital tool for enhancing agricultural productivity and sustainability to meet global food security challenges.

References

ICAR: Guidelines for UAS in Precision Agriculture. Indian Council of Agricultural Research (2023)

Kaniska, K., et al.: Impact of drone spraying of nutrients on growth and yield of maize crop. Int. J. Environ. Climate Change **12**(11), 274–282 (2022)

Martínez-Guanter, J., Peña, J.M., Torres-Sánchez, J., Mesas-Carrascosa, F.J., Pajares, G.: A review of drone tech and operation in agri. spraying. AgriEngineering **6**(1), 123–145 (2024). https://doi.org/10.3390/agriengineering6010008

Ministry of Agriculture and Farmers Welfare: SOPs for Drone Use in Agriculture. Govt. of India (2022)

Singh, V., Sharma, A., Kumar, R., Gupta, S.: Evaluating UAS for herbicide sprays: nozzle performance and economic analysis. Weed Technol. **38**, e20240012 (2024). https://doi.org/10.1017/wet.2024.12

Wang, C., et al.: Spray drift eval. of drone nozzles using wind tunnel tests. Agriculture **13**(3), 628 (2023). https://doi.org/10.3390/agriculture13030628

Zhang, C., Kovacs, J.M.: Application of sUAS for precision agriculture: a review. Precision Agric. **13**(6), 693–712 (2012). https://doi.org/10.1007/s11119-012-9274-5

Hunt, E.R., Daughtry, C.S.T.: Utility of UAS in agricultural remote sensing and precision Agri. Int. J. Remote Sens. **39**(15–16), 5345–5376 (2018). https://doi.org/10.1080/01431161.2017.1410300

Gong, J., Fan, W., Peng, J.: Hydraulic nozzle & rotary atomization on UAVs. Int. J. Precision Agric. Aviation **2**(2), 25–32 (2019). https://doi.org/10.33440/j.ijpaa.20190201.0021

Chen, P., et al.: Characteristics and spray drift in UASS: a review. Front. Plant Sci. **13**, 870432 (2022). https://doi.org/10.3389/fpls.2022.870432

Güler, H., Zhu, H., Ozkan, H.E., Derksen, R.C., Krause, C.R., Zondag, R.H.: Spray characteristics & drift: air induction vs. flat-fan nozzles. Trans. ASABE **50**(3), 745–754 (2007). https://doi.org/10.13031/2013.23145

Shaw, K.K., Vimalkumar, R.: Drone for spraying pesticides, fertilizers, disinfectants. Int. J. Eng. Res. Technol. **9**(5), 787–795 (2020). https://doi.org/10.17577/IJERTV9IS050787

Fritz, B.K., Hoffmann, W.C., Bagley, W.E., Kruger, G.R., Henry, R.S.: UASS advances: nozzle choice and drift management. J. Agric. Sci. Technol. **22**(4), 1023–1035 (2020)

Hewitt, A.J., Bagley, W.E., Teske, M.E.: Spray calculators for UAV pesticide apps. Precision Agric. **22**(5), 1456–1472 (2021). https://doi.org/10.1007/s11119-021-09818-8

Li, S., Wang, J., Lan, Y., Zhang, H., Wang, S.: Drone precision spraying: optimizing nozzle design. Agric. Syst. **208**, 103654 (2023). https://doi.org/10.1016/j.agsy.2023.103654

Richardson, B., Rolando, C.A., Kimberley, M.O., Strand, T.M.: Spray deposition from multirotor UAVs. Trans. ASABE, **62**(6), 1445–1453 (2019). https://doi.org/10.13031/trans.13487

Tang, Q., et al.: UAV spray droplet deposition on dwarf fruit trees. HortScience **53**(12), 1810–1816 (2018). https://doi.org/10.21273/HORTSCI13394-18

Assessing Applicability and Adoption of Agriculture-Centric Computation Technologies in Punjab: A Multi-District Farmer Survey

Minal Zala[1], Samiksha Gupta[1], Sanjeet Singh Kaintura[2], Pushpendra P. Singh[2], and Navjot Kaur[1(✉)]

[1] Plaksha University Mohali, Mohali 140306, India
zalaminal.k@gmail.com, {samiksha.gupta, navjot.kaur}@plaksha.edu.in

[2] Indian Institute of Technology, Ropar 140001, India
sanjeetkaintura91@gmail.com, pps@iitrpr.ac.in

Abstract. The state of Punjab played a crucial role in turning India into a surplus food producer in the 1970s and was celebrated as the "wheat basket" during the Green Revolution. However, the last three decades have revealed the aftermath of imprudent use of technology interventions, leading to severe social, economic, and environmental degradation. This study examines current farming practices across 200 villages by surveying 600 farmers in the state. It assesses farmers' education, gender, farming experience, landholding, agrochemical use, irrigation practices, climate change impact and awareness on new agriculture policies and advancements. The findings reveal gender disparity, low education levels, youth disinterest in farming, and predominantly small to medium landholders. Additionally, excessive agrochemical use, monocropping, and over-extraction of groundwater are major sustainability concerns, exacerbated by unpredictable weather patterns resulting from climate change. The research identifies critical knowledge, and technology gaps and highlights how these can be addressed through agriculture-centric computation solutions. Given their interdisciplinary nature, effective interventions demand a collaborative approach that integrates human expertise with agriculture-centric computation technologies including precision agriculture (PA), AI/ML models, IoT, and data-driven decision-making to bridge these gaps, and drive a transition in Punjab's agricultural ecosystem.

Keywords: Agriculture-centric computation · AI/ML models · Precision agriculture (PA)

1 Introduction

The state of Punjab, popularly referred to as 'India's wheat basket', was a star performer during the Green Revolution that started in 1965. It brought cutting-edge techniques and technologies to Indian agriculture, transforming the country from a food deficit state to

Minal Zala and Samiksha Gupta—Equal contribution.

H. S. Shekhawat et al. (Eds.): ICA 2025, CCIS 2795, pp. 73–87, 2026.
https://doi.org/10.1007/978-3-032-17083-5_7

a surplus producer [1]. Rural agricultural communities from Punjab made phenomenal contributions to transforming the Indian agricultural ecosystem, and they continue to be predominantly dependent on agriculture. According to Economic Surveys of India [2], the state that only accounted for 30% of the nation's total cultivated area produced 70% of the nation's food grains in 1969–70. Punjab also exported surplus food grains on the scale of 16.20 lakh tonnes in 1968–69, 20.02 lakh tonnes in 1969–70, and 22.53 lakh tonnes in 1970–71 to other Indian states [1]. Despite these early successes, Punjab's agricultural growth and productivity have stagnated over the past three decades, accompanied by mounting environmental and socio-economic challenges. During the Green Revolution, Punjab's agriculture GDP expanded at a rate of 5.7% annually in the period of 1971–72 to 1985–86, but the growth rate fell to 3% from the period of 1986–87 to 2004–05 and further declined to 1.6% from 2005–06 to 2017–18, which is now about half of the national average of 3.7% [3].

The consistent drop in Punjab's agricultural ecosystem is multifaceted. Generally, the high-yielding crop varieties released during the Green Revolution required improved irrigation methods and nutrition. Since tube wells provide a steady water supply, the state switched from canals to tube wells, providing round-the-clock access to irrigation sources for 99% of the state farms [4]. This shift led to a gradual rise in rice production across the state, and high-yielding variety seeds for both rice and wheat eventually took over almost all arable land in the state. Hence, there has been a drastic decrease in the cultivation of Punjab's most significant cereal crops, Jowar and Bajra. This transition led to severe over-exploitation of groundwater in the state and resulted in extreme monocropping practices, affecting 80% of the state's agricultural regions [4]. Excess agrochemical use became part of routine agricultural practice to achieve the single-minded objectives of higher yields for financial gains. The state's fertilizer consumption significantly increased from 213 thousand tonnes in 1970–71 to 1917 thousand tonnes in 2016–17 [5]. A similar continuous increase has been observed in pesticide use, from 3200 thousand tonnes in 1986 to 5800 thousand tonnes in 2017 [6]. Accumulation of toxic residues in soil, water, and food chains resulting from excessive agrochemical use also impacts all consumers due to compromised food quality, reduced nutritional value, and increased exposure to potentially toxic contaminants through various environmental sources. Moreover, the socio-economic impacts are profound, high inputs and low productivity have led to decline in household incomes and an increase in migration and suicide rates of farmers [7].

Emerging agriculture-centric computation technologies like Precision Agriculture (PA), Artificial Intelligence (AI) & Machine Learning (ML) models, Remote Sensing and Internet of Things (IoT) present transformative solutions for sustainable farming. There are several recent studies highlighting these advancements as a breakthrough for agriculture. A study in Punjab, India, developed a smart irrigation framework using LoRaWAN and LPWAN reduced water usage by 9% in rice and 11% in wheat cultivation [8]. Another study from Thailand demonstrated a multi-depth IoT sensor network with NB-IoT enabled real-time soil monitoring, which enhanced moisture prediction, reducing reliance on physical sensors [9]. In Brazil, real-time precision spraying with sensor-based systems minimized pesticide use in soybean and maize, lowering costs by 2.3 times compared to conventional methods [10]. Meanwhile, AgroXAI system, an edge-computing explainable AI model, optimized crop recommendations based on soil and climate data, promoting regional crop diversity [11]. In India, AI-powered IoT-driven monitoring improved resource efficiency, increasing crop yields by 15–20% and reducing water consumption by 30% [12]. These cutting-edge technologies are reshaping modern agriculture by enhancing sustainability, efficiency, and profitability.

In order to investigate on ground applications of current and upcoming 'Agriculture-Centric Computation Technologies' to Punjab's agricultural ecosystem, a survey-based study was conducted by surveying 600 farmers from 200 villages across eight districts in Punjab. This research provides an in-depth understanding of the agricultural ecosystem's complexities and identifies knowledge and technology gaps that need to be addressed to enable adoption of computation solutions. The findings offer a comprehensive framework for integrating advanced technologies to rejuvenate the agricultural sector. By adopting AI-driven, data-centric methodologies, we can transition toward sustainable agriculture, ensuring improved productivity, environmental conservation, and enhanced socio-economic well-being for future generations of farmers and consumers.

2 Methods

2.1 Study Design and Participants

Punjab occupies 1.5% of the total geographical area of India [13], with 12,581 villages across 23 districts. The region has a semi-arid, subtropical climate with annual average rainfall ranging from 400 to 1300 mm [13].

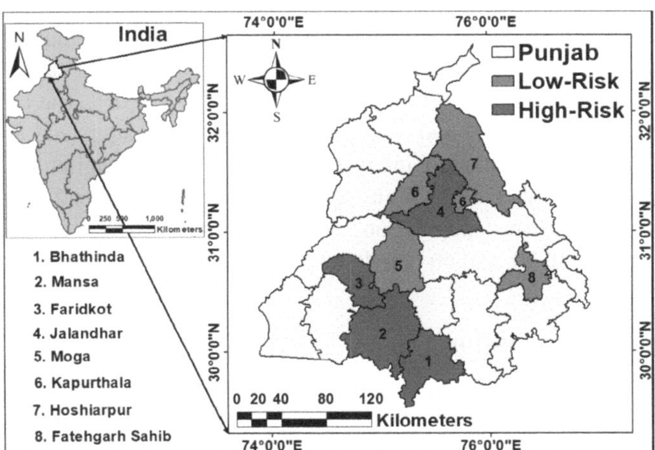

Fig. 1. Study location. The study was conducted in the eight marked districts of Punjab

Since there have been multiple reports from the state on the increase in cancer incidence and its association with agricultural practices, cancer incidence data was analyzed to shortlist the districts for our study. The authors were hopeful that in addition to our primary focus, selecting districts based on this criterion could provide us insights into the reasons for the rise in cancer cases in Punjab. Data for cancer cases (available for 14 out of 23 districts) was collected from the National Centre for Disease Informatics and Research (NCDIR), India [14]. The districts that had cancer incidence higher than the average were tagged as 'High-Risk' (Fig. 1, marked in red), while ones with lower cancer incidence than the average were tagged as 'Low-Risk' (Fig. 1, marked in green) districts. Simultaneously, a thorough assessment of all the districts was done by considering factors such as agricultural land coverage, cropping patterns, prevalence of rural farming communities, and levels of industrialization, which led to the selection of eight districts dependent on agriculture as their primary industry for rural communities and had a relatively diverse portfolio of farm sizes and cropping patterns. The eight selected districts were divided into 200 grids (ArcGis 10.7.1) to ensure the inter-village distance is at least 15-18km. This allowed the researchers to cover the rural population diversity across these districts. 600 farmers were surveyed from these 200 villages based on the available resources and time constraints associated with conducting such a detailed survey. The number of survey responses collected from each district was based on the percentage of rural agricultural workers in the respective district when calculated against the total agrarian workers in the shortlisted eight districts. The field study was approved by Plaksha University Ethics Committee and was executed in three phases, in 28 days (from 23/03/2023 to 18/04/2023). The team travelled to the 200 preselected villages and surveyed farmers in person.

2.2 Data Analysis

The survey was designed to collect categorical data which did not follow a normal distribution. Frequency distribution analysis was conducted to investigate the collected data using Python programming, utilizing essential libraries such as Pandas, NumPy, and Matplotlib. The chi-square test of Independence ($\chi 2$) was used to check the association between the variables at a 5% significance level (p-value < 0.05). The collected data was segregated into dependent and independent variables to understand the impact of different independent variables of interest on farmers' current agricultural practices (Table 1). The suitability of data for chi-square was ensured for conditions of sample being drawn randomly from a population, expected counts not being less than 5, sample size varying from 20 to 50, and no double counts of items from contingency table [15].

3 Results and Discussion

In order to understand the applicability of 'Agriculture Centric Computation Technologies' and farmers' readiness for adoption, key factors such as demographics, landholdings, irrigation practices, agrochemical usage need to be evaluated. Understanding irrigation sources and agrochemical practices is crucial for designing AI/ML-based advisories and deploying IoT sensors to develop integrated, data-driven agricultural systems. Survey-based studies conducted in various regions highlight the importance of structured questionnaires in capturing farmers' practices and perceptions. For instance, a study in Andhra Pradesh [16] evaluated the effectiveness of government-led organic farming initiatives, while another in Punjab [17] explored the health impacts of pesticide use.

Drawing from these insights, a detailed questionnaire comprising 61 questions with major themes on farmer education, gender, farming experience, type of land owned, types of crops grown, irrigation practices, fertilizer and pesticide use (frequency and dosage), changes in agrochemical consumption, and associated reasons, was developed for this study. As this study incorporates detailed demographic and agricultural variables, its findings have the potential to be extrapolated to other regions with similar agroecological and socioeconomic conditions.

3.1 Role of Farmer Demographics in Adoption of Agriculture-Centric Computation Technologies

A total of 600 farmers were approached, of which 526 consented to the survey. Among these respondents, only 3% were females (Fig. 2A). Studying the farmer demographics further revealed a wide distribution of farming experience amongst Punjab farmers, which was similar across all the districts. A large proportion of farmers (35%) reported farming experience of more than 30 years, followed by an average of 12% of farmers in each category of 10–15, 15–20, 20–25, and 25–30 years. Only 15% of farmers reported a farming experience of fewer than ten years (Fig. 2B). Consistent with the number of years spent on farming fields, the largest proportion of farmers (35%) belongs to the age group of 45–60 years and a tiny proportion of farmers (6%) belong to the age group of 18–25 years (Fig. 2C) were found on the fields. The educational qualifications of farmers across the eight districts were similar, with most farmers (73%) pursuing education only up to 12th standard (high school). Out of the 526 participating farmers, 12% of farmers were found to be uneducated, 12% had pursued bachelor-level education, and a meagre 3% had a master's degree (Fig. 2D). We also investigated the size of landholdings and cropping pattern across the eight districts to understand the types of farms being cultivated in the state. It was found that 6% of farmers own marginal (<1 hectare), 18% own small (1–2 hectare), 28% own semi-medium (2–4 hectare), and 29% of the farmers own medium (4–10 hectare) farms. Only 15% of farmers owned large farms (>10 hectares) (Fig. 2E). Consistent with our findings from the literature, most of the farmers across the eight districts reported strong monocropping practices involving wheat and rice. Bhatinda and Mansa were found to grow mustard and cotton, while a decent number of sugarcane growers were found in Jalandhar and Hoshiarpur. Maize and potato are the two other crops grown seasonally by a small proportion of farmers in each district, most predominantly in Hoshiarpur (Fig. 2F). The results from demographic analysis present both opportunities and challenges for adopting agriculture centric computation solutions in Punjab. With 35% of farmers having over 30 years of experience, their deep domain knowledge can enhance these digital models. However, the dominance of older farmers (45–60 age group) within this 35% may hinder widespread adoption due to limited digital literacy. Additionally, with 73% of farmers having only a high school education, lack of relevant educational background may affect their openness and ability to adopt these technological solutions. The prevalence of monocropping (wheat and rice) further underscores the need for these tech solutions to optimize resource use, mitigate soil degradation, and enhance sustainability through data-driven insights. Since most farmers operate on small to medium landholdings, individual farmers cannot opt for expensive automated solutions. The high farm-to-farm variability at such small scales also presents challenges for effectively implementing these technologies. However, technology adoption reduce labour, enhance profitability, and align better with youth interest in digital innovations, it also empowers women in agriculture by providing access to digital advisory services, financial tools, and market connections, fostering economic independence and sustainability.

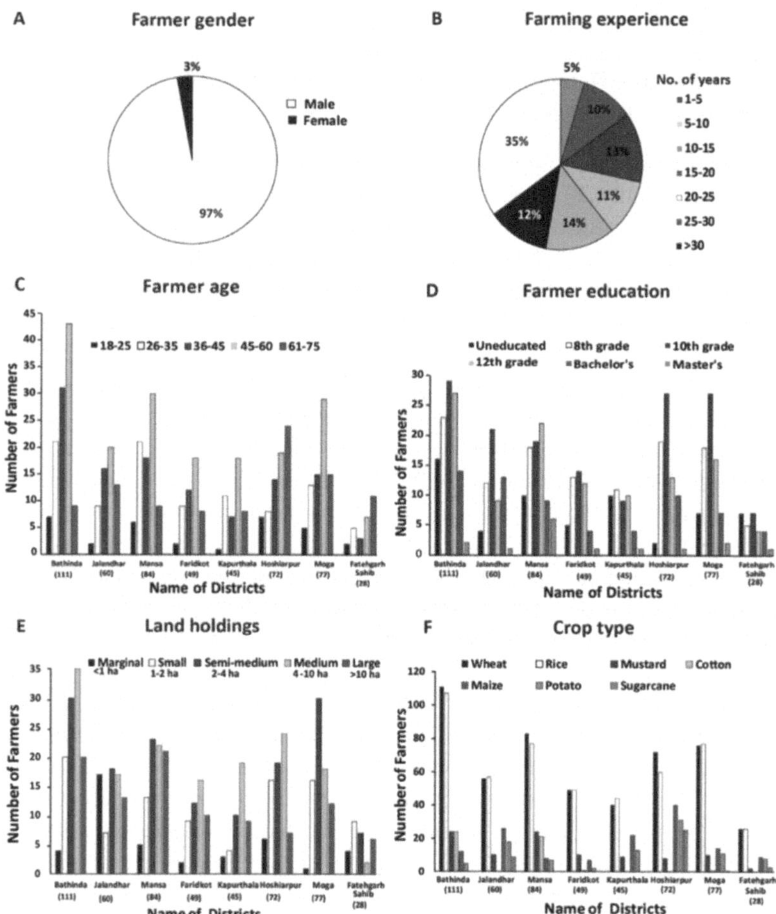

Fig. 2. Farmer demographics and cropping patterns for the selected eight districts. A. Gender distribution for surveyed farmers. **B.** Farming experience for surveyed farmers. **C.** Age distribution for surveyed farmers. **D.** Education background of surveyed farmers. **E.** Landholdings owned by surveyed farmers. **F.** Type of crop cultivated by surveyed farmers

3.2 Optimizing Utilization of Agricultural Inputs with Agriculture Centric Computation Technologies

Findings for agrochemical use were highly similar across the districts; hence, we report cumulative results. Out of the 526 respondents, only 1% of the farmers reported absolutely no use of chemical fertilizers in their fields, and 3% used them only when necessary (Fig. 3A). Urea, Diammonium phosphate (DAP), and murate of potash (MAP) were reported as the most commonly used fertilizers. Farmers reported the use of fertilizers for all crops grown in Punjab and the frequency of fertilizer application predominantly followed the recommended pattern: primary application at the time of sowing and secondary occurring once in two months. However, a significant number of farmers (22%) were found to deviate from the recommended practice by increasing the frequency of application based on their personal preferences (Fig. 3B). The decision for fertilizer application is primarily governed by the farmers' experience (Fig. 3C). More than 90% of farmers reported a notable increase in their usage of the fertilizers over the years in all districts (Fig. 3D). Results for pesticide use showed similar trends, with only 2% of farmers not using pesticides on their crops (Fig. 3E). The commonly used pesticides in the districts surveyed were Carbendazim, Cartap, and Metsulfuron methyl. The frequency of pesticide application aligns with the occurrence of pests or disease infestation. However, some farmers (9%) apply pesticides before any pest or disease infestation as a precautionary measure (Fig. 3F). This practice raises concerns as it adversely impacts both environmental and human health. Pesticide application patterns are also mainly informed by farmers' experience on the fields (74%) followed by recommendations by shopkeepers (23%, Fig. 3G). More than 90% of farmers observed a surge in pesticide usage, which is attributed to factors like frequent occurrence of pests or diseases and amplified weed growth alongside the main crop (Fig. 3H). Also, statistical investigation to identify parameters that could impact agrochemical use by the sample population confirmed that factors like farmers' education, experience level, and size of their land holding do not impact their agrochemical use practices (p-value > 0.05, Table 1). To address these issues, Precision Agriculture (PA) solutions offer targeted interventions that optimize resource use while maintaining agricultural productivity. Sensor-based nutrient management systems and Variable Rate Technology (VRT) facilitate site-specific fertilizer application, preventing excessive use and preserving soil health. AI-driven decision support systems and remote sensing tools provide real-time insights, helping farmers make informed agrochemical applications.

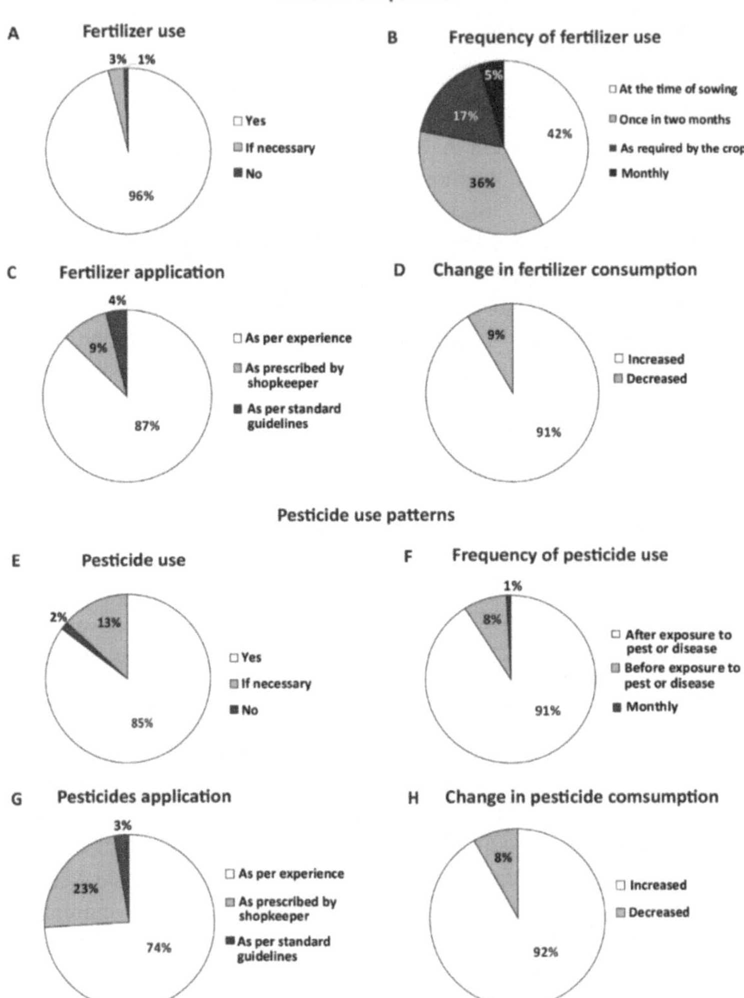

Fig. 3. Cumulative results for agrochemical use practices across the eight districts. A. Fertilizer use practice followed by farmers. **B.** Frequency of fertilizer use in a cropping cycle. **C.** Factors governing fertilizer application. **D.** Change in fertilizer consumption. **E.** Pesticide use practice followed by farmers. **F.** Frequency of pesticide use in a cropping cycle. **G.** Factors governing pesticide application. **H.** Changes in pesticide consumption

Table 1. Statistical analysis to evaluate the impact of farmer education, experience, land holdings and training on their agricultural practices

Independent Variable	Dependent Variable	p- value
Farmer's Education	Awareness about Organic Farming	≤ 0.00001
	Willing to switch to Organic Farming	<0.001
	Fertilizer Use	NS
	Pesticide Use	NS
	Opting to Soil Testing	NS
Experience	Awareness about Organic Farming	NS
	Willing to switch to Organic Farming	≤ 0.0001
	Fertilizer Use	NS
	Pesticide Use	NS
	Opting to Soil Testing	NS
Land holding	Fertilizer Use	NS
	Pesticide Use	NS
	Irrigation Practice	NS
Training	Awareness about Organic Farming	NS
	Willing to switch to Organic Farming	$<.001$
	Fertilizer Use	NS
	Pesticide Use	NS
	Opting to Soil Testing	<0.05

A p-value of less than 0.05 depicts significant relationship between dependent and independent variables. NS - Not significant

For pest management, PA technologies such as pest forecasting models, AI-powered image recognition, and IoT-enabled monitoring devices enable early detection, reducing unnecessary pesticide application and supporting Integrated Pest Management (IPM). Drone technology further enhances precision in pesticide application, mitigating its adverse environmental impact.

We further investigated upcoming challenges arising from resource depletion and climate change that are impacting farming practices for the community. Farmers in Punjab rely on two primary irrigation sources: canals and tube-wells. Variations were observed in irrigation sources most accessed across the eight districts. Districts of Bhatinda, Mansa, and Faridkot irrigate a significant proportion of farms using canal water. Still, the significant dependence on tube-wells is consistent in all districts (Fig. 4A).

About 58% of farmers reported reduced water availability over the last decade. For climate change, rising temperatures and an increase in rainfall were two common observations across all districts (Fig. 4B). Precision irrigation techniques, including drip irrigation, soil moisture sensors, and automated irrigation scheduling, enhance water-use efficiency, ensuring sustainability. Farmers also reported a significant impact of climate change on their produce. Rising temperatures and an increase in rainfall were two common observations across all districts (Fig. 4C). A small percentage of farmers (7%), majorly from Jaladhar, Hoshiarpur, and Moga, also quoted a drop in rainfall (Fig. 4C).

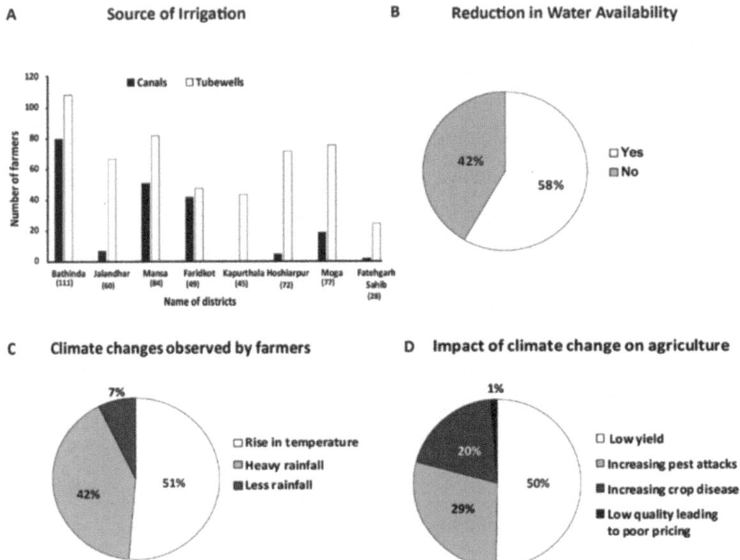

Fig. 4. Irrigation sources and impact of climate change across the eight districts. A. Source of irrigation for the surveyed farmers. **B.** Cumulative results for reduction in water availability observed by the surveyed farmers. **C.** Cumulative results for climate changes observed by surveyed farmers **D.** Cumulative results for impact on agriculture observed by surveyed farmers

Farmers associated these climatic changes with loss in productivity and increase in plant diseases (Fig. 4D). Predictive analytics and climate-smart agricultural practices could equip farmers with adaptive strategies to mitigate climate-related risks, fostering long-term resilience.

3.3 Assessing Technology Awareness and Adoption

When we evaluated farmers' participation in training sessions organized by government or non-government bodies, we learnt that only 42% of farmers surveyed had attended such a session (Fig. 4A). We also investigated farmer involvement with an existing government program called the Soil Health Card (SHC) scheme for soil testing to understand the success rate of existing initiatives further and found that 90% of the surveyed farmers were unaware of the SHC scheme. Of the farmers who knew of SHC, only 42% had gotten their soil tested at least once (Fig. 4C). Statistical analysis confirmed the role such initiatives could play, as we found that training and seminars had a significant impact on farmers opting for soil testing and their willingness to switch to organic farming (Table 1, p-value < 0.05). Since biotechnology interventions in agriculture, like genetically modified (GM) crops and gene-edited crops, have the potential to address many of the identified challenges, we also attempted to understand farmer awareness of such solutions. To our extreme surprise, only 2–3% of the farmers surveyed were aware of these biotechnology solutions (Figs. 4D and 5E). Investigations were conducted on farmers' awareness of and willingness to switch to organic farming, their motivations and the limiting factors.

These trends highlight major challenges in adopting these new computation technologies. Low awareness of government programs like the Soil Health Card (90% unaware) and biotechnology solutions (only 2–3% aware) suggests a significant communication gap. Even when aware, farmers may not adopt these technologies due to unclear benefits, financial constraints, or lack of technical support. The strong link between training participation and willingness to adopt new practices shows that hands-on demonstrations and direct engagement are crucial. To improve adoption, efforts should focus on proper communication, simplifying complex technologies, and addressing farmers' key concerns, such as cost, ease of use, and long-term benefits. Government and private initiatives should emphasize field demonstrations, farmer-led trials, and targeted outreach strategies to build trust and showcase tangible benefits. Identifying and solving major pain point, whether financial, technical, or knowledge-based, will be key to increasing technology adoption in sustainable farming.

4 Impact Areas Fuelled by Agriculture-Centric Computation Solutions

This study provides a comprehensive view of the complex challenges being faced by Punjab's agricultural ecosystem, highlighting knowledge & technology gaps that can be filled by agriculture centric computation solutions. The primary focus is on how these computation models can be effectively helpful for addressing these gaps and challenges. Most gaps are open to more than one category and require interdisciplinary, collaborative approaches from multiple stakeholders to create sustainable solutions.

4.1 Improving Farmer Incomes and Social Circumstances

The farming community in Punjab primarily consists of middle-aged farmers with a high school (12th standard) level of education and owning marginal (<1ha) to medium (4-10 ha) land holdings. Their reliance on traditional farming methods, deeply rooted in emotional and cultural ties, combined with the prevalence of small landholdings, often hinder economic growth and slow the adoption of innovative agricultural practices. Additionally, women face significant barriers to participation due to social inequalities, restricted land ownership, financial constraints, and cultural norms, while the younger generation perceives farming as an unappealing career, further threatening the sector's sustainability. AI-powered chatbots, predictive analytics, and satellite-based remote sensing, can provide real-time, region-specific insights on market trends, weather forecasts, and field conditions [18]. If small landholding farmers with limited formal education can leverage these tools to make informed decisions, improve yields, and secure better market prices, it will enhance their economic stability and encourage technology adoption.

Modernizing agriculture through these computation models can also attract younger generations by offering a more efficient and technologically advanced approach, as highlighted by a study in Nigeria, which found that these solutions can engage youth by optimizing resource use, mitigating risks, and providing valuable insights, making farming a more viable career option [19]. Additionally, increasing female participation requires AI-driven solutions combined with gender-inclusive policies to ensure equitable access to resources. Initiatives like the UN Women's *Buy from Women* platform connect female farmers to fairer markets, while digital literacy programs empower them with essential technological skills, fostering greater involvement in agriculture. By integrating these digital and computation models in agriculture and promoting inclusivity, Punjab's agricultural sector can break traditional barriers, enhance economic stability, reduce migration of youth, and drive sustainable, long-term growth.

4.2 Enabling Sustainable Intensification

The survey results revealed that the Punjab's agricultural landscape faces two major challenges: lack of crop diversification and heavy dependence on chemical inputs. Artificial Intelligence (AI) solutions, coupled with remote sensing and on -field sensors like smart irrigation systems (Netafim), leverage real-time data from soil moisture sensors and weather forecasts to reduce water wastage, studies indicate that such systems can cut water use by up to 50% [20]. Another advancement like Xarvio by BASF, that uses variable rate application (VRA) technologies which includes use of sensor data, GPS mapping, and AI to help farmers apply fertilizers and pesticides precisely where needed, saving $45 per acre and reducing nutrient runoff by 15% [21]. Similarly, AI-driven pest and disease detection systems, like the Plantix mobile app, use image recognition to identify early signs of infestations with 99.35% accuracy [22] Another case study from California demonstrated PA model a smart irrigation system leveraged by AI in a 500-acre vineyard, reducing water usage by 40% [22]. Climate FieldView, developed by The Climate Corporation (Bayer), optimizes crop monitoring and resource use through IoT, drones, and satellites. Used on 180 million acres across 23 countries, it enhances decision-making (92%), increases yields (3–5%), reduces water use (15%), and lowers fertilizer application (7–12%), making precision farming more efficient [23]. These examples highlight the potential of the digital solutions to revolutionize agricultural sector, making it more sustainable, efficient, and resilient to climate change.

4.3 Accelerating Farmers Awareness and Improving Access to Tech Solutions.

Our findings reveal that the majority of farmers lack awareness of government initiatives and policies. Additionally, many remain uninformed about advancements in the agricultural ecosystem, preventing them from leveraging new technologies. Although numerous agricultural workshops and seminars are organized, farmers often miss them due to a lack of timely information or find them unengaging and irrelevant. Addressing these challenges, many countries are shifting to AI-driven e-governance to create more transparent, efficient, and service-oriented agricultural governance. AI-powered e-governance aim to optimize agricultural policies, enhance service scalability, and strengthen digital public services [24]. One such innovation is KisanQRS, an AI-driven query-response system tested on 300,000 samples from India's Kisan Call Centre, where it outperformed traditional methods with a 96.58% top F1-score, ensuring fast and accurate responses to farmer queries [25]. Similarly, FarmerChat, a generative AI-powered chatbot, addresses language barriers. It has been deployed in four countries, and assisted over 15,000 farmers with 300,000 queries. KissanAI Dhenu 1.0 provides voice-based, multilingual support, making AI-driven agricultural assistance more inclusive for farmers across different regions [26]. These AI initiatives enhances decision-making by clustering and ranking farmer queries using deep learning techniques, thereby improving engagement and service scalability.

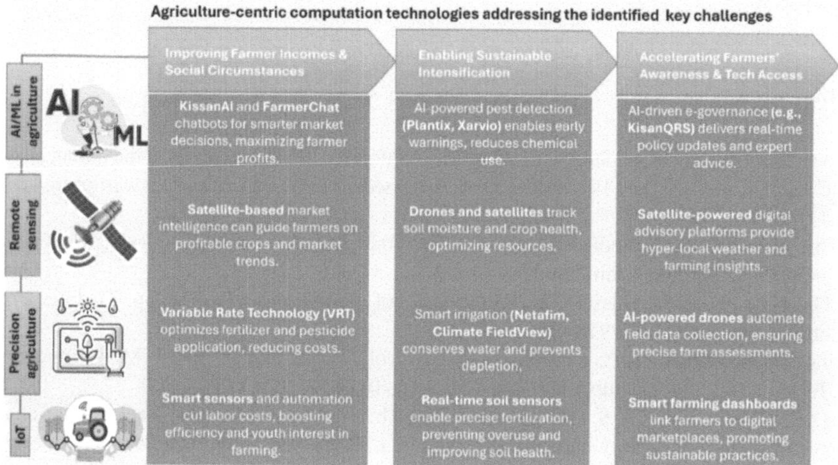

Fig. 5. Applications of Agriculture-centric computation technologies

5 Conclusion and Limitations

The study highlights Punjab's agricultural challenges, including gender disparities, an aging and undereducated farming population, youth disinterest and migration, overuse of agrochemicals, monocropping, groundwater depletion, and climate-induced productivity and economic losses. Existing government initiatives have had limited impact and require advanced reformative measures for greater effectiveness. Agriculture centric computation offers transformative solutions to enhance farmer incomes, enable sustainable intensification, and improve tech access. AI-driven tools, IoT monitoring, and digital platforms can empower small farmers, optimize resource use, and boost economic viability. Integrating these innovations with gender-inclusive policies and e-governance can bridge knowledge gaps and drive sustainable growth, ensuring a resilient future for Punjab's agriculture. The findings of this study may be limited by the lack of geographic and demographic diversity in the sample, as a broader range of participants from different regions and farming practices could provide more generalized insights. Additionally, focus group discussions with forward-thinking or younger farmers are needed to further assess their willingness to adopt ag-centric computational technologies within Punjab's agricultural ecosystem.

Acknowledgments. We want to thank Mr. Ravi Saini for his support in conducting the field study and Ms. Anshika Singh and Mr. Vikas Kumar for their support in the formal analysis. We are also grateful to Prof. Frances Ligler for providing critical feedback for improving the quality of this manuscript.

Role of Funding Source. This work was supported by the Startup Research Grant (SRG-006) from Plaksha University, Mohali, Punjab, India, and an external research grant (AWaDH-SpINe/IITRPR/2022/P100017) from IIT Ropar Technology and Innovation Foundation for Agriculture and Water Technology Development Hub (AWaDH), awarded by to Dr. Navjot Kaur.

Declaration of Competing Interests. The authors declare that they have no known competing financial interests or personal relationships that could have appeared to influence the work reported in this paper.

References

1. Sandhu, J.S.: Green revolution: a case study of Punjab… In: Proceedings of the Indian History Congress, vol. 75, pp. 1192–1199, 2014. Accessed 17 Feb 2024. http://www.jstor.org/stable/44158509
2. "Economic Survey of India." Accessed 17 Feb 2024. https://www.indiabudget.gov.in/budget_archive/es1985-86/esmain.htm
3. Gulati, A., Roy, R., Hussain, S.: Performance of agriculture in Punjab, pp. 77–112 (2021). https://doi.org/10.1007/978-981-15-9335-2_4
4. Jaiswal Shalini, J.P.B.P.: Case study-green revolution in Punjab. Int. J. Mod. Trends Sci. Technol. **6**(7), 61–68, July 2020. https://doi.org/10.46501/IJMTST060709
5. Dhillon, B.S., Bains, N.S., Mohapatra, T.: Punjab agriculture in the post green revolution era: perceptions and a reality check. Agricult. Res. J. **58**(5), 934–940 (2021). https://doi.org/10.5958/2395-146X.2021.00134.4
6. Singh Lovepreet, B.S.S.I.: Sustainability of agriculture systems: a case study of Punjab. Ind. J. Econ. Dev. 225–231, June 2020. https://doi.org/10.35716/ijed/NS20-046
7. Singh, H.: Socio-economic prospects of Punjab farmers. J. Emerg. Technol. Innov. Res. **6**(1), 321–328 (2019)
8. Mishra, A., et al.: An autonomous irrigation system framework for precision agriculture in Punjab, India. Irrig. Sci. (2025). https://doi.org/10.1007/s00271-025-01000-5
9. Wattanapanich, C., Imjai, T., Kefyalew, F., Aosai, P., Garcia, R., Vapppangi, S.: Integration of Internet of Things (IoT) and machine learning for management of ground water banks in drought-prone areas: a case study from Imjai organic garden, Thailand. Eng. Sci. (2024). https://doi.org/10.30919/es1248
10. Zanin, A.R.A., et al.: Reduction of pesticide application via real-time precision spraying. Sci. Rep. **12**(1), 5638 (2022). https://doi.org/10.1038/s41598-022-09607-w
11. "Turgut, Ö., Kök, İ., Özdemir, S.: AgroXAI: explainable AI-Driven crop recommendation system for agriculture 4.0. In: 2024 IEEE International Conference on Big Data (BigData), pp. 7208–7217. IEEE, December 2024
12. Madrewar, S., Khadkikar, N., Suryawanshi, O., Mulani, J., Sagar, A.: Digital agriculture: impact of IoT and AI on Indian agribusiness. Int. J. Appl. Econ. Account. Manag. (IJAEAM), **2**, 327–350 (2024). https://doi.org/10.59890/ijaeam.v2i4.2346
13. Kaur, P., 'Sandhu Singh, S., 'Singh, H., 'Kaur, N., 'Singh, S., 'Kaur, A.: Climatic features and their variability in Punjab (2016). https://doi.org/10.13140/RG.2.2.29571.91688
14. Development of an Atlas of Cancer in Punjab State. Accessed 06 Jan 2024. https://ncdirindia.org/ncrp/pca/RPT_2013_13/District_Wise.htm

15. Singhal, R., Rana, R.: Chi-square test and its application in hypothesis testing. J. Pract. Cardiovas. Sci. **1**(1), 69 (2015). https://doi.org/10.4103/2395-5414.157577
16. Jaacks, L.M., et al.: Impact of large-scale, government legislated and funded organic farming training on pesticide use in Andhra Pradesh, India: a cross-sectional study. Lancet Planet Health **6**(4), e310–e319 (2022). https://doi.org/10.1016/S2542-5196(22)00062-6
17. Singh, B.: Cancer deaths in agricultural heartland a study in Malwa region of Indian Punjab. Cancer Deaths in Agricultural Heartland a Study in Malwa Region of Indian Punjab (2008)
18. Mgendi, G.: Unlocking the potential of precision agriculture for sustainable farming. Discov. Agricult. **2**(1), 87 (2024). https://doi.org/10.1007/s44279-024-00078-3
19. Olagunju, O.O.: Harnessing artificial intelligence for youth engagement in agriculture: lessons from global practices and prospects for Nigeria. Int. J. Adv. Soc. Sci. Educ. (IJASSE) **2**(2), 83–94 (2024). https://doi.org/10.59890/ijasse.v2i2.1490
20. Touil, S., Richa, A., Fizir, M., Argente García, J.E., Skarmeta Gómez, A.F.: A review on smart irrigation management strategies and their effect on water savings and crop yield. Irrigation Drain. **71**(5), 1396–1416, December 2022. https://doi.org/10.1002/ird.2735
21. European Commission, Joint Research Centre, Precision agriculture and the future of farming in Europe (2021). https://www.europarl.europa.eu/RegData/etudes/STUD/2016/581892/EPRS_STU(2016)581892_EN.pdf
22. Basa, R.: AI in agriculture: revolutionizing precision farming and sustainable crop management. Int. J. Sci. Res. Comput. Sci. Eng. Inform. Technol. **10**(5), 535–543 (2024). https://doi.org/10.32628/CSEIT241051040
23. Chlingaryan, A., Sukkarieh, S., Whelan, B.: Machine learning approaches for crop yield prediction and nitrogen status estimation in precision agriculture: a review. Comput. Electron. Agric. **151**, 61–69 (2018). https://doi.org/10.1016/j.compag.2018.05.012
24. Gkikas, D.C., Theodoridis, P.K., Gkikas, M.C.: Artificial Intelligence (AI) use for e-Governance in agriculture: exploring the bioeconomy landscape, 141–172 (2023). https://doi.org/10.1007/978-3-031-22408-9_7
25. Rehman, M.Z.U., Raghuvanshi, D., Kumar, N.: KisanQRS: a deep learning-based automated query-response system for agricultural decision-making. Comput. Electron. Agric. **213**, 108180 (2023). https://doi.org/10.1016/j.compag.2023.108180
26. Mohanty, S.P., Hughes, D.P., Salathe´, M.: Using deep learning for image-based plant disease detection. Front. PlantSci. **7**, 1419 (2016)

A Model Based Comparative Study for Conventional and Drone Based Farming: Case Studies from India

Kondaka Shiva Sankar, Ajay Dashora(✉), and Pradip K. Das

Indian Institute of Technology Guwahati, Guwahati, Assam 781039, India
k.shivasankar729@gmail.com, {abd,pkdas}@iitg.ac.in

Abstract. Agriculture in India largely depend on conventional farming methods, exemplified by high labor costs, inefficient resource utilization, and suboptimal yields. This study explores the transformative potential of precision agriculture using drones through a detailed cost comparison with conventional practices. Focusing on two Indian farmers, the analysis examines labor costs, cultivation yields, and chemical usage. For farm area 1 (wheat, Uttar Pradesh), conventional labor costs per season are reduced by 29.8% with drones, while chemical usage decreases by approximately 40% and yield improves from 1 ton to 1.2 tons per acre. For farm area 2 (paddy, Andhra Pradesh), labor costs drop by 19.1%, with similar reductions in chemical inputs and a yield increase from 2 tons to 2.5 tons per acre. A drone based farming estimate shows how this technology can reduce costs by 19–29%, minimize chemical wastage by 40%, and enhance yields by 20–25%, offering a sustainable and profitable alternative for Indian agriculture.

Keywords: Precision Agriculture · Unmanned Aerial Vehicle · Conventional Farming · Drone Based Farming

1 Introduction

Agriculture is a cornerstone of India's economy, contributing 18.2% to GDP and supporting 42.3% of the population (GoI, 2024). However, conventional farming methods are labor-intensive, resource-inefficient, and environmentally unsustainable, resulting in high costs and inconsistent yields (Bruno et al., 2014). Precision agriculture, particularly through drones, offers a solution by enhancing efficiency and sustainability. Drones monitor crop health, assess soil conditions, and apply inputs like fertilizers and pesticides with precision, reducing chemical use and environmental impact (Abderahman et al., 2022). Studies show drones can cut pesticide application time by up to 90%, saving labor and boosting productivity (Tsouros et al., 2019; Yallappa et al., 2017).

India's drone market is expanding rapidly, with a projected 25% CAGR from 2022 to 2028, driven by government initiatives like the 'Drone Didi' program (BWC, 2022; MPower, 2024). Drones enable early detection of pests and nutrient issues, improving crop management and yields. This study compares conventional and drone-based farming in India, using mathematical models to analyze labor, chemical use, time, and

H. S. Shekhawat et al. (Eds.): ICA 2025, CCIS 2795, pp. 88–99, 2026.
https://doi.org/10.1007/978-3-032-17083-5_8

yields across two farms. The results highlight drones potential to make agriculture more profitable and sustainable, addressing key challenges in cost and environmental impact.

2 Methodology

This study employs an algebraic model based approach to evaluate economic viability of conventional and drone based farming in India. The models enable a systematic comparison accounting for the fundamental differences in operational paradigms of conventional and drone-based farming while maintaining consistent evaluation metrics.

Conventional farming has several costs factors, namely, labor costs, equipment costs, chemical costs and market costs. The cost in conventional farming (C_{conv}) per unit area is modeled as a linear combination of key cost factors, which are expressed in Eq. 1 as:

$$C_{conv} = C_L + C_E + C_C + C_M \quad (1)$$

Where, C_L, C_E, C_C and C_M represent labor cost, equipment cost, chemical cost, and market or freight cost, respectively. Labor cost (C_L) depends on the number of days the laborers work which varies on the size of the farm, crop type, and daily wages rate which varies according to geographical location in the country. The cost of labor is expressed in the Eq. 2 as

$$C_L = N_L \times N_D \times W_D \quad (2)$$

Where, N_L is number of laborers, N_D is number of working days and W_D is wage per day.

Equipment Cost (C_E) is sum of spraying equipment cost along with its seasonal maintenance and repair. Common practices among farmers are to hire heavy equipment on rent (e.g. tractor and ploughing machine) and to procure and own spraying machine which works for an average of 10 years. The equation for equipment cost is mentioned in below.

$$C_E = C_{Se} + C_{MR} + C_R \quad (3)$$

Where, C_{Se} is the average cost of spraying equipment per season, which can be calculated as average cost taking 10 years of life cycle for a spraying equipment and average of 2 crop seasons per year. C_{MR} is seasonal cost of maintenance and repair for spraying equipment, and C_R is cost of rent for an equipment (or its accessories) such as tractor or ploughing machine.

Chemical Cost (C_C) consists of fertilizer or nutrients that needs to be supplied to the plants, pesticide which is used for regulating crops infected with pests, or herbicide is for removing unwanted plants or weed plants. The chemical cost can be written as:

$$C_C = C_F + C_P + C_H \quad (4)$$

Where, C_F is fertilizer cost per unit area, C_P is pesticide cost per unit area, and C_H is herbicide cost per unit area.

Market cost (C_M) includes cost for transportation or freight charges to local market, labor charges for loading at farm and unloading of harvested crop at market, unforeseen contingency charges which includes renting and local taxes. These charges are expressed as:

$$C_M = C_T + C_{LU} + C_{UC} \tag{5}$$

Where, C_T is transportation charges, C_{LU} is labor charges for loading/unloading harvested crop, C_{UC} unforeseen contingency charges (i.e. charges for hiring space in local market for safekeeping of crops, hiring local manpower for various purposes, insurance of crops, and charges for other unforeseen events).

On the other hand, drone based farming cost (C_{drone}) consists of service cost of drone used for spraying operations, labor cost for various farm activities, which are not automatized using drones, rent for an equipment, reduced chemical usage of 40% due to drone based spraying (Adoni, et al., 2024) and market cost. This is modeled as a linear combination of above key cost factors, which are expressed as:

$$C_{drone} = (C_S + C_L^{'}) + C_R + \alpha C_C + C_M \tag{6}$$

Where, drone service cost (C_S) is fee for spraying using drone per unit area multiplied number of days. Labor Cost ($C_{L'}$) consists of plowing, sowing, harvesting done by labor as these activities are not automatized by drones. Rent Cost (C_R) of an item such as tractor or ploughing machine. Chemical Cost (αC_C) is adjusted by reduction factor α (α = 0.6 for 40% savings). Market cost (C_M) is assumed to be same as in Eq. 5.

Two cost models (expressed by Eqs. 1–6) provide a framework for comparing the economic viability of conventional and drone based farming. In next section, we demonstrate these models to two distinct farming scenarios (farmlands and crops) in India, utilizing real world data to quantify the cost differences of two methods of agriculture lands.

3 Case Study

This section applies and demonstrates the two cost models (C_{conv} and C_{drone}) to two distinct farming scenarios in India. The selected farmlands for this paper are situated in Shekhvapur village of Hardoi district (Uttar Pradesh), and Narsipatnam village in Anakapalle district (Andhra Pradesh). Locations of the two farmlands are described by the latitudes and longitudes as: (27°06′01.9"N 80°09′28.5"E), and (17°39′48"N 82°35′56"E), respectively. Also, the two farmlands grow wheat and paddy (rice) over 5 acre and 4.5 acre areas, respectively. Google Earth images showing locations and detailed maps of the two farmlands are shown by Figs. 1, 2 and 3 below.

For evaluating cost models for conventional farming and drone farming, data were collected through farmer interviews.

3.1 Cost Analysis for Conventional Farming

Conventional farming costs for the wheat farm in Uttar Pradesh and the paddy farm in Andhra Pradesh share four key components: labor, equipment, chemical inputs, and

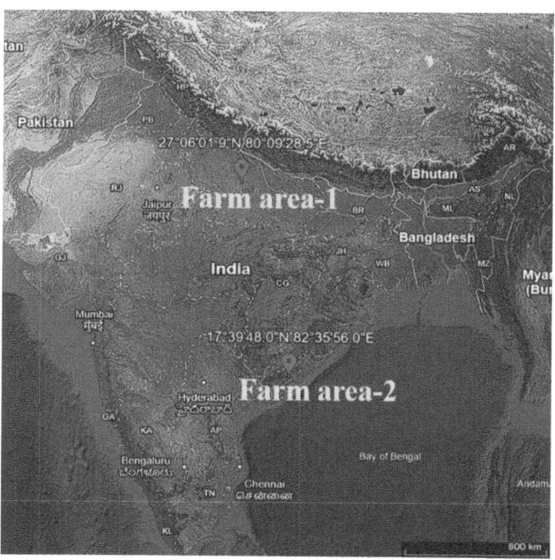

Fig. 1. Google Earth image showing both locations on one map of India

Fig. 2. Image of Shekhvapur Farm Area (scale: 1:13333)

market expenses. These common factors provide a consistent basis for comparing costs across regions and crops.

The wheat farm yields 1 ton per acre, with laborers paid ₹500 per day. It requires 2 L of pesticide, 1 L of herbicide, and 100 kg of fertilizer per acre, with spraying taking 10 days using two workers. The paddy farm yields 2 tons per acre, with labor costs at ₹400 per day per worker. It uses 0.8 L of pesticide, 1.5 L of herbicide, and 30 kg of fertilizer per acre, with spraying completed in 5 days by four workers.

The detailed analysis of these shared components for both the farm areas are discussed as follows:

Fig. 3. Image of Narsipatnam Farm Area (scale: 1:13333)

i. Labor Cost (C_L): Total labor cost is the sum of costs of plowing, sowing, spraying, weeding and harvesting. Cost calculation of these factors according to services of daily waged workers for farm area 1 and 2 are calculated using Eq. 2 and tabulated below.

Table 1. Labor cost for farm area 1 and farm area 2

	Farm Area 1				Farm Area 2			
Activity	No. of Laborers	No. of Days	Daily Wage (₹)	Total Cost (₹)	No. of Laborers	No. of Days	Daily Wage (₹)	Total Cost (₹)
Plowing	2	3	500	3000	2	4	400	3200
Sowing	3	2	500	3000	4	2	400	3200
Spraying	2	10	500	10000	4	5	400	8000
Weeding	4	5	500	10000	4	2	400	3200
Harvest	6	7	500	21000	8	4	400	12800
Total				**47000**	**Total**			**30400**

According to Table 1, the total labor cost (C_L) of farm area 1 and farm area 2 for conventional farming are ₹47,000 and ₹30,400 respectively.

ii. Equipment Cost (C_E) includes owned knapsack sprayer cost of ₹15,000 which is averaged based on the life cycle of over 10 years taking 2 crop seasons per year i.e. ₹750 per season with additional maintenance and repair costs for every season. Equipment rent expenses are for renting a tractor for 10 h at ₹1,000 per hour for farm area 1 and 8 h at ₹1,200 per hour for farm area 2. These costs contribute as input to Eq. 3 for calculations, which are mentioned in Table 2.

As shown in the above table the equipment cost for farm area 1 and farm area 2 are ₹11,200 and ₹10,850 respectively.

Table 2. Equipment cost for farm area 1 and farm area 2

Type of Cost	Farm Area 1 (₹)	Farm Area 2 (₹)
Spraying Equipment	750	750
Maintenance and Repair	450	500
Equipment Rent	10,000	9,600
Total	**11,200**	**10,850**

iii. Chemical Cost (C_C) is calculated as sum of fertilizer, pesticide, and herbicide costs for each crop season in Eq. 4. Details of chemical costs for farm area 1 and 2 are tabulated in Table 3.

Table 3. Chemical cost for farm area 1 and farm area 2

	Farm Area 1				Farm Area 2			
Application	Area (acre)	Qty	Price/ acre (₹)	Total (₹)	Area (acre)	Qty	Price/ acre (₹)	Total (₹)
Fertilizer	5	100 kg	25	12500	4.5	30 kg	120	16200
Pesticide	5	2 L	800	8000	4.5	0.8 L	2000	7200
Herbicide	5	1 L	900	4500	4.5	1.5 L	800	5400
	Total			**25000**	**Total**			**28800**

As per the Table 3, the total chemical cost (C_C) of farm area 1 and farm area 2 are ₹25,000 and ₹28,800 respectively.

iv. Expenses related to market cost covers transportation costs of labor costs for loading/unloading and unforeseen contingency charges, which are calculated according to Eq. 5, are mentioned in the Table 4.

Table 4. Market cost for farm area 1 and farm area 2

Expenditure Head	Farm Area 1 (₹)	Farm Area 2 (₹)
Transportation	2000	8000
Loading/Unloading	2000	0
Unforeseen Contingency	4500	1800
Total	**8500**	**9800**

Table 4 shows the total market cost (C_M) as ₹8,500 for farm area 1 and ₹9,800 for farm area 2.

Cost of conventional farming is calculated using Eq. 1 and cost for both the farms are shown in Table 5 below.

Table 5. Conventional farming cost for farm area 1 and farm area 2

Type of Cost	Farm Area 1 (₹)	Farm Area 2 (₹)
Labor Cost	47000	30400
Chemical Cost	25000	28800
Equipment Cost	11200	10850
Market Cost	8500	9800
Total	**91700**	**79850**

Total costs of conventional farming for farm area 1 and farm area 2 according to Table 5 are ₹91,700 and ₹79,850, respectively. It should be noted that compared to agricultural based farming, drone based farming increased the production of wheat and paddy by 20% (from 1.0 to 1.2 tons) and 25% (2.0 to 2.5 tons), respectively. However, we have ignored the likely increase in market costs for both wheat and paddy as the increased costs will be compensated by likely higher gains obtained by selling the crops in local markets.

3.2 Cost Analysis for Drone Based Farming

This section analyzes the projected operational costs associated with adopting drone based precision agriculture for the two case study farms. Utilizing the drone cost model defined in the methodology (Eq. 6), this analysis estimates the economic landscape when key conventional practices, particularly manual chemical spraying, are replaced by drone technology.

This calculation incorporates the cost of drone services (C_S) alongside the costs for remaining essential manual labor ($C_{L'}$) such as plowing, sowing, and harvesting. It also accounts for the anticipated reduction in chemical expenses due to precision application, represented by the factor α applied to the conventional chemical cost (αC_C), reflecting a 40% saving ($\alpha = 0.6$) in this study. Costs for persistent inputs like tractor rental (C_R) for land preparation and standard market costs (C_M) are also included to provide a comprehensive estimate.

The following distribution as presented in Table 6 details these projected costs for both farm area 1 and farm area 2, providing a basis for comparison with the conventional farming costs analyzed previously.

The total drone based farming cost for farm area 1 and 2 are ₹66,500 and ₹61,280, respectively. To assess the economic viability of drone based precision agriculture, the costs incurred in conventional farming are compared with the projected costs associated with utilizing drones in the results section.

Table 6. Drone based farming cost for farm area 1 and farm area 2

Type of Cost	Farm Area 1 (₹)	Farm Area 2 (₹)
Drone Service + Labor Cost ($C_S + C_L'$)	33000 (6000 + 27000)	24600 (5400 + 19200)
Equipment Rent (C_R)	10000	9600
Chemical (αC_C)	15000	17280
Market (C_M)	8500	9800
Total	**66500**	**61280**

4 Results

This section presents a detailed cost comparison between costs incurred by conventional and drone based farming for the two case studies.

4.1 Farm Area 1: Wheat Cultivation in Uttar Pradesh

Comparative cost analysis on individual cost factors:

Labor and Drone Service: As can be seen, a reduction from ₹47,000 (C_L) under conventional farming to ₹33,000 (C_S + $C_{L'}$) with drone farming adoption. This saving of ₹14,000 (~29.8%) reflects the replacement of intensive manual spraying labor (₹10,000) and potentially other efficiencies with the drone services (costing ₹6,000), while accounting for essential labor for remaining farm activities (₹27,000).

Equipment Costs: The equipment cost decreased slightly from ₹11,200 (C_E) to ₹10,000 (C_R). This reduction of ₹1,200 is solely due to the elimination of the cost associated with owning/maintaining a manual sprayer, as the tractor rental cost remained same for the conventional and drone based practices.

Chemical Costs: A substantial saving was observed with costs dropping from ₹25,000 (C_C) to ₹15,000 (αC_C). This reduction of ₹10,000 is attributed to the assumed 40% increase in chemical efficiency ($\alpha = 0.6$) enabled by precision application via drones.

Market Costs: Marketing costs (C_M) were assumed to remain same at ₹8,500 in both scenarios for this analysis.

Conventional farming incurred a total seasonal cost (C_{conv}) of ₹91,700. However, the cost using drone based methods (C_{drone} was calculated as ₹66,500. This represents a significant total cost reduction of ₹25,200, equivalent to approximately 27.5% savings per crop season of wheat. Table 7 summarizes the cost reductions of various cost components by drone based farming.

For the wheat farm in Uttar Pradesh, drone adoption demonstrates considerable cost savings potential (~27.5%). Among all costs components, maximum saving is from reduced chemical usage (40%) followed by reduction in labor/service costs contributing to the overall economic benefit. The elimination of manual sprayer and consequent maintenance costs also provide a substantial and additional saving (10%) (Table 8).

Table 7. Summary of cost components for Conventional vs. Drone based farming

Cost Component	Conventional (₹)	Drone Based (₹)	Savings (₹)	Reduction
Labor	47,000	33,000*	14,000	29.80%
Equipment	11,200	10,000#	1,200	10.70%
Chemicals	25,000	15,000	10,000	40.00%
Market	8,500	8,500	0	-
Total	**91,700**	**66,500**	**25,200**	**27.50%**

* Includes ₹6,000 (drone service) + ₹27,000 (labor costs for plowing/harvesting).
# Tractor rental cost for plowing

4.2 Farm Area 2: Paddy Cultivation in Andhra Pradesh

The analysis for Farm Area 2 (4.5 acre paddy farm) also indicates economic advantages with drone technology.

Comparative analysis on individual cost factors.

Labor and Drone Service: This component decreased from ₹30,400 (C_L) for conventional farming to ₹24,600 ($C_S + C_{L'}$) with drone based farming, showing a reduction of ₹5,800 (~19.1%). Similar to Farm area 1, this reflects the drone service cost (₹5,400) is replacing manual spraying labor (₹8,000) alongside essential labor costs (₹19,200) for remaining farm activities.

Equipment Costs: Equipment costs reduced slightly from ₹10,850 (C_E) to ₹9,600 (C_R). The saving of ₹1250 corresponds to the elimination of the manual sprayer cost, while tractor rental cost remains as the primary equipment expense.

Chemical Costs: This component showed the largest absolute saving for Farm area 2. Costs decreased from ₹28,800 (C_C) to ₹17,280 (αC_C), resulting in a reduction of ₹11,520 (40%) due to precision application.

Market Costs: Marketing costs (C_M) were assumed equal at ₹9,800 for both conventional and drone based farming practices.

Table 8. Summary of cost factors for Conventional vs. Drone based farming

Cost Component	Conventional (₹)	Drone Based (₹)	Savings (₹)	% Reduction
Labor	30,400	24,600*	5,800	19.1%
Equipment	10,850	9,600#	1250	11.5%
Chemicals	28,800	17,280	11,520	40.0%
Market	9,800	9,800	-	-
Total	**79,850**	**61,280**	**18,570**	**23.2%**

* Includes ₹5,400 (drone service) + ₹19,200 (labor costs for plowing/harvesting).
# Tractor rental for plowing

Total cost analysis: Conventional farming costs (C_{conv}) totaled ₹79,850 per season. The projected drone based farming cost (C_{drone}) was ₹61,280. This results in a total cost saving of ₹18,570, or approximately 23.2%. Also, in this case of farm area 2, the reduction in chemical costs (40%) represents the highest single component of savings, exceeding the savings in the labor/service component (19%). The equipment cost saving is minimal for all cost components.

4.3 Inter Crop Comparison of Drone Based Costs

This section compares drone based operational costs between the Uttar Pradesh wheat farm and the Andhra Pradesh paddy farm to assess if drone farming economics are consistent across different crops and regions. The comparison aims to identify key cost drivers and understand how factors like crop type, regional wages, and other needs influence costs even with drone adoption. The analysis utilizes total drone costs (C_{drone}) per acre (or cost per unit area) for normalization. It specifically examines the per acre cost distributions of major components, namely, combined labor/drone service cost, equipment rent cost, and chemical cost to reveal the sources of variation between the two distinct farming systems. However, unlike other components, market cost cannot be normalized by unit area criteria for two reasons:

Freight charges (transportation and loading/unloading charges) are less for 1.2 ton/acre wheat than that of 2.5 ton/acre of paddy.

Minimum support price (MSP) for the wheat and paddy crops are similar with minimum difference. Thus, commercial gains to a farmer by selling 1.2 ton/acre wheat is significantly less than that of selling the paddy (2.5 ton/acre) in respective local markets.

Therefore, considering both factors listed above, we calculate normalized market cost by dividing the market cost by area and total weight of produced crop. The resulting unit of normalized market cost is "per acre per ton". Table 9 presents and compares various normalized cost components of drone based farming for two farm areas growing wheat and paddy.

Table 9. Inter-comparison of Drone based farming costs for wheat and paddy crops

Cost Component	Farm Area 1	Farm Area 2
Drone Service	1200 (₹/acre)	1200 (₹/acre)
Labor Cost	6600 (₹/acre)	5467 (₹/acre)
Chemical Cost	3000 (₹/acre)	3840 (₹/acre)
Market Cost	1416.7 (₹/acre/ton)	871.2 (₹/acre/ton)
Total	**12216.7 (₹)**	**11378.2 (₹)**

Similar to total drone cost (calculated in previous sub-section), the total normalized drone cost is marginally lower for the farm area 2 than that of farm area 1. This difference is primarily driven by higher paddy production; though chemical cost is higher for the paddy crop. Farm 2 is also benefitted by lower labor cost potentially due to regional wage differences influencing the labor cost component for farm activities, which are performed by drones. Farm 1 shows higher normalized drone based cost (₹12,217 vs ₹11,378), which stems from higher expenses for labor work force and market expenses, despite lower chemical cost. This suggests that crop type (paddy vs. wheat) and yield volume influence cost structures, with paddy's higher chemical intensity and transport needs offsetting market expenses.

5 Discussions

Drone based farming outperforms conventional methods, cutting total costs by 27.5% (Farm 1) and 23.2% (Farm 2). Key savings stem from labor reductions (29.8% and 19.1%) and 40% chemical cuts (Yallappa et al., 2017; Tsouros et al., 2019). Variations arise from crop and location differences: Farm 1 (wheat) saves more due to high spraying costs, while Farm 2 (paddy) benefits from scaled production but faces higher chemical expenses. The 40% chemical reduction boosts sustainability by lowering runoff (Abderahman et al., 2022). However, this study assumes constant non spraying labor and market costs, limiting its scope. Future research could explore drones for sowing and monitoring. Overall, drones offer a cost-effective, eco-friendly solution for Indian agriculture, with benefits varying by crop and scale.

6 Conclusion

This study demonstrates that drone based precision agriculture offers a transformative pathway for Indian farming, addressing critical challenges of labor inefficiency, resource wastage, and stagnant yields. Through a two case study of wheat and paddy cultivation in different geographical locations and climate of Uttar Pradesh and Andhra Pradesh, respectively, the research validates the economic and environmental superiority of drone technology over conventional practices. Operational costs were substantially reduced, by approximately 27.5% for the wheat farm and 23.2% for the paddy farm, compared to conventional methods. These savings were primarily driven by two key factors: a notable reduction in labor and associated service costs (ranging from 19% to 30% savings) due to the efficiency of drone operations replacing manual field tasks like spraying, and a significant decrease in chemical input costs (40% savings observed based on the precision application factor) which also carries positive environmental implications. While the specific cost structures and magnitudes of savings varied between the wheat and paddy farms, reflecting differences in crop needs and regional factors, the overall trend of improved economic efficiency was consistent.

The findings underscore the capacity of drone technology to address critical issues facing Indian agriculture, including optimizing resource use, mitigating labor shortages, reducing the environmental footprint of chemical inputs, and ultimately boosting the economic resilience of farming operations. Policymakers should consider subsidies and training programs to facilitate adoption, while awareness campaigns and investment in research and development can further tailor drone solutions to Indian farming conditions.

6.1 Limitations and Future Work

- Manual labor for plowing/harvesting remains a bottleneck, necessitating integrated mechanization.
- Using drones for sowing will help in more optimized plant spreading across the farm land and reduces labor dependence.

References

Rejeb, A., Abdollahi, A., Rejeb, K., Treiblmaier, H.: Drones in agriculture: a review and bibliometric analysis. Comput. Electron. Agricult. **198** (2022)

Sreenivas, A.G., Ranganath, Wazid, Vijayalakshmi: Unmanned Aerial Vehicle (UAVs): a novel spraying technique for management of sucking insect pests in cotton ecosystem. J. Biosyst. Eng. **49**(2), 193–201 (2024)

BWC: Blue Weave Consulting, "India Agriculture Drones Market - Industry Trends & Forecast Report 2028 (2022). https://www.blueweaveconsulting.com/report/india-agriculture-drones-market. Accessed 6 Apr 2025

Faiçal, B.S., et al.: The use of unmanned aerial vehicles and wireless sensor networks for spraying pesticides. J. Syst. Architect. **60**(4), 393–404 (2014)

Tsouros, D.C., Bibi, S., Sarigiannidis, P.G.: A review on UAV based applications for precision agriculture. Information **10**(11), 349 (2019)

Yallappa, D., Veerangouda, M., Maski, D., Palled, V., Bheemanna, M.: Development and evaluation of drone mounted sprayer for pesticide applications to crops. In: 2017 IEEE Global Humanitarian Technology Conference (GHTC), pp. 1–7 (2017)

Wang, G., et al.: Comparison of spray deposition, control efficacy on wheat aphids and working efficiency in the wheat field of the unmanned aerial vehicle with boom sprayer and two conventional knapsack sprayers. Appl. Sci. **9**(2), 218 (2019)

GoI: Government of India, "Economic Survey 2023–24" (2024). https://static.pib.gov.in/WriteReadData/specificdocs/documents/2024/jul/doc2024722351601.pdf. Accessed 6 Apr 2025

Indian Council of Agricultural Research (ICAR): Revolutionizing Agriculture: The Digital Transformation of Farming (2024). http://www.cazri.res.in/marquee/icarreport-2024.pdf. Accessed 6 Apr 2025

MPower: Agricultural Drone Subsidy & Government Policy in India (2024). https://mpowerlithium.com/blogs/blog/agricultural-drone-subsidy-government-policy-in-india. Accessed 6 Apr 2025

Yanliang, Z., Qi, L., Wei, Z.: Design and test of a six-rotor unmanned aerial vehicle (UAV) electrostatic spraying system for crop protection. Int. J. Agricult. Biol. Eng. **10**(6), 68–76 (2017)

Enhancing Large Language Model Performance for Agricultural Domain Translation via Specialised Dictionaries and Embeddings

Shyama Wilson[1,2](✉), Kishan Fernando Warnakulasuriya[3], Dilum Bandara Wijesundara[4], and Athula Ginige[1]

[1] Western Sydney University, Parramatta, NSW 2150, Australia
22085197@student.westernsydney.edu.au, shyama@uwu.ac.lk, A.Ginige@westernsydney.edu.au

[2] Uva Wellassa University, Passara Road, Badulla 90000, Sri Lanka

[3] Deakin University, Locked Bag 20000, Geelong VIC 3220, Australia
krishanfernando129@gmail.com

[4] University of Ruhuna, Matara 81000, Sri Lanka
dilumbandara095@gmail.com

Abstract. Translating agricultural Packages of Practices (PoPs) into local languages is crucial for knowledge dissemination but poses challenges in terms of cost, time, and accuracy, particularly for domain-specific terminology. This paper investigates automating PoP translation using Large Language Models (LLMs), specifically Gemini-1.5-Flash. Initial experiments revealed inaccuracies in translating agricultural terms. To address this, we propose a novel dictionary-based approach, integrating a specialised agricultural dictionary (approx. 10,000 terms) with the LLM. The method employs embeddings ("all-MiniLM-L6-v2") and K-Means clustering to efficiently retrieve and provide contextually relevant dictionary terms to the LLM during translation. This significantly reduces computational overhead compared to a linear search. Results demonstrate a substantial improvement in translation accuracy for Sinhala and Tamil, validated by domain experts. This optimised dictionary-enhanced LLM approach offers an effective solution for accurate and efficient automated translation of specialised agricultural content.

Keywords: Package of Practices · Sinhala and Tamil Languages · Multilingual Dictionary · Embedding · LLMs

1 Introduction

Decision Support Systems (DSS) play a vital role in agriculture by providing timely information that enables farmers to make informed decisions. However, the effectiveness of a DSS depends critically on its ability to deliver contextual information in local languages [1, 2]. Without this capability, farmer adoption remains limited. Providing agricultural information in local languages is challenging due to the time-consuming, costly, and expertise-driven nature of manual translation, particularly regarding specialised agricultural terminology.

H. S. Shekhawat et al. (Eds.): ICA 2025, CCIS 2795, pp. 100–111, 2026.
https://doi.org/10.1007/978-3-032-17083-5_9

These challenges are evident in our DSS, Govinena, developed for Sri Lankan and Indian farmers [3–5]. Govinena delivers information based on farmer context and language preferences, using content known as Packages of Practices (PoPs). A dedicated team creates these PoPs and manually translates them into local languages like Sinhala and Tamil, followed by expert validation before DSS deployment.

Recognising the inefficiencies of manual translation, this study investigated automating the process using advanced LLMs [6, 7]. Initial evaluations of models like GPT and Gemini indicated that Gemini-1.5-Flash offered superior translation performance for the target languages and was freely available. Consequently, Gemini-1.5-Flash was selected for automating PoP translation.

Early experiments using zero-shot and few-shot prompting with Gemini-1.5-Flash revealed significant limitations: the translations, while generally coherent, and contained numerous inaccurate terms, especially for domain-specific agricultural vocabulary. This motivated the development of a more robust method. We integrated a specialised agricultural dictionary (containing approx. 10,000 terms) with the LLM. However, directly inputting the entire dictionary proved infeasible due to LLM token limitations.

To overcome this, we implemented a dictionary-based approach utilising embeddings ("all-MiniLM-L6-v2″) to represent terms numerically. This allowed the system to retrieve contextually relevant terms from the dictionary based on semantic similarity to the input text. Furthermore, to address the computational inefficiency of searching the large, embedded dictionary, we employed K-Means clustering to group terms, significantly reducing the search space and improving processing speed. This optimised system retrieves relevant dictionary terms (English and local language equivalents) and incorporates them into the prompt provided to the LLM.

This dictionary-enhanced, optimised LLM approach resulted in significantly improved translation accuracy, particularly for agricultural terms, and garnered positive feedback during expert verification. This paper details the implementation of this dictionary-based LLM translation system, highlighting the techniques used to overcome initial LLM limitations and improve accuracy while mitigating hallucinations.

The paper is structured as follows: Sect. 2 discusses related work that utilises LLMs for agriculture and content translation. Section 3 explains the translation process before and after integrating a dictionary with an LLM. Section 4 explains the challenges encountered during the implementation of the dictionary-based approach and how they were effectively solved. Section 5 discusses the results. Finally, Sect. 6 concludes the author's work.

2 Literature Review

The generative power of large language models (LLMs) is increasingly being applied across various domains, including agriculture. Numerous applications of LLMs can be seen in addressing agricultural challenges, particularly in areas such as content generation, image creation, and communication (e.g., chatbots) [8–14]. While many studies demonstrate the potential of LLMs, discussions on their use for translating agricultural content remain limited, and practical evidence is scarce. Thus, we analysed the capability of LLMs in language translation for low-resource languages rather than thoroughly sticking to the agricultural context.

Researchers in [11, 15] have highlighted that many LLM models, such as mT5 [16], XLMs [17], BLOOM [18], Gemini [19] and GPT [6] are capable of understanding and translating multiple languages. However, when applied them in practice, the quality of their language translation is often low [11, 15]. For this reason, agriculture applications use different software or strategies to translate agriculture content into different local languages [11, 20, 21]. For example, Kisaan GPT is a chatbot that allows farmers to access crucial agricultural information in English and Hindi [11, 20, 21]. This app utilises the Dhenu 1.0 model, an agriculture-specific LLM developed by KissanAI [11, 20, 21], to support two languages. Dhenu 1.0 is trained on 300,000 instruction sets in both English and Hindi, enabling it to process and translate agricultural queries effectively between these languages.

In the context of low-resource languages, the scholars in [22] examined the capability of existing LLMs to translate text into a low-resource language: Sinhala. They found that Claude and GPT-4o performed better in translation than the fine-tuned models of LLaMA 3 and Mistral. Additionally, an experiment on English and Indian language (22 languages) translation was conducted in [23]. This experiment demonstrated that fine-tuned LLaMA-2-based models outperformed other models (i.e., opt-6.7b, Bloom-7B, Mistral-7B, MPT-7B, Falcon) in both zero-shot and in-context example-based learning. The researchers in [24] experimented with translating English into the Mambai language, evaluating the efficacy of few-shot LLM prompting. In this experiment, they showed that translation accuracy could be significantly improved by using dictionary entries (i.e., a set of 1,187 parallel sentences in English and Mambai) with the prompt. Moreover, SingRAG [25] employed a Translation-Augmented framework to improve the accuracy of the language translation: English-Singlish (Sinhala-English code-mixed language). This approach shows higher accuracy than the based model (LLaMA-2 7B) that they used.

Based on existing work in language translation, it can be observed that advanced LLMs now have significant capability to translate even low-resource languages. However, their accuracy is still relatively low, and researchers have shown that it can be improved by applying various strategies, such as fine-tuning, integrating external dictionaries, and providing more contextual information in the prompts.

3 Automation of PoP Translation

This section first presents the manual PoP preparation process (see Fig. 1). Thereafter, the section explains how an advanced LLM is utilised to automate the PoP translation accurately.

3.1 Existing Approach of PoP Preparation and Translation

The Agri team in our DSS, i.e., Govinena, first develops Packages of Practices (PoPs) for specific crops (see Sample PoP in [26]). A PoP is a workbook or spreadsheet created based on various factors, conditions, parameters, and constraints that influence farmers' decisions. Additionally, it includes a set of guidelines to be followed in doing different activities throughout the farming life cycle [27]. Usually, PoPs are prepared in English.

However, these PoPs are translated into local languages based on the countries where the DSS is used. For example, PoPs are currently translated into two languages, Sinhala and Tamil, as Govinena is actively used in Sri Lanka. Therefore, this study conducted all experiments using Sinhala and Tamil to demonstrate the effectiveness of integrating a specialised dictionary with LLMs.

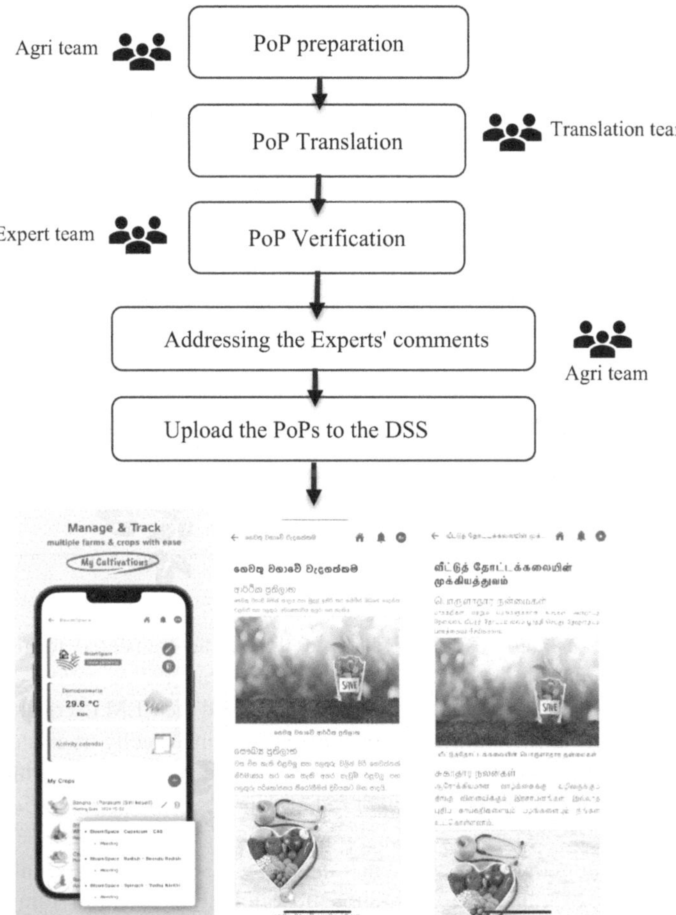

Fig. 1. The flow of the PoP creation and some mobile interfaces of the DSS, i.e., Govinena

Once the PoPs are translated, they are sent to relevant experts in the Department of Agriculture for verification. After expert review, the verified sheets are returned to the internal Agri team, and they refine the PoPs based on the domain experts' feedback. Once finalised, the PoPs are uploaded to the DSS to make the information accessible to farmers [4, 5].

3.2 AI-Based PoP Content Translation

The most time-consuming and labour-intensive step of the PoP preparation is *translating* it into different languages. Hiring labour for these tasks and paying them over an extended period is neither practical nor cost-effective. Therefore, this study began with the aim of investigating an approach to automating language translation in the context of agriculture. Moreover, as mentioned above, the study considered the two languages used in Sri Lanka. The experiments were initiated with LLM models, as advanced LLMs now offer several capabilities for processing natural language. Based on recent literature and practical experience, we also identified GPT and the Gemini model [6, 7] as potential LLM models for translation. However, the study selected a Gemini model for the experiments as our preliminary experiments demonstrated that the Gemini-1.5-Flash model provides more accurate translations compared to GPT models. Consequently, Gemini-1.5-Flash was selected for automating the translation process.

In the experiment with Gemini-1.5-Flash, translations were performed by directly prompting the model to translate PoP data. At this stage, the PoP data was provided in JSON format to Gemini, along with Prompt 01 (Table 1).

Table 1. Example for a prompt

Prompt 01: Zero-Shot Strategy
Translate this json text data to <<target_language>>. Do not translate keys. Translate only values and give me a JSON-type translated answer. Consider the agricultural term in the translation. Give me exactly json format without any other notes, headers or texts. Here is the JSON data to be translated: <<content>>

However, the translated responses showed that some translated terms were not relevant to the agricultural domain and presented significant hallucinations.

For example, although the prompt explicitly instructed the LLM to consider agricultural terminology, the term "Nursery Management" was incorrectly translated into Sinhala as "ළදරු පාසල් කළමනාකරණය(ladharu pasal kalamanakaranaya)" which refers to a *preschool or nursery school management* rather than " තවාන් කළමනාකරණය(thawan kalamanakaranaya)".

To address the hallucination problem in translation, we evaluated different prompting strategies, including examples (i.e., few-shot strategies). However, providing examples within the prompt is not practical, as it requires crafting prompts with multiple examples to explain various agricultural terms. To address this challenge, the study introduces a dictionary-based approach for the LLM. This approach utilises a dictionary containing nearly 10,000 agricultural terms along with their meanings in Sinhala and Tamil (see Agriculture Glossary Sri Lanka in [26]). Then, the LLM was first instructed to refer to the dictionary for relevant terms before processing translation. However, due to the input token limitations of LLMs, it is not possible to consider the entire dictionary for translation. Instead, the LLM processes terms only up to the token limit and incorporates those terms during translation. Consequently, this constraint reduces translation quality, as relevant terms from the dictionary may not be included.

The following section provides a detailed explanation of how the study addressed these challenges and modified the algorithms to achieve significant accuracy in translation.

4 Dictionary-Based LLM Local Language Translation

As explained in the previous section, providing the dictionary in its original form to the LLM does not enhance translation accuracy. Therefore, we conducted additional experiments based on theories such as information retrieval and augmentation [28]. Subsequently, an embedding mechanism was employed to assign embedding values to dictionary terms, facilitating the identification of the most similar terms for a given text content. For example, the dictionary terms were assigned embedding values using the "*all-MiniLM-L6-v2*", which is from the sentence-transformers library by Hugging Face [29]. This embedded dictionary is integrated into the translation systems (see Fig. 2).

For example, the system first receives the PoP content to be translated in JSON format (Step 1 in Fig. 2). Then, the system pre-processes the content by removing propositions, stop words, and other words that should not be included in the dictionary search. Then, these pre-processed terms are sent to the searching algorithm to identify agricultural terms from the dictionary and their corresponding equivalents in local languages. At this stage, the algorithm retrieves relevant records (i.e., dictionary terms) comparing the embedding values of the dictionary for each pre-processed term (Step 2 in Fig. 2). Then, the algorithm provides the relevant agriculture terms associated with them (Step 3 in Fig. 2). These terms (i.e., English and its local agriculture terms) are then incorporated into the LLM prompt alongside the original input content, instructing the LLM to consider them during translation (Step 4 in Fig. 2). The LLM subsequently utilises the recommended terms while translating the content into the target language and returns the translated response in JSON format (Step 5 in Fig. 2). Finally, the translated output is structured into the appropriate PoP format and delivered to the user (Step 6 in Fig. 2).

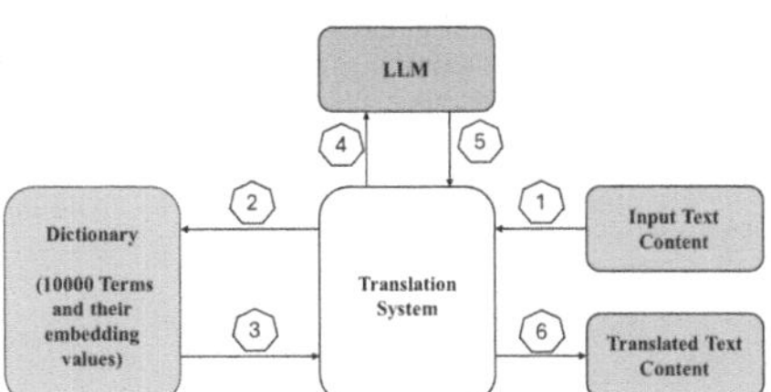

Fig. 2. Initial version of the dictionary-based LLM translation process

However, we observed that this method consumes a significant amount of time. Upon analysis, it was found that since the dictionary contains 10,000 records, the algorithm must perform 10,000 calculations to find the most similar record for a single word. Consequently, translating an entire dataset requires a substantial number of calculations. To optimise this process, clustering was employed [30]. Specifically, K-Means clustering [30] was used to group the 10,000 words into clusters, with each cluster containing

20 words, resulting in nearly 500 clusters (see Fig. 3). Moreover, each cluster has an average embedding value, which was calculated based on the embedding values of the records/terms in a cluster.

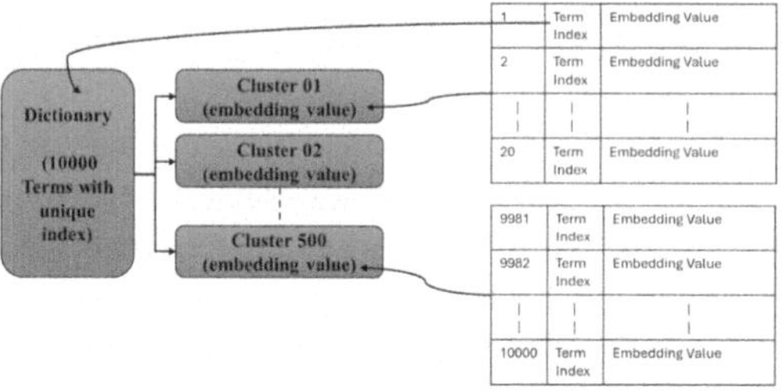

Fig. 3. Clustering of dictionary records

With this approach, the system first pre-processes the given content, as mentioned previously and sends the required terms to the search algorithm. It then performs 500 calculations on the clusters to identify the most similar cluster associated with a term at a time. Thereafter, 20 additional calculations were performed to find the most relevant records (i.e., a record contains an embedding value and the term index) within that cluster. This reduces the total number of calculations to be performed for a term from 10,000 to 520, significantly improving computational efficiency.

Then, the most relevant records for a term are sent to the system. The system then finds the corresponding dictionary terms using the unique index of the records. Finally, the retrieved dictionary terms, along with the original input (i.e., sentences/text content), are sent to the LLM for translation. Then, the LLM returns the translated content to the system in JSON format, and the system provides the PoP to the user after structuring it into the appropriate format. The described process is summarised in Fig. 4.

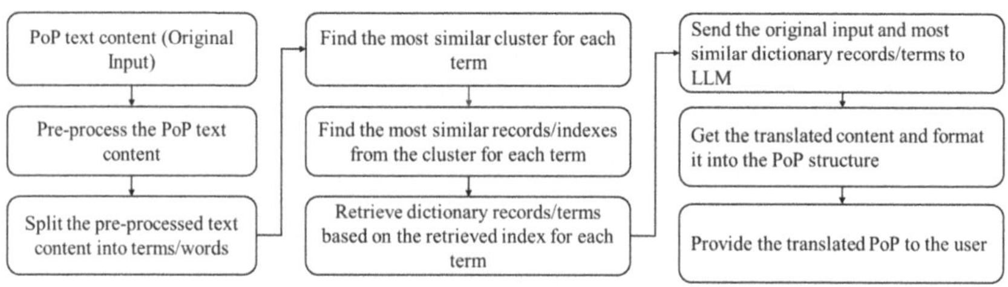

Fig. 4. Improved dictionary-based LLM translation process

5 Results and Discussion

As explained in Sects. 3 and 4, various translation approaches were tested using the Gemini-1.5 Flash model. The experiments primarily focused on two languages: Sinhala and Tamil. Initially, the translation of PoP content was tested by directly providing the content to the LLM. While the resulting translations demonstrated considerable accuracy, they included inaccuracies in the translation of agricultural terms (see Table 2).

After introducing the dictionary-based approach, the study observed a significant increase in the accuracy when translating English into local languages. Table 3 presents how the example given in Table 2 was improved through the dictionary-based LLM translation.

The developed system was tested and deployed as a cloud-based system in Digital Ocean [31] and the LLM model is accessed via API calls.

Table 2. Example - the inaccurate terms included in the LLM-based Sinhala translation

Sentence	LLM - Wrong Sinhala translation	Expected translation
The control methods of rice pest	සහල් කළුඹුහානියේ ක්‍රම (sahal kalubhuhaaniye krama)	වී වගාවේ පළිබෝධ පාලන ක්‍රම(vee wagawe palibodha palana krama)

Table 3. Example - dictionary-based LLM translation

Sentence	Extracted Dictionary Terms	LLM response
The control methods of rice pest	[0] {"English": "Pest control", "Sinhala": "පළිබෝධ පාලනය", "Tamil": "பீடைக் கட்டுப்பாடு"}, [1] {"English": "Chemical control", "Sinhala": "රසායනික මර්ධනය; රසායනික පාලනය", "Tamil": "இரசாயனவாட்சி, இரசாயனக் கட்டுப்பாடு" }, [2] {"English": "Weed control", "Sinhala": "වල් පැළෑටි පාලනය", "Tamil": "களை கட்டுப்படுத்தல்" } Indexes: [0] => 7863 [1] => 1259 [2] => 10612	වී වගාවේ පළිබෝධ පාලන ක්‍රම(vee wagawe palibodha palana krama)

Furthermore, we employed two methods to verify the accuracy of the AI-generated content. The first was a quantitative evaluation using similarity metrics such as BLEU (Bilingual Evaluation Understudy) [32], METEOR [33], TER (Translation Edit Rate) [34], and BERTScore [35]. The second method involved manual verification by domain experts. In the quantitative evaluation, our initial objective was to identify a similarity metric that performs well for Sinhala and Tamil languages. However, preliminary experiments revealed that commonly used metrics like BLEU, METEOR, and TER did not adequately capture semantic similarity in these languages. For example, when testing semantically equivalent sentence pairs in Sinhala or Tamil, the metrics produced similarity scores close to zero (i.e., indicating dissimilarity - see Fig. 5), despite the sentences conveying nearly identical meanings.

Further experimentation showed that BERTScore [35] performed better in evaluating semantic similarity for these languages. However, we have not yet verified its reliability across a wide range of linguistic scenarios. Therefore, its applicability remains uncertain at this stage, and further testing is ongoing.

Given these limitations, we selected manual verification, an expert-based evaluation approach, as our primary method for assessing the accuracy of the AI-generated content within PoPs. An additional reason for this decision is the real-world nature of the project, where the content serves real users. In such cases, ensuring 100% accuracy is essential.

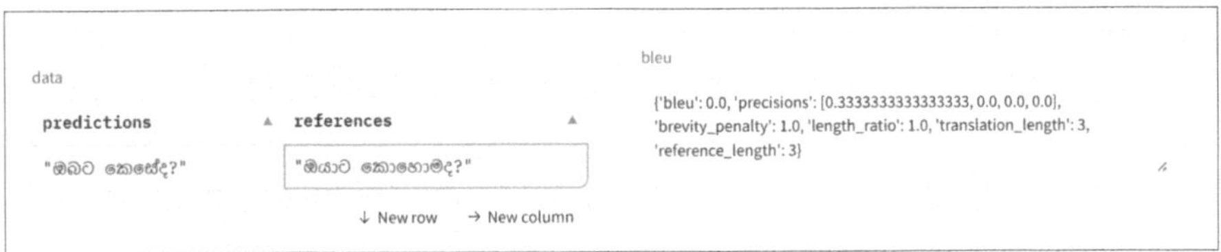

Fig. 5. BLUE score generated for the semantically equivalent sentences in the Sinhala Language

Then, the accuracy of the translated PoPs was verified manually with the support of the agriculture experts. For that, we developed a system (see Fig. 6) that allows users to upload an English PoP for translation. Once they upload a PoP, they can select the local language to be considered in the translation and submit the data to the system. Then, the system provides the translated PoP as an outcome [26]. Each PoP is then submitted for expert verification. The experts review the content for accuracy and make modifications if necessary. So far, we have received constructive and positive feedback on the PoPs generated using the dictionary-based LLM approach.

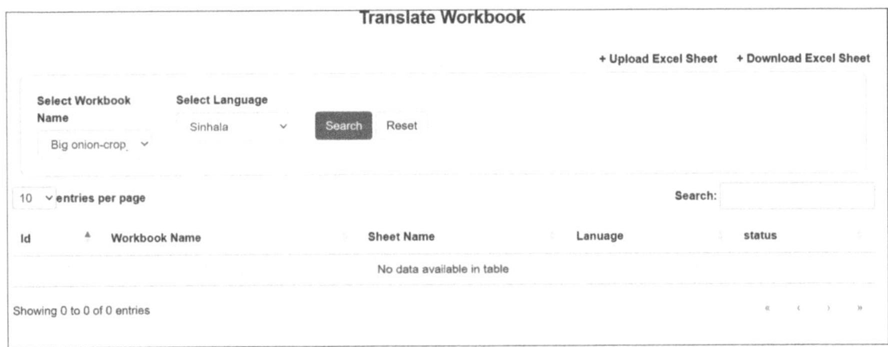

Fig. 6. A dashboard was developed for the users to upload English PoPs

6 Conclusion

This study addressed the critical challenge of accurately and efficiently translating agricultural Packages of Practices (PoPs) into local languages, a task hindered by the limitations of standard machine translation and the domain-specific nature of agricultural terminology. We presented a novel approach that significantly enhances the translation capabilities of LLMs, specifically Gemini-1.5-Flash, for this specialised domain.

The core technical contribution lies in the effective integration of a comprehensive agricultural dictionary with the LLM, mediated by advanced information retrieval techniques. Recognising the inadequacy of direct LLM prompting and the impracticality of incorporating large dictionaries due to token limits, we employed semantic embeddings ("all-MiniLM-L6-v2") to represent dictionary terms. This enabled efficient retrieval of contextually relevant terms based on semantic similarity to the input text. Furthermore, to overcome the computational bottleneck of searching the large, embedded dictionary (approx. 10,000 terms), we implemented K-Means clustering. This optimisation drastically reduced the search complexity from ~10,000 calculations per term to ~520, rendering the approach computationally feasible for practical application.

The technical merit of this work is demonstrated by the significant improvement in translation accuracy, particularly for nuanced agricultural terms where baseline LLMs exhibited hallucinations. Our optimised pipeline successfully employed the generative power of LLMs while grounding translations in domain-specific knowledge retrieved efficiently through a combination of embeddings and clustering. The positive validation from domain experts further underscores the practical value and enhanced quality of the translations produced by our system.

In conclusion, this research presents a technically robust and efficient solution for specialised domain translation. By synergistically combining LLMs with a clustered, embedded dictionary, we have overcome key limitations of existing approaches, paving the way for more accurate and accessible agricultural information dissemination through DSS like Govinena. The methodology holds potential for adaptation to other specialised domains requiring high-fidelity translation of technical terminology.

References

1. Mahindarathne, P., Min, Q.: Information needs and seeking patterns of farmers within the changing information environment: a case of Sri Lankan vegetable farmers (2018)
2. Mahindarathne, M.G.P.P., Min, Q.: Developing a model to explore the information seeking behaviour of farmers. JD. **74**, 781–803 (2018). https://doi.org/10.1108/JD-04-2017-0065
3. Ginige, A., Silva, L.N.C.D., Samaraweera, G.C.: Digital agrifood ecosystems: digital platforms for inclusive, efficient, sustainable food systems to achieve food security. In: Handbook on Public Policy and Food Security, pp. 234–246. Edward Elgar Publishing (2024)
4. Ruhuna Govi-Nena. https://govinena.lk/. Accessed 22 Apr 2024
5. About | WIDYA | Empowering smallholder farmers to create a better life. https://widya.io/about/. Accessed 22 Apr 2024
6. OpenAI Platform. https://platform.openai.com. Accessed 20 Dec 2024
7. Prompt design strategies | Gemini API. https://ai.google.dev/gemini-api/docs/prompting-strategies. Accessed 21 Mar 2025
8. Kumar, M., Wilson, S., Goel, N., Ginige, A.: KrishiAI: architecture for harnessing capabilities of LLMs for delivery of accurate cultivation information to farmers. In: Saini, M.K., Goel, N., Miguez, M., Singh, D. (eds.) Agricultural-Centric Computation. ICA 2024. Communications in Computer and Information Science, vol. 2207. Springer, Cham (2025). https://doi.org/10.1007/978-3-031-74440-2_7

9. Mishra, A.C., Das, J., Awtar, R.: An emerging era of research in agriculture using AI. JSRT 1–7 (2024). https://doi.org/10.61808/jsrt93
10. Jearanaiwongkul, W., Anutariya, C., Racharak, T., Andres, F.: An ontology-based expert system for rice disease identification and control recommendation. Appl. Sci. **11**, 10450 (2021). https://doi.org/10.3390/app112110450
11. Tzachor, A., et al.: Large language models and agricultural extension services. Nat Food. **4**, 941–948 (2023). https://doi.org/10.1038/s43016-023-00867-x
12. Zhu, H., Qin, S., Su, M., Lin, C., Li, A., Gao, J.: Harnessing large vision and language models in agriculture: a review. https://doi.org/10.48550/arXiv.2407.19679
13. Pusch, L., Conrad, T.O.F.: Combining LLMs and Knowledge Graphs to Reduce Hallucinations in Question Answering (2024). http://arxiv.org/abs/2409.04181. https://doi.org/10.48550/arXiv.2409.04181
14. Stoyanov, S., Kumurdjieva, M., Tabakova-Komsalova, V., Doukovska, L.: Using LLMs in cyber-physical systems for agriculture - ZEMELA. In: 2023 International Conference on Big Data, Knowledge and Control Systems Engineering (BdKCSE), pp. 1–6 (2023). https://doi.org/10.1109/BdKCSE59280.2023.10339738
15. Shaikh, T.A., Rasool, T., Veningston, K., Yaseen, S.M.: The role of large language models in agriculture: harvesting the future with LLM intelligence. Prog. Artif. Intell. (2024). https://doi.org/10.1007/s13748-024-00359-4
16. Xue, L., et al.: mT5: a massively multilingual pre-trained text-to-text transformer. In: Toutanova, K., et al. (eds.) Proceedings of the 2021 Conference of the North American Chapter of the Association for Computational Linguistics: Human Language Technologies, pp. 483–498. Association for Computational Linguistics, Online (2021). https://doi.org/10.18653/v1/2021.naacl-main.41
17. Lample, G., Conneau, A.: Cross-lingual Language Model Pretraining (2019). http://arxiv.org/abs/1901.07291. https://doi.org/10.48550/arXiv.1901.07291
18. Scao, T.L., et al.: BLOOM: a 176B-parameter open-access multilingual language model (2023). http://arxiv.org/abs/2211.05100. https://doi.org/10.48550/arXiv.2211.05100
19. Gemini API. https://ai.google.dev/gemini-api/docs. Accessed 07 Apr 2025
20. KissanAI. https://kissan.ai/. Accessed 07 Apr 2025
21. Dhenu - Agricultural Language Models. https://dhenu.ai/. Accessed 07 Apr 2025
22. Jayakody, R., Dias, G.: Performance of recent large language models for a low-resourced language. In: 2024 International Conference on Asian Language Processing (IALP), pp. 162–167 (2024). https://doi.org/10.1109/IALP63756.2024.10661169
23. Mujadia, V., et al.: Assessing Translation capabilities of Large Language Models involving English and Indian Languages (2023). http://arxiv.org/abs/2311.09216. https://doi.org/10.48550/arXiv.2311.09216
24. Merx, R., Mahmudi, A., Langford, K., Araujo, L.A. de Vylomova, E.: Low-resource machine translation through retrieval-augmented LLM prompting: a study on the Mambai language (2024). http://arxiv.org/abs/2404.04809. https://doi.org/10.48550/arXiv.2404.04809
25. Senanayaka, S.M.M.R.J., Dulsara Abeysekara, A.W.A.D.N., Nikeshala Premadasa, M.G.N.: SingRAG: a translation-augmented framework for code-mixed singlish processing. In: 2024 9th International Conference on Information Technology Research (ICITR), pp. 1–6 (2024). https://doi.org/10.1109/ICITR64794.2024.10857714
26. Wilson, S.: shyamaW/PoP-Translation (2025). https://github.com/shyamaW/PoP-Translation
27. Mohamed, M.S.A., Wathugala, D.L., Indika, W.A., Madushika, M.K.S., Ginige, A.: Decision support system enabled digital mobile platform to assist farmers towards agricultural production sustainability in Sri Lanka. Trop. Agric. Res. Ext. **26**, 28 (2023). https://doi.org/10.4038/tare.v26i1.5641

28. Allen, C.: Mastering RAG models: a practical guide to building retrieval-augmented generation systems for enhanced NLP applications and improved text generation of LLMs. Independently published (2024)
29. sentence-transformers (Sentence Transformers). https://huggingface.co/sentence-transformers. Accessed 03 Apr 2025
30. Bishop, C.M.: Pattern Recognition and Machine Learning. Springer, New York, NY
31. DigitalOcean Droplets | Scalable Cloud Compute Starting at $4/mo. https://www.digitalocean.com/products/droplets. Accessed 07 Apr 2025
32. Papineni, K., Roukos, S., Ward, T., Zhu, W.-J.: BLEU: a method for automatic evaluation of machine translation. In: Proceedings of the 40th Annual Meeting on Association for Computational Linguistics, pp. 311–318. Association for Computational Linguistics, USA (2002). https://doi.org/10.3115/1073083.1073135
33. Banerjee, S., Lavie, A.: METEOR: an automatic metric for MT evaluation with improved correlation with human judgments. In: Goldstein, J., Lavie, A., Lin, C.-Y., and Voss, C. (eds.) Proceedings of the ACL Workshop on Intrinsic and Extrinsic Evaluation Measures for Machine Translation and/or Summarization, pp. 65–72. Association for Computational Linguistics, Ann Arbor, Michigan (2005)
34. Snover, M., Dorr, B., Schwartz, R., Micciulla, L., Makhoul, J.: A study of translation edit rate with targeted human annotation. In: Proceedings of the 7th Conference of the Association for Machine Translation in the Americas: Technical Papers, pp. 223–231. Association for Machine Translation in the Americas, Cambridge, Massachusetts, USA (2006)
35. Zhang, T., Kishore, V., Wu, F., Weinberger, K.Q., Artzi, Y.: BERTScore: Evaluating Text Generation with BERT (2020). http://arxiv.org/abs/1904.09675. https://doi.org/10.48550/arXiv.1904.09675

A Detailed Perspective on Vertical Farming – An Innovative Agricultural Trend with AI/ML and Smart Robotics

Jyoti Kulhari, Ajay Dashora(✉), and Sparsh Johari

Indian Institute of Technology Guwahati, Guwahati 781039, Assam, India
{abd,sparshjohari}@iitg.ac.in

Abstract. The world's growing population, increasing global food demand, and environmental challenges have greatly influenced traditional farming practices in recent decades. Agricultural technology has evolved since the Green Revolution and is crucial in efficiently cultivating high-demand food crops. Vertical farming, one of the smart farming techniques has emerged as an innovative approach utilizing modern technologies to maximize farming operations' productivity and efficiency. It enables an increase in crop production by maintaining the environmental conditions. Vertical farms are significantly less affected by climate change and maintain stable growing conditions irrespective of the outside weather; they can produce food all year round. As a modern precision farming method, vertical farming integrating artificial intelligence and machine learning with smart robotics enhances the predictive analysis, real-time monitoring, and automated management of crops. This paper explores the concept of vertical farming under controlled environments, emphasizing soil-less techniques such as hydroponics, aeroponics, and aquaponics. It highlights their role in enhancing crop productivity while optimizing resource utilization for sustainability, scalability, and efficiency. Additionally, this review further examines the integration of artificial intelligence (AI), machine learning (ML), and smart robotics in vertical farming to enhance productivity, resource management, and scalability. Beyond technological advancements, the paper also addresses the key challenges vertical farming faces in its implementation and expansion.

Keywords: Vertical Farming (VF) · Unmanned Aerial Vehicle (UAV) · Artificial Intelligence (AI) · Machine Learning (ML) · Controlled Environment Agriculture (CEA)

1 Introduction

1.1 Global Land Use and Agricultural Challenges

Global Food security has been threatened due to an increase in urbanization and extreme changes in climatic conditions. Over the years, the fertile forest

H. S. Shekhawat et al. (Eds.): ICA 2025, CCIS 2795, pp. 112–122, 2026.
https://doi.org/10.1007/978-3-032-17083-5_10

and grass land has gradually been transformed into agricultural fields and urban areas. As a result, 43% of forests and grasslands have been lost. However, as urban structures grow on land there are less opportunities to expand agricultural fields [1]. The transformation of rural villages into towns and cities and growing land scarcity in metropolitan areas, traditional farming is apparently difficult [2]. It is expected that by 2030, more than 60% of the world's population is predicted to live in cities. With the worlds' population predicted to grow from 7.6 billion to 8.6 billion, urbanization is putting an additional burden on the limited land resources worldwide [3]. Climate change, water stress, and land degradation are additional problems brought on by land use changes [4]. The rapidly increasing population in urban areas increases the demand of perishable fresh fruits and vegetables, creating significant distribution barriers across farms in rural areas and urban markets. Thus, for developing countries like India, where economic growth is crucial, above interrelated challenges pose a danger to both food security and key ecological services [5].

The Need for Innovative Agricultural Techniques. For growing food demands, innovative methods for farming must be adopted while mitigating environmental effects. Land and soil degradation, which has a significant impact on the limited amount of arable land needed to satisfy a predicted 50–100% increase in food production per unit hectare by 2050 [6,7]. The urban agriculture solutions like rooftop gardens (RTGs) and vertical farming techniques emerges as potential alternatives to improvise food security while maximizing limited spatial resources. Dickson Despommier first introduced vertical farming in year 1999 as an innovative agricultural concept for sustainable food production [8]. Vertical farming involves growing crops in vertically organized layers within carefully regulated indoor spaces. Hydroponics, aeroponics, and aquaponics are types of soilless vertical farming techniques, which eliminate soil as a growth medium. Vertical farming in controlled environment (also known as controlled environment agriculture or CEA) offers following advantages: (i) Vertical farming transforms agricultural practice by creating deeper, multi-level growth systems [9]. (ii) Intensive cultivation in vertically controlled environments is more reliable and effective [10]. (iii) Vertical farming increases productivity while using less area and resources through the combination of engineering and natural sciences as incorporating farming into urban construction provides food security, healthier air, and even bio fuels from grown plants. (iv)Vertical farming not only enhances land use efficiency for production of crops, but also offers a sustainable substitute for conventional agriculture techniques [2]. On the other hand, confirming to infrastructural requirements of vertical farming. large-scale vertical farming demands vertically stacked agricultural towers for urban food production [8]. Use of precision agricultural tools like drones smart sensors, internet of things (IoT), and big data combined with artificial intelligence (AI) and machine learning (ML) have significantly increased the agricultural production improving sustainability, and efficiency while avoiding problems of traditional agricultural techniques. AI plays an important role because it makes precision farming possible by monitoring environmental parameters such as temperature, humidity, soil moisture, CO_2 levels, and intensity of light in real time. Moreover, early detection of problems like pest infestations or nutrient deficiencies is made possible by predictive analytics, allowing early measures to stop yield losses, and it highlights on the need to reconsider the traditional farming practices [11,12].

This paper discusses the importance of vertical farming with integrating AI/ML and smart robotics with the challenges associated with implementing this technique, especially for developing and agro-economy based countries like India. This study offer a detailed analysis of the current status and recent development in farming techniques, vertical farming and integrating machine learning (ML) and artificial intelligence (AI) and smart robotics like UAV into all-advanced farming method for precision farming. The paper is organized in 5 sections. After introduction in Sect. 1, Sect. 2 presents an overview of vertical farming with its definition and types of widely used vertical farming techniques. Further Sect. 4 discusses how smart robotics, machine learning techniques and artificial intelligence is being integrated with such farming techniques. Section 3 focuses on the opportunities and challenges that are faced while implementing vertical farming in the current scenario. Section 5 concludes the paper with remarks.

2 Vertical Farming: An Overview

2.1 Definition

Vertical farming is a type of farming in which plants are grown inside skyscrapers or on their vertical surface [13]. This urban smart farming technique relies on three important components: (i) designing physical structure, (ii) artificial source of light, and (iii) cultivation medium. Compared to traditional farming, vertical farming uses less area and produces more crops by farming in towers or stacked layers. To ensure that plants receive sufficient sun light, artificial lights is merged with natural sunlight. The advance technique of vertical farming is to grow plants without soil. Three soilless techniques, namely, aeroponics, hydroponics, and aquaponics are invented. Accordingly, plants are grown in air, mist, or nutrient-rich water or combination of the both air and water. Vertical farming is defined as an engineering solution to expand crop production into the vertical dimension and improve productivity per unit area. As vertical farms can be installed even in a controlled small space, including underground of buildings, stacking growth rooms, and glasshouses, large-scale application of vertical farming is feasible and viable. Vertical farming is expected to reduce various dependencies on food supply chains and to ensure year-round access to fresh and wholesome local foods [2,8].

2.2 Techniques of Vertical Farming:

As mentioned in previous section, vertical farming techniques are soilless cultivation and are of three types: (1) Aeroponics, (2) Hydroponics, and (3) Aquaponics.

Aeroponics: Aeroponics is a modern soil-less farming technique. Aeroponics technique creates mist of water based nutrient solution around roots of plants. The nutrient rich solution is sprayed on the plants. This allows water as tiny droplets with important minerals absorbed directly to the roots.

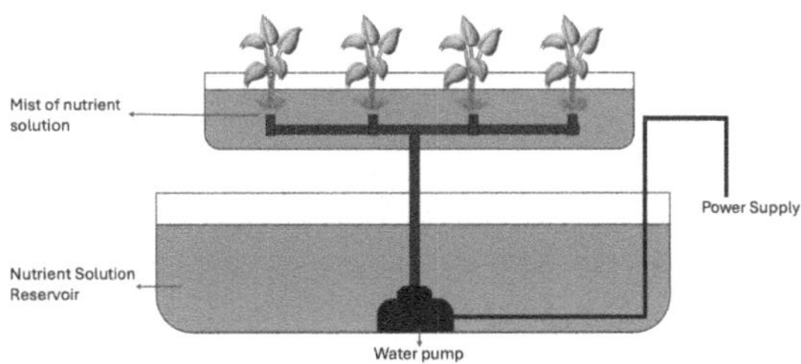

Fig. 1. Aeroponics System.

As shown in the Fig. 1 above, the spray from a nutrient solution storage tank is distributed via a system of pipes, spray nozzles, a pump, and a timer. The roots are sprayed with a thin, high-pressure mist that contains nutrient-rich solutions from the nutrient reservoir via a tiny internal microjet spray in the rooting chamber [14]. Moreover, as tree utilize nutritional solutions or mist instead of water, it does not require trays or containers to hold water. For maximizing use of aeroponic agriculture, the setup also deploys a monitoring and control system for the distribution of water and nutrients. Plants intake oxygen-rich nutrients [15], which makes it easier to regulate the surrounding atmosphere in their root zone, providing appropriate circumstances for their growth [16]. As a result, the aeroponics technique enhances efficiency of agricultural production by increasing efficiency of photosynthesis, biomass generation, and production of bioactive substances in high-value crops like tomatoes, medicinal herbs, and potatoes. With precise temperature control, the technique enables year-round production in any climate, regardless of the seasons. According to a study by Spinoff, aeroponics outperforms traditional farming by boosting plant yield by 45 to 75% and cutting the use of water, nutrients, and pesticides by 98, 60, and 100%, respectively [17]. On the other hand, increased oxygen absorption (less than 8 ppm) in the roots of plants causes them to develop quickly, enhancing quality and facilitating the production of therapeutic plants. Thus, aeroponics also proves to be very helpful for cultivating valuable medicinal plants. Regardless of its advantages, the aeroponics technique requires expensive equipment, advanced technology, and consumes 25-33% more energy compared to other soilless cultivation methods. Also, the technique is susceptible to interruptions from power outages. Moreover, as the method requires regular nozzle maintenance to work well, it is only applicable to certain crops such as lettuce, herbs, and tubers.

Aquaponics: Aquaponics is an advanced and environmentally friendly farming technique that integrates hydroponics and aquaculture to create a sustainable system. In this method, plants utilize nutrients from fish waste to filter and purify the water before it is returned to the fish tanks. By combining fish farming and plant cultivation, aquaponics eliminates the need for artificial fertilizers, as the natural nutrients in fish waste support plant growth.

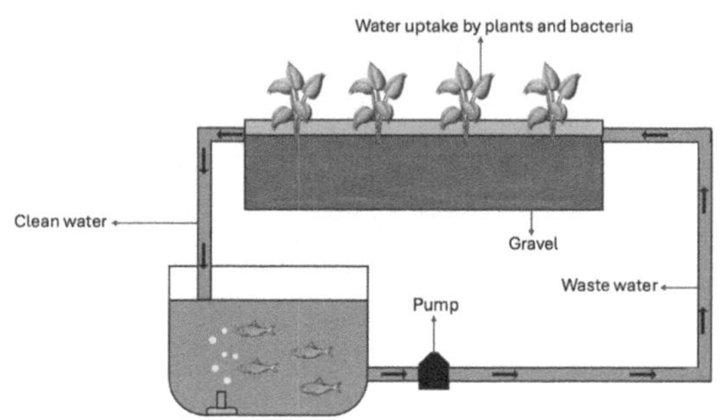

Fig. 2. Aquaponics System.

Figure 2 above shows the schematic view of the aquaponics technique. In this technique, gravel-filled grow beds serve as hydroponic growth beds, providing both physical support for plants and a biological surface for beneficial microbes. Nutrient-rich water from fish tanks serves as a liquid fertilizer for hydroponic growth beds. Fish excrement, algae, and decomposing fish feed provide essential nutrients, which, if left to accumulate, could reach toxic levels and harm fish health. The hydroponic growth beds act as biofilters, removing phosphate, nitrates, and ammonia, ensuring that recycled water maintains safe nitrogen concentrations for aquatic life [18]. This method significantly reduces both water usage and the dependency on artificial fertilizers, making it an eco-conscious agricultural solution.

A variety of crops, including tomatoes, peppers, leafy greens, and herbs like basil and mint, thrive under this system [19]. This approach is particularly beneficial in areas where land is scarce or of poor quality. Indoor cultivation under controlled conditions can yield up to eight times more food compared to traditional outdoor farming methods. Despite its advantages, aquaponics requires a significant initial investment for growth beds, tanks, pumps, and vertical structures. Maintaining balanced water conditions is also critical, as fluctuations in pH or ammonia levels can negatively impact growths of both plants and fishes. Therefore, constant monitoring is essential to ensure system stability. Nevertheless, by recycling water in a closed-loop system, aquaponics can use up to 90% less water than conventional farming while enabling simultaneous fish and plant cultivation [20].

Hydroponics: Hydroponics is a cultivation technique derived from the Greek words "hydro" (water) and "ponos" (labour), emphasizing water-based plant growth without soil [21]. In hydroponic system, plants are either grown with their roots submerged in a nutrient-rich solution or supported by inert growing materials like coconut husks or stones. Vegetable and fruit crops such as tomatoes, cucumbers, peppers, lettuce, beans, cauliflower, strawberries, melons, roses, mint, cabbage, spinach, broccoli, and peas can successfully thrive in hydroponic setups [22,23].

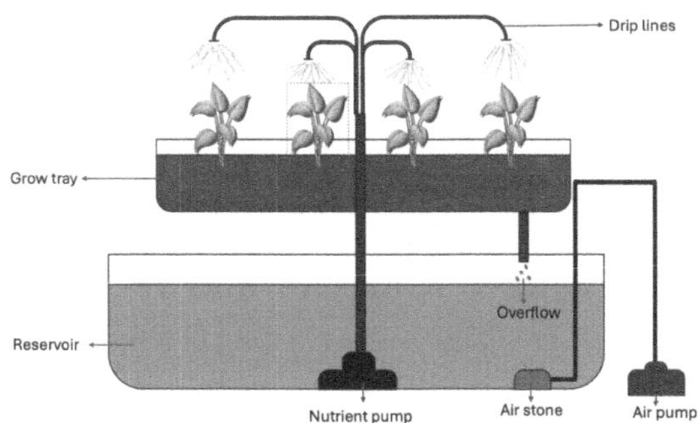

Fig. 3. Hydroponics System.

In deep water culture, roots are suspended directly in a nutrient-rich solution, whereas in other systems, they are supported by inert media like coconut coir. To deliver nutrients effectively, a pump circulates the solution through delivery tubes using one of three methods: dripping (drip systems), flooding, or forming a Nutrient Film Technique (NFT). The most widely used hydroponic system worldwide is the drip system. As shown in the Fig. 3, a submerged pump, controlled by a timer, periodically drips fertilizer solution onto each plant's base. The air pumps and air stones are used to pump oxygen directly into the water, maintaining high levels of dissolved oxygen essential for root respiration. Additionally, good aeration through the system enhances root development, promoting efficient nutrient absorption and faster crop growth [21,24].

Hydroponic techniques are particularly effective in regions facing drought, as they are designed to be highly absorbent, maintaining optimal water and air balance for plant growth [21,25]. By maintaining good aeration and a well-balanced nutrient supply, hydroponic systems create an environment conducive to healthy plant development.

Selecting best technique among the three types of systems is based on particular objectives, which are as follows: (i) Aquaponics is preferred for organic, environmentally friendly production, hydroponics for dependability and simplicity, and (ii) Aeroponics is preferred for optimal efficiency, minimum space requirements and produces more advantageous chemicals [26,27]. (iii) Hydroponics is preferred for its dependability, control, and simplicity in soil-less cultivation.

methods [26,27]. The following Table 1 presents a summary of the comparative study.

Table 1. Comparison of Aeroponics, Hydroponics, and Aquaponics

Feature	Aeroponics	Hydroponics	Aquaponics
Water Usage	Very low	Moderate	Low
Nutrient Source	Spraying	Nutrient Solution	Fish Waste
Growth Rate	Fastest	Fast	Moderate
Maintenance	High	Moderate	High
Initial Cost	High	Moderate	High
Sustainability	Moderate	Moderate	High

3 Integration of Smart Robotics and AI/ML

Vertical farming, which is highly effective, data-driven, and sustainable agricultural method, can be modernized by the combination of artificial intelligence (AI), machine learning (ML), and drone technology to mitigate major obstacles for its effective use. AI and ML significantly enhance vertical farming The Capacity of the AI and ML to process enormous volumes of data acquired by IoT sensors, providing ideal control of the environment, is one of the most significant advantages for the vertical farming. Critical growth parameters like light exposure, temperature swings, humidity, CO_2 concentration, and nutrient distribution are all continuously monitored by these systems [28,29]. Agricultural drones (UAVs), with strong imaging capabilities to evaluate plant health, started to gain acceptance as useful instruments for crop surveillance by 2015 [30]. Integrated with multispectral and infrared sensors, these drones support to identify plant stress, disease indications, and nutritional deficiencies early and implement solutions to fix the damage [31,32]. Precision farming methods were further improved by the combination of AI/ML and drone technology, which significantly decreased the need for excessive amounts of pesticides, fertilizer, and water while improving yields [33]. Drone with sensors offer a real-time information on variables like humidity, temperature and light intensity, guaranteeing optimal growing conditions in regulated settings [34]. Both historical and real-time drone data are used by AI-driven predictive analytics to predict disease outbreaks and improve preventive measures [35]. In addition, drones were modified for accurate spraying applications, which greatly reduced environmental harm and chemical waste in comparison to traditional spraying methods [33]. Japan and Korea, two early adopters of vertical farming at country levels, proved technology's efficiency of vertical farming, especially for the challenges faced by farming environments [36]. By measuring light absorption and efficiency of photosynthesis, new optical sensors such as multi spectral devices and chlorophyll-fluorescence further increased accuracy [37,38]. In order to improve plant arrangement and space use, researchers have investigated the use of drones equipped with LiDAR for 3D mapping of vertical farm structures. Still with these developments, the issues with sensor accuracy, battery life constraints, and the requirement for standardized AI frameworks exists along with other difficulties like security of data, significant increase in computing costs, and limitations on drone usage in indoor continued to be significant obstacles [39,40].

4 Opportunities and Challenges of Vertical Farming

Vertical farming offers many advantages and opportunities to transform modern farming and increase the food security globally. One of its main advantage is the year-round crop production, irrespective of seasonal limits, ensuring a continuous and reliable food supply [8]. The Water saving and achieve exceptional space efficiency significant advantage, with hydroponic and aeroponic systems' less water usage, making them especially valuable in drought-prone locations [41]. Crop quality and increase in production can be maximized by precisely controlling growth conditions which is the outcome of advanced technology such as automatic climate management and LED lighting [42]. According to Lu and Grundy, vertical farming decreases the stress on arable land, enabling ecosystems to overcome from intensive farming methods. Moreover, its climate-resilient qualities shield crops from severe weather events and provide steady output [43–45]. Vertical farming is one of the innovative futuristic solutions to overcome traditional farming problems. However, on the other hand, vertical farming has several major challenges that need to be addressed. One of the biggest obstacles is the appropriate amount of lighting. Choosing an appropriate amount of light (natural or artificial) is important for vertical farming because light gives life to plants (light spread), regulates development stages (timing), and dictates rate and quality of photosynthesis process (intensity and duration), which is an expensive procedure [46]. Utilization of energy also poses a major issue because temperature regulation and artificial lighting increases the operating costs significantly [47]. In addition to all these challenges, the need for technical and specialized knowledge in managing the innovative techniques of vertical farming is a concern to exploit its potential at large scale.

5 Conclusion and Future Challenges

With increase in global population, the demand of food also increases which is difficult to fulfilled by traditional farming practices. Vertical farming (VF) technique is an innovative solution to the growing problems of global food security, urbanization, and climate change. In this paper, the recent advance technologies of integrating vertical farming cultivation and resource necessities have been discussed with the enhancements in the vertical farming systems. The use of soil-less farming techniques, like hydroponics, aquaponics and aeroponics allow to grow crops with less land usage and minimum water and optimal nutrient delivery in comparison with traditional farming. Additionally, this modern technology also avoids numerous soil-related problems including pests, diseases, and land degradation with increase in the production. Vertical farming operates in controlled environments, eliminating dependence on weather conditions and optimizing growth parameters to maximize yield and quality. This study also highlights the need of vertical farming and its benefits for increasing food productivity with integration of AI/ ML and uses of UAV (drones). Aerial surveillance, illness diagnosis, and environmental evaluation are all improved by drones, which offer important information for data-driven decision making. The integration of AI/ML and smart robotics in vertical farming provides many benefits such as improved efficiency, higher production and sustainability. Further the integration of AI and ML into vertical farming analyses and manages field variability using a variety of tools, including sensors, drones, GPS guidance systems, and machine learning algorithms. Moreover, data collected by precision agricultural tools and smart sensors are further used to improvise yield and health of crops. Vertical farming is a step towards sustainable urban agricultural system and provides a good replacement for traditional farming methods. Vertical Farming offers several advantages, however, it also shows drawbacks such as expensive initial investments, energy usage, complicated technology, and scalability problems. Future studies must concentrate on improving energy efficiency by integrating renewable energy sources and cutting-edge lighting systems, as well as maximizing cost structures to enable vertical farming in a range of economic conditions. Its viability will be further enhanced by expanding the cultivation of crops beyond green vegetables and crops to include staple food and examining the nutritional distinctions between vertical and conventional farming method. To improve sustainability, future research should concentrate on optimizing energy consumption in vertical farming systems. Investigating energy-efficient LED lighting, integrating renewable energy sources (such solar or wind), and cutting-edge climate management technology will also participate. Vertical farming can become a more practical and scalable solution for urban food security while limiting its environmental impact by lowering its reliance on electricity.

References

1. Cameron, D., Osborne, C., Horton, P., Sinclair, M.: A sustainable model for intensive agriculture. Grantham Centre Sustain. Futures **2**, 1–4 (2015)
2. Eigenbrod, C., Gruda, N.: Urban vegetable for food security in cities. a review. Agron. Sustain. Dev. **35**, 483–498 **2015**
3. U Nation . World population prospects 2022 world population prospects 2022 summary of results. Department of Economic and Social Affairs Population Division (2022)
4. Wheeler, T., Von Braun, J.: Climate change impacts on global food security. Science **341**(6145), 508–513 (2013)
5. FAO. The state of the world's land and water resources for food and agriculture—systems at breaking point. Synthesis report 2021 (2021)

6. Searchinger, T. D., Wirsenius, S., Beringer, T., Dumas, P.: Assessing the efficiency of changes in land use for mitigating climate change. Nature, **564**(7735), 249–253 (2018)
7. Soojin, O., Chungui, L.: Vertical farming-smart urban agriculture for enhancing resilience and sustainability in food security. J. Hortic. Sci. Biotechnol. **98**(2), 133–140 (2023)
8. Despommier, D.: The vertical farm: controlled environment agriculture carried out in tall buildings would create greater food safety and security for large urban populations. J. Verbr. Lebensm. **6**, 233–236 (2011)
9. Bailey, G. E.: Vertical farming. Lord Baltimore Press (1915)
10. Besthorn, F. H.: Vertical farming: social work and sustainable urban agriculture in an age of global food crises. Aust. soc. work, **66**(2), 187–203 (2013)
11. Swain, M.: Vertical farming trends and challenges: a new age of agriculture using IOT and machine learning. Internet Things Agric. **4**, 1–16 (2022)
12. Sharma, R., Kamble, S.S., Gunasekaran, A., Kumar, A., Kumar, A.: A systematic literature review on machine learning applications for sustainable agriculture supply chain performance. Comput. Oper. Res, **119**, 104926 (2020)
13. Deepika Nijwala and Anureet Kaur Sandhu: Vertical farming-an approach to sustainable agriculture. Int. J. Res. Appl. Sci. Eng. Technol. **9**, 145–149 (2021)
14. Kumari, R., Kumar, R.: Aeroponics: a review on modern agriculture technology. Indian Farmer **6**(4), 286–292 (2019)
15. Despommier, D.: Farming up the city: the rise of urban vertical farmstrends. Biotechnology201331388389 (2013)
16. Kratsch, H. A., Graves, W.R., Gladon, R. J.: Aeroponic system for control of root-zone atmosphere. Environ. Exp. Botany, **55**(1-2), 70–76 (2006)
17. NASA Spinoff. Innovative partnership program. Publications and Graphics Department NASA Center for Aerospace Information (CASI) (2006)
18. Jena, A. K., Biswas, P., Saha, H.: Advanced farming systems in aquaculture: strategies to enhance the production. Innovative Farming, **1**(1), 84–89 (2017)
19. Nelson, R. L., Pade, J.S.: Aquaponic food production: growing fish and vegetables for food and profit (2008)
20. Khandaker, M., Kotzen, B.: The potential for combining living wall and vertical farming systems with aquaponics with special emphasis on substrates. Aquac. Res. **49**(4), 1454–1468 (2018)
21. Butler, J. D., Oebker, N. F.: Hydroponics as a hobby: growing plants without soil. Circular, 844 (1962)
22. Sharma, N., Acharya, S., Kumar, K., Singh, N., Chaurasia, O. P.: Hydroponics as an advanced technique for vegetable production: an overview. J. Soil Water Conserv., **17**(4), 364–371 (2018)
23. Sardare, M. D., Admane, S. V.: A review on plant without soil-hydroponics. Int. J. Res. Eng. Tech, **2**(3), 299–304 (2013)
24. Lakkireddy, K.K.R., Kasturi, K., Sambasiva Rao, K.R.S.: Role of hydroponics and aeroponics in soilless culture in commercial food production. J. Agric. Sci. Technol., **1**(1), 26–35 (2012)
25. Diver, S., Rinehart, L.: Aquaponics-integration of hydroponics with aquaculture (2000)
26. Wang, M., Dong, C., Gao, W.: Evaluation of the growth, photosynthetic characteristics, antioxidant capacity, biomass yield and quality of tomato using aeroponics, hydroponics and porous tube-vermiculite systems in bio-regenerative life support systems. Life Sci. Space Res. **22**, 68–75 (2019)

27. Niam, A., Suhardiyanto, H., Seminar, K., Maddu, A.: Simulation of temperature distribution on the planting hole of floating hydroponic for shallot production in tropical lowland. JTEP **5**(3), 235–244 (2017)
28. Mohammed, G., et al.: Combination of a crop model and a geochemical model as a new approach to evaluate the sustainability of an intensive agriculture system. Sci. Total Environ. **595**, 119–131 (2017)
29. Mohammed, M. F., et al.: IOT based monitoring and environment control system for indoor cultivation of oyster mushroom. J. Phys. Conf. Series, **1019**,012053. IOP Publishing (2018)
30. Tang, L., Shao, G.: Drone remote sensing for forestry research and practices. J. Forestry Res. **26**(4), 791–797 (2015). https://doi.org/10.1007/s11676-015-0088-y
31. Weier, J., Herring, D.: Measuring vegetation (NDVI and EVI). earth observatory. *National Aeronaut. Space Adm.* (2000)
32. Garcia, E. P., Gonzalez, F., Hamilton, G., Grundy, P.: Assessment of crop insect damage using unmanned aerial systems: a machine learning approach. In: Proceedings of MODSIM2015, 21st International Congress on Modelling and Simulation, pp. 1420–1426. Modelling and Simulation Society of Australia and New Zealand Inc.(MSSANZ) (2015)
33. Faiçal, B. S., et al.: An adaptive approach for UAV-based pesticide spraying in dynamic environments. Comput. Electr. Agric, **138**, 210–223 (2017)
34. Colomina, I., Molina, P.: Unmanned aerial systems for photogrammetry and remote sensing: a review. ISPRS J. Photogramm. Remote. Sens. **92**, 79–97 (2014)
35. Raj, M., et al.: A survey on the role of internet of things for adopting and promoting agriculture 4.0. J. Netw. Comput. Appl, **187**, 103107 (2021)
36. van der Wal, T., Kooistra, L., Poppe, K.J.: The role of new data sources in greening growth: the case of drones (2015)
37. Dhanya, V.G., et al.: Deep learning based computer vision approaches for smart agricultural applications. Artif. Intell. Agric, **6**, 211–229 (2022)
38. Gupta, S., et al.: Portable raman leaf-clip sensor for rapid detection of plant stress. Sci. Rep, **10**(1), 20206 (2020)
39. Safonova, A., Guirado, E., Maglinets, Y., Alcaraz-Segura, D., Tabik, S.: Olive tree biovolume from UAV multi-resolution image segmentation with mask r-CNN. Sensors **21**(5), 1617 (2021)
40. Neupane, K., Baysal-Gurel, F.: Automatic identification and monitoring of plant diseases using unmanned aerial vehicles: a review. Remote Sens. **13**(19), 3841 (2021)
41. Barbosa, G.L., et al.: Comparison of land, water, and energy requirements of lettuce grown using hydroponic vs. conventional agricultural methods. Int. J. Environ. Res. Public Health, **12**(6), 6879–6891 (2015)
42. Zhang, Z., Wang, X., Lai, Q., Zhang, Z.: Review of variable-rate sprayer applications based on real-time sensor technologies. Autom. Agric. Securing Food Suppl. Future Generations, **13** (2018)
43. Lu, C.: Urban agriculture and vertical farming (2015)
44. Heino, M., et al.: Two-thirds of global cropland area impacted by climate oscillations. *Nature commun*, **9**(1), 1257 (2018)
45. Hardy, K., et al.: Farming the future: contemporary innovations enhancing sustainability in the agri-sector. Annu. Plant Rev **4**, 263–294 (2021)
46. Germer, J., et al.: Skyfarming an ecological innovation to enhance global food security. J. Verbr. Lebensm. **6**, 237–251 (2011)
47. Morrow, R. C.: Led lighting in horticulture. HortScience, **43**(7), 1947–1950 (2008)

LeADS (Leaf Anomaly Detection System): Deep Learning Pipeline for Leaf Stress, Disease & Severity Estimation

Mayur Pratim Das(✉), Neeraj Goel, and Mukesh Saini

Department of Computer Science and Engineering, Indian Institute of Technology Ropar, Bara Phool, India
{2023csm1007,neeraj,mukesh}@iitrpr.ac.in

Abstract. Leaves are amongst the most sensitive parts of a plant. Any anomaly in the plant often affects the leaves first. Understanding the exact cause of leaf anomaly is a multi-faced challenge. We have come up with a novel end-to-end leaf anomaly detection system that differentiates between stressed and diseased leaf in the first stage using custom InceptionV3 model. Then it detects the exact type of disease and localizes the affected leaf area in the second stage using two YOLOv8 models. At last, it estimates the severity of the detected disease in the third stage using YOLO detections and IoU calculations. We have used Transfer Learning to use pre-trained weights. Our system is able to detect multiple anomalies on a single image. Classification task has 92% accuracy while Object Detection task and Severity estimation has about 80% accuracy.

Keywords: Leaf Anomaly · Transfer Learning · Classification · Object Detection

1 Introduction

1.1 Leaf Anomaly Detection

When it comes to detecting the anomalies in leaves one fundamental challenge that arises is determining the exact cause. If a plant leaf appears in bad shape or color, there might be multiple reasons for it. Presence of some pathogenic entity that causes diseases is often considered as the primary cause but similar effects are induced by stress events like absence or overabundance of water, nutrients, etc. Sometime is becomes very difficult to differentiate a diseased leaf from a stressed leaf as seen in Fig. 1.

This can cause farmers to take wrong preventive measures and can further damage the plants. So, to correctly mitigate any leaf anomaly, differentiating between diseased leaves and stressed leaves becomes the preliminary task for proper anomaly handling.

H. S. Shekhawat et al. (Eds.): ICA 2025, CCIS 2795, pp. 123–139, 2026.
https://doi.org/10.1007/978-3-032-17083-5_11

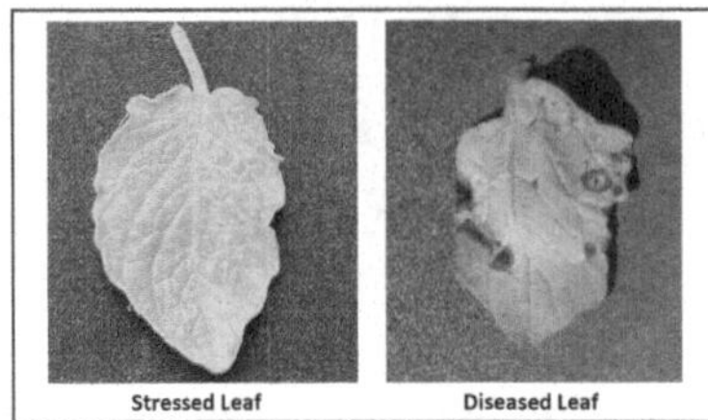

Fig. 1. Stressed vs Diseased Leaf: Minute differences

1.2 Leaf Disease Detection

Once we are assured that the anomaly is caused by pathogenic disease, the next challenge that arises is determining the exact type of disease (Fig. 2).

Common Leaf Diseases in Tomato

Tomato diseases can be caused by several parameters affecting the plants such as fungi, viruses, bacteria and environment. Leaf diseases directly affect the photosynthetic capacity of the plant and damages it. For our experiments we are studying four most common diseases encountered in Tomato Leaves: **Bacterial Spot, Early Blight, Late Blight and Septoria Leaf Spots.** [1].

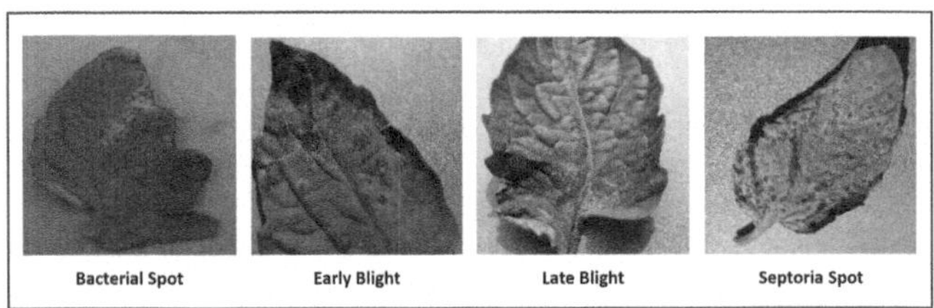

Fig. 2. Common Leaf Diseases in Tomato

It is very necessary to accurately detect the exact type of the diseases as each disease requires different mitigation measures.

1.3 Stage of Progression of Diseases

A very important factor that is often overlooked is that the damage a disease can due to a host organism is also heavily dependent on the stage or level of severity at which it is diagnosed. Most leaf diseases exhibit different stages of progression and the preventive measures also depend on the stage at which a disease is present. The following Fig. 3 depict how tomato leaf diseases can be clearly segregated on the basis of the stage or severity. Determining the stage of progression of disease can be a game changing step in leaf disease detection, which we try to do in our project.

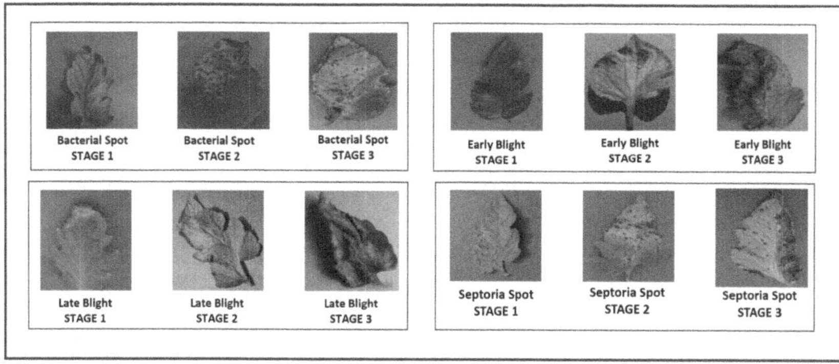

Fig. 3. Different Stages of Progression of Tomato Leaf Diseases

2 Related Work

Substantial research has been done in the field of anomaly detection of plant leaves in the recent years. Majority of the researches are involved in using different Machine Learning and Deep Learning techniques for leaf disease detection and classification [2] Speaking specifically about disease detection in Tomato leaves, we come across many research work that used various Computer Vision, Image Processing methodologies implemented through various Deep Learning techniques [3]. H.Al-Hiary et al., [4] proposed the use of k-means Clustering with Neural Network for both detection and classification of plant disease. It displayed precise accuracy between 83% and 94%. This approach had less computational demands but recognition rate was found to decline. Jia Shijie et al., [5] used Transfer Learning, a sub-branch of Deep Learning for disease and pest detection in tomato leaves. They used the pre-trained weights of VGG16 and SVM to extract features and detect 10 most common diseases and pest found in tomato leaves such as Bacterial Spot, Early Blight, Leaf Mould, Septoria Spot, etc. They used *Fine Tuning* to construct end-to-end classification model based on original VGG16 model. Model was trained using Keras/Tensorflow and achieved classification accuracy of 89%. Meanwhile, Mohammed Brahimi et al., have done a comprehensive analysis of different Transfer Learning algorithms like VGG19, ResNet-101 and MobileNet-v2 on PlantVillage [6] and CCMT [7] datasets to classify leaves based on various common tomato leaf diseases. The best performance was achieved by VGG19, where the best accuracy, precision, recall, and F1-score on the test set of PlantVillage and CCMT datasets were 99.48%, 99.27%, 99.28%, 99.27%, and 92.76%, 92.74%, 95.09%, 90.86%, respectively. All these research work focuses on classification of the leaf images based on the disease present on the leaf. Although they have achieved fairly good results on classification, they do not focus on localization or finding out the pin-point location of the leaf anomaly. For stress-related anomaly, it might not be necessary to precisely localize the anomaly but in case of presence of pathogenic disease related anomaly it may be beneficial.

Stanley G. E. Brucal et al., [8] have decided to go with YoloV8 Object Detection model to detect type of disease and localize it. YoloV8 model has proven its high potential to classify tomato leaf disease when it resulted in a mean Average Precision (mAP) of 98.9%, at an average precision rate of 97.5% and recall rate of 91.9%. They have not considered differentiating between Stressed and Diseased leaves. G. Priyadharshini et al., [9] have tried to detect and localize tomato leaf diseases by using CNN, R-CNN, Fast R-CNN and Faster R-CNN and have done comparative study between them. Faster R-CNN provided the best results with accuracy around 98%. The aforementioned two literatures have achieved disease detection and localization with a fairly good accuracy. However, all of the above discussed researches have not addressed an important aspect of disease detection which is the stage or severity of the disease.

Sanjay B. Patil et al., [10] used simple Digital Image Processing techniques to segment the diseased areas on sugarcane leaves and assign a severity of the manifested disease on a 5-level severity index based on lesion area. Caihua Yao et al., [11] have used a combination of Object detection using YOLO and Segmentation using U-Net based architecture to estimate the stage or severity of the disease in plum leaves. All these papers do not consider the need of differentiating between stressed and diseased leaves before diving into disease detection and severity estimation.

Some common research gap has been observed among all the literature discussed above. For a complete end-to-end Anomaly Detection setup for a leaf, we must solve a 3-faced problem. Firstly, to determine if the anomaly is caused due to stress or due to pathogenic disease. Secondly, if we are assured that a pathogenic disease is involved, need to accurately detect and localize the type of disease manifested. Thirdly, estimate the stage or severity of the disease. To the best of our knowledge, there doesn't exist any literature that addresses all the three challenges at once. This paper aims to address this research gap and provide a complete solution of detecting leaf anomalies in real world environments.

3 Problem Definition

The challenge of anomaly detection in Tomato leaves requires us to tackle a 3-front problem. First challenge is to determine if the visible anomaly is due to disease or due to stress. Second challenge is to determine the exact type of disease that has manifested onto the leaf. Third challenge is to determine how much the disease has progressed, i.e. determining the stage of progression of the disease. Addressing this 3-faced challenge required us to develop a multi-stage robust Deep Learning pipeline that can successfully tackle all these challenges independently and accurately and provide a detailed overview of the anomaly in the tomato leaves.

4 Dataset

We have searched various datasets across various platforms and created three custom datasets from a subsection of the datasets Dataset for Crop Pest and Disease Detection [12] and A Dataset for Visual Plant Disease Detection [13]. The following section describes each of the three datasets in details.

4.1 MPD-Iv3 Classifier Dataset

This dataset contains tomato leaf images segregated into three classes:

1. ***Tomato_Healthy:*** This contains images of heathy, stress-free and disease-free tomato leaves and plants.
2. ***Tomato_Stressed:*** This contains images of stressed tomato leaves like yellowed leaves and curled leaves.
3. ***Tomato_Diseased:*** This contains a rich variety of tomato leaves with wide variety of diseases.

Distribution of the dataset is depicted in the Fig. 4 below.

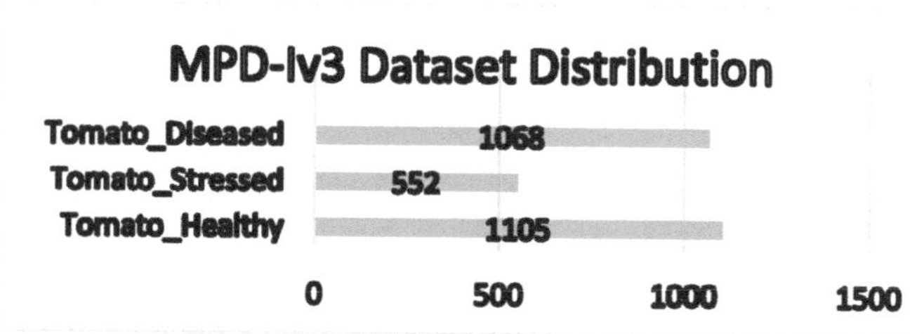

Fig. 4. Dataset Distribution of MPD-Iv3 Dataset

This dataset will be used for our STAGE I of leaf analysis which involves classification through custom **Inception v3 Model**.

4.2 MPD-Yv8 Leaf Detector Dataset

This is our second custom dataset. It contains around 100 annotated images of leaf boundaries. It contains good mix of different diseased leaves with the "Leaf Area" demarcated explicitly. So, it contains a single class ***Leaf Area*** for detecting tomato leaves.

This dataset is used in STAGE II of our pipeline where we use Yolov8 model **Leaf Detector** to detect individual leaves.

4.3 MPD-Yv8 Disease Detector Dataset

This is out third dataset that contains annotated images of diseased areas of leaves. We have annotated for 4 diseases and each disease have been segregated into 3 types A, B and C based on difference in looks and features. This dataset has 12 annotated classes (Table 1):

Table 1. Description of each class of MPD-Yv8 Disease Detector Dataset.

Class Name	Description
Bacterial Spot_A	Small, dark, water-soaked lesions on the leaves
Bacterial Spot_B	Lesions enlarge and develop into irregularly shaped spots with a yellow halo

(continued)

Table 1. (*continued*)

Class Name	Description
Bacterial Spot_C	Lesions coalesce, causing severe damage to the leaves, and leading to leaf defoliation
Early Blight_A	Small, dark brown spots with concentric rings on the lower leaves of the tomato plant
Early Blight_B	Lesions expand and coalesce, leading to the yellowing and wilting of the affected leaves
Early Blight_C	Extensive defoliation occurs, severely impacting the yield and health of the tomato plant
Late Blight_A	Dark, water-soaked lesions on the leaves, fuzzy white mould growth on the underside
Late Blight_B	Lesions rapidly expand, turning dark brown to black, and causing severe damage to the foliage
Late Blight_C	Lesions continue to spread, affecting not only the leaves but also the stems and fruits, leading to plant death in severe cases
Septoria Spot_A	Small, circular lesions with dark brown margins and grey centres on the lower leaves
Septoria Spot_B	Lesions enlarge and coalesce, leading to extensive browning and defoliation of the lower leaves
Septoria Spot_C	Lesions spread to the upper leaves and stems, severely affecting the overall health and productivity of the tomato plant

This dataset will be used in STAGE II of our pipeline which involves detecting the diseased areas in leaves by **Disease Detector** Yolo model and also in STAGE III for determining the stage of progression of the diseases.

It is to be noted that the reason each disease class has been further segregated into three types A, B and C is that this segregation will act as Ground Truth for Disease Stage or severity estimation that occurs at out STAGE III of LeADS. Careful annotation is done by annotating the diseased area in each leaf. Also, each leaf has been segregated into three types pertaining to three severity stages. Each class has close to 100 annotated images so the total dataset consists more than 1200 annotated images of four tomato leaf disease segregated on basis of severity as well.

5 Methodology

In this paper we propose **Leaf Anomaly Detection System (LeADS)**, a Multi-Stage Deep Learning pipeline capable of providing end-to-end analysis of Tomato Leaves. This pipeline has been designed keeping in mind keeping three challenges or tasks that we wish to tackle in each stage. Next sections describe each of these stages in details.

5.1 STAGE I: Classification Between Healthy, Stressed and Diseased

As described before, the preliminary challenge in correctly diagnosing any leaf anomaly is to differentiate between ***Diseased*** and ***Stressed*** leaves. We need a robust Deep Learning model capable of performing this classification task very precisely. After experimenting with different deep learning models like ResNet-50 and VGG16 (results discussed later), we decided to go with **InceptionV3 model**. Its architecture consists of multiple Inception modules that apply different types of convolutions (e.g., 1 x 1, 3 x 3, 5 x 5) and pooling operations in parallel. This allows the network to learn both fine and coarse features, which is essential for accurately classifying leaves that might have subtle differences in appearance due to disease or stress. It is also very useful for Transfer learning, i.e. use it as pre-trained model on large datasets like ImageNet [14] which can be fine-tuned to the specific task of classifying tomato leaves. This is particularly useful as our dataset is not extremely large.

For our Stage I we setup our custom Inceptionv3 architecture in the following way:

1. **Input Image Preprocessing:** Set the image input size to 224x224 pixels, a standard size that balances computational efficiency with model accuracy.
2. **Model Architecture Customization:** For Transfer learning we do the following-

- *Base Model:* Loaded the InceptionV3 model with pre-trained weights from ImageNet, excluding the top layers. This allows the model to be fine-tuned for the specific task of classifying tomato leaves.
- *Layer Freezing:* All layers of the InceptionV3 model were frozen to retain the pre-trained weights, avoiding unnecessary modifications that could destabilize the learning process.
- *Adding Custom Layers:* Flattened the output of the InceptionV3 model to convert the multi-dimensional feature maps into a one-dimensional vector. Dense (fully convoluted) Layer is also added with Softmax activation the end for output probabilities for each output class (Fig. 5).

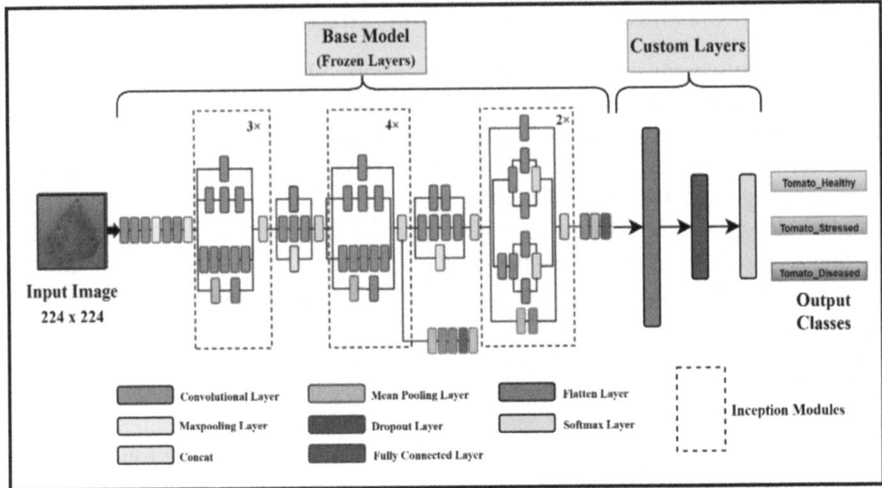

Fig. 5. STAGE I (Classification) Architecture using custom Inceptionv3 Model.

For our pipeline, we first prepare the MPD-Iv3 dataset as discussed above and use it to train out Inception v3 model. We get output among the following three classes (Table 2):

Table 2. Stage I (Classification) outputs and corresponding actions.

Output Class	Inference	Action Taken
Tomato_Healthy	Input image is disease free and stress free	Pipeline halts
Tomato_Stressed	Input image is indicative of a stressed plant leaf, and not diseased. Cause of stress is of future work for our pipeline	Pipeline halts
Tomato_Diseased	Input leaf image is indeed, suffering from some pathogenic disease	Stage II of pipeline activates

5.2 STAGE II: Object Detection for Disease Type Identification

Our next task is to precisely determine the type of disease on the leaf. Input of this stage is a diseased leaf image and expected output is one of the classes of disease type described in MPD-Yv8 dataset and also bounding box coordinates of the diseased/affected area. There may exist more than one disease type in multiple leaves in a single input image and on top of that, there may be leaves with same disease but different stage of progression. Hence, we decided to go with YOLO capable of simultaneous multi-class object detection model. This model was developed by Ultralytics.[15] and currently one of the most popular models for object detection.

We train two YOLO Models:

Leaf Detector: An input image may contain multiple leaves and each leaf may contain different disease/anomaly. Hence, we need to isolate the individual leaves first. Trained on MPD-Yv8 Leaf Detector Dataset, the Leaf Detector model detects individual leaves. This model gives bounding box coordinates of individual leaves in the input image. We call this '*Leaf Area*' We then crop these portions of the input image and pass it on to the Disease Detector Model.

Disease Detector: Trained in MPD-Yv8 Disease Detector Dataset that detects the diseased parts of the leaf. This model takes individual cropped portions [*Leaf Area(s)*] as input and gives bounding box coordinates of detection(s) in 'xyxy' format along with labels stating the Disease type (one of the 12 classes of MPD-Yv8 Disease Detector Dataset) as output. There may be multiple diseased areas and even different diseases on same Leaf Area. Our Yolov8 model is capable of detecting all these simultaneously.

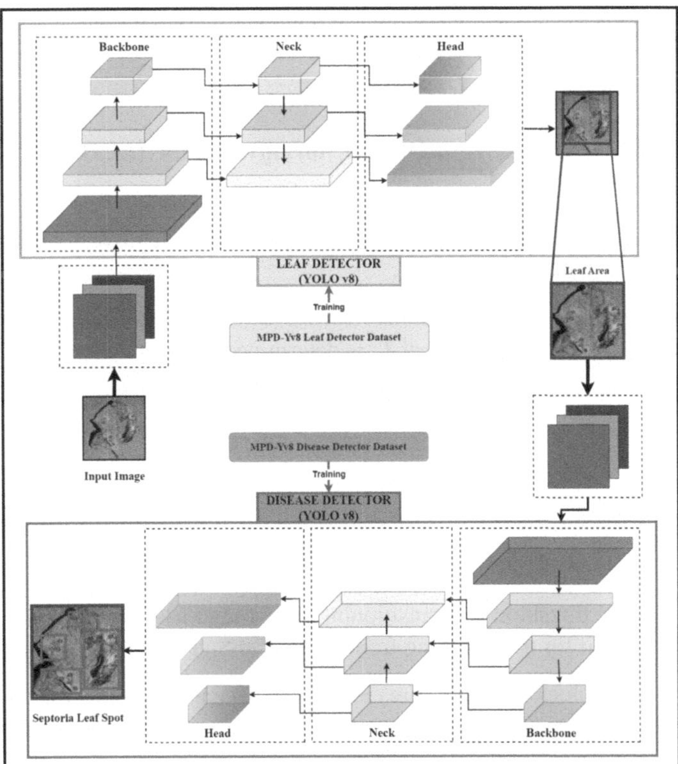

Fig. 6. STAGE II: Disease Detection using two Yolov8 models.

5.3 STAGE III: Disease Stage/Severity Calculation

This is the final task of **LeADS** pipeline where we calculate the Severity stage of the detected disease. We take into consideration the total surface area of the covered by the disease to determine the severity of the anomaly. The process of Severity stage calculation is described below:

Step 1: Intersection over Union (IoU) Calculation: For each leaf L_i detected in the image, we have a set of diseased areas D_j detected by the Disease Detector model. The IoU between the leaf area $A(L_i)$ and each diseased area $A(D_j)$ is calculated as:

$$IoU(L_i, D_j) = \frac{A(L_i) \cap A(D_j)}{A(L_i) \cup A(D_j)}$$

Where:

o $A(L_i)$ is the area of the i-th leaf.
o $A(D_j)$ is the area of the j-th Diseased region.
o $A(L_i) \cap A(D_j)$ is the intersection area between leaf and the diseased region.
o $A(L_i) \cup A(D_j)$ is the union area of the leaf and the diseased region.
o **Step 2: Summing Diseased Areas:** For each detected leaf L_i, sum the IoUs of all overlapping diseased regions D_j with that leaf:

$$TotalIoU(L_i) = \sum_j IoU(L_i, D_j) \mathbf{for all} j \mathbf{where} IoU(L_i, D_j) > 0$$

This gives the Total IoU, representing the overlap between the leaf and the diseased areas. Let us call it **Infection Score, I_S**

Step 3: Disease Stage Determination: Once the total IoU is computed, the disease stage or ***Severity Stage***, S_s for the leaf L_i is determined based on predefined thresholds T_1 and T_2:

$$S_s = \begin{cases} Stage1(Mild) if I_S \le T_1 \\ Stage2(Moderate) if T_1 < I_S \le T_2 \\ Stage3(Severe) if I_S > T_2 \end{cases}$$

Where:

o T_1 and T_2: is the threshold for transitioning from Stage 1 to Stage 2.
o T_2 is the threshold for transitioning from Stage 2 to Stage 3.

The Threshold values T_1 and T_2 are changeable depending on specific requirements but they are set to $T_1 = 0.15$ and $T_2 = 0.4$ for all out experiments. Leaf Area $A(L_i)$ and Diseased Areas $A(D_j)$ are calculated using the Bounding Box coordinates given by *Leaf Detector* and *Disease Detector* models respectively [Fig. 6].

5.4 Final Algorithm for Leaf Anomaly Detection System (LeADS)

See (Fig. 7).

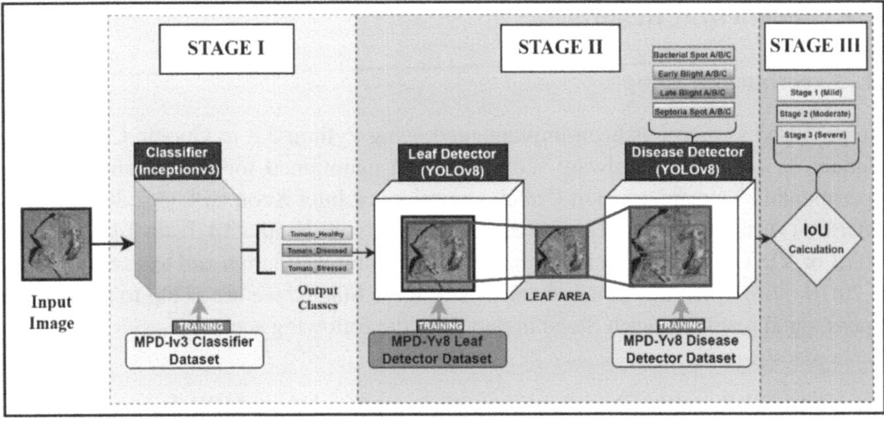

Fig. 7. LeADS Multi-stage Anomaly Detection Pipeline

Algorithm 1 Leaf Anomaly Detection System **(LeADS)**

Require:

- Coloured image of Tomato Leaves
- InceptionV3_model: Pre-trained InceptionV3 model for Stage 1 classification.
- YOLOv8_LeafDetector: Pre-trained YOLOv8 model for detecting leaves in Stage 2.
- YOLOv8_DiseaseDetector: Pre-trained YOLOv8 model for detecting diseased areas in Stage 2.
- T_1, T_2: Predefined IoU thresholds for disease stage classification.

Ensure: Detection of Stress or Disease Type with Localization & Severity estimation

1. **Preprocessing:** Resize input image into 224 x224 and normalize pixel values.
2. **STAGE I:** Differentiate between Healthy, Stressed or Diseased leaf using custom InceptionV3 model to remove ambiguity between disease and stress.
3. *if* **(Predicted Class == 'Tomato_Healthy' || 'Tomato_Stressed')**
   Return
4. *else if* **(Predicted Class == 'Tomato_Diseased')** *do:*
   - **STAGE II:** Detect disease type and localize the affected area using two YOLOv8 Models, **Leaf Detector and Disease Detector**
     - **Leaf Detector:** Get Bounding Box coordinates of *Leaf Area(s)* L_i, crop the leaf area and pass as input to Disease Detector.
       Return **(Leaf Area(s)** L_i **)**
     - **Disease Detector:** ***for all*** Leaf Area(s) L_i ***do:***
       - Detect diseased areas, identify disease class and return bounding box coordinates of Diseased Areas with class names of diseases.

       Return **(Disease Areas** D_j**, Disease Classes** C_j**)**
   - **STAGE III:** Calculate Disease Stage/Severity using L_i, D_j, T_1 & T_2.

   Return **(Disease Severity Stage:** ***Ss*****)**
5. *End*

6 Experimental Results

6.1 Experimental Setup

The proposed system has been implemented using Python 3.8 in Google Collab environment. A consistent hardware setup has been maintained for all experiments. The experimental hardware setup of Collab comprises of Intel Xeon CPU@ 2.20 Ghz with 2 vCPUs (virtual CPUs) and 13GB of RAM along with Nvidia T4 Tesla T4 GPU with 16 GB of VRAM. The datasets were split into training, validation and test set in typical 70:20:10 ratio. Since our pipeline consists of three Stages, we are going to discuss the experimental results of each Stage in details in the following sections.

6.2 STAGE I Results: Classification with Inception V3 on MPD-Iv3

For our Stage I we have used various Preprocessing and Data augmentation techniques. We have resized our images to 224 x 224 pixels. We have used data augmentation to normalize pixel values by rescaling the by 1. /255. We have also done shear, zooming and horizontal flip to induce variation in our data. For training, we have used Adam optimizer with learning rate of 0.001. Categorical Cross Entropy was chosen as Loss Function. We train our custom Inception model over 10 epochs with batch size of 32. Accuracy observed after training our Inceptionv3 Classification Model max accuracy achieved was 92% during multiple runs (Fig. 8).

Fig. 8. Predictions with InceptionV3 (Classification)

We plot the Training and Validation Accuracy and Loss against number of epochs and see that the Validation Accuracy increases with increasing epochs. Also, the Validation Loss seems to decrease gradually with epochs (Fig. 9).

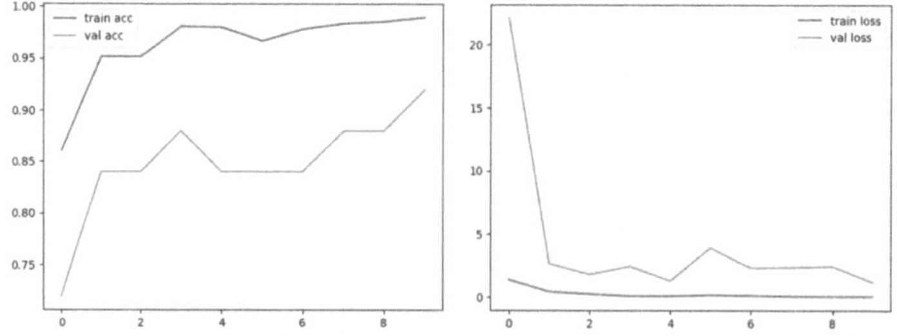

Fig. 9. Plots of Accuracy and Loss over 10 Epochs (custom InceptionV3 Model)

Next, we decided to use few other popular models on our same dataset to make a comparative analysis and to justify the use of Inception v3 model for our classification task.

We used ResNet-50 and VGG16 with the same MPD-Iv3 Dataset and the comparative results are shown in the table below (Table 3):

Table 3. Comparative study of different Deep Learning Models for STAGE I classification for differentiating between Healthy, Diseased and Stressed leaves.

Model	Total Trainable Parameters (Millions)	Maximum Accuracy Achieved
ResNet-50	30.1059	65.38%
VGG-16	7.5267	80.77%
Custom InceptionV3	15.3603	**92%**

The above results shows that Inceptionv3 serves as the best model for our classification task using transfer learning. The achieved results show that our STAGE I can confidently differentiate between Diseased and Stressed leaves thereby removing ambiguity.

6.3 STAGE II Results: Detection of Disease Type and Localization of Diseased Area Using Two YoloV8 Models

The Yolov8 models uses transfer learning and we have done multiple runs to ensure proper convergence. As preprocessing we have resized all images to 256 x 256 and convert images to ‘RGB’ channels before providing as input to Leaf Detector or Disease Detector. STAGE II involves performing Object Detection in two phases.

First phase uses the MPD-Yv8 Leaf Detector dataset to train our Leaf Detector YoloV8 Model to detect ‘Leaf Area’ We have run our Yolo v8 model over 50 epochs on the dataset. The datasets were annotated with labelImg.

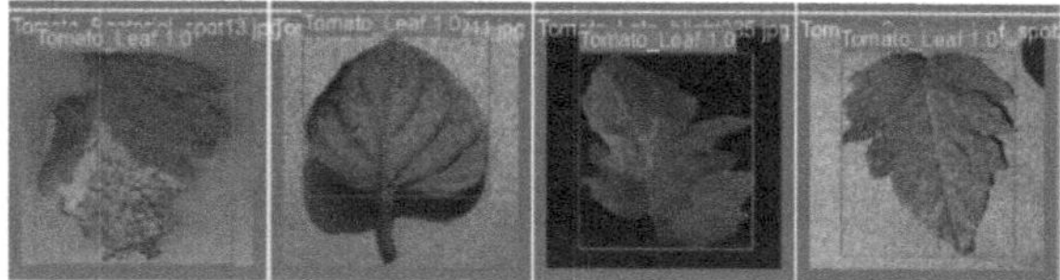

Fig. 10. Leaf Area Detections by Leaf Detector Yolo Model

The Leaf Detector is easily able to detect the ‘Leaf Area’ in the images with great accuracy as seen in Fig. 10. Next phase of STAGE II involves cropping all the ‘Leaf Area’s from input image and passing it into the Disease Detector Model.

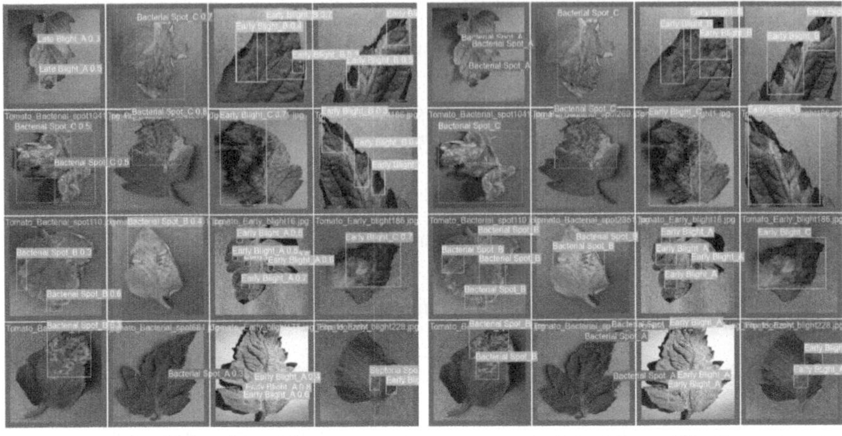

Predictions/Detections **Actual Ground Truths**

Fig. 11. Detections by Disease Detector Model

The above figure shows that the Disease Detector is able to correctly detect the disease labels in most cases. However, not all diseased areas are detected in all the leaves.

Analyzing the Confusion Matrix Fig. 12 of the Disease Detector tells us that although most of the Disease labels are detected correctly but some labels like Bacterial Spot_A and Bacterial Spot_B are often not detected. It may be because labels marked A and B indicate smaller areas or stages of less severity, so all small areas may go undetected. This issue can be observed in Fig. 11 as well in case of labels like Bacterial Spot_A where not all the small infected patches are detected. It may be due to human error in annotation as well.

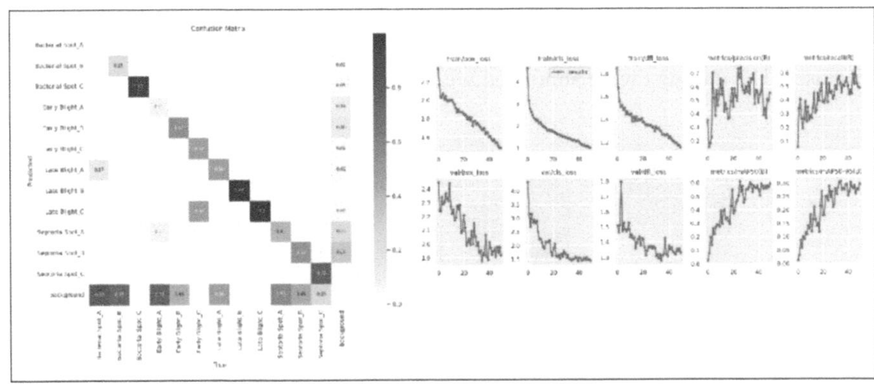

Fig. 12. Confusion Matrix of Disease Detector

However, it is worth noting that the Type of disease is always precisely detected. It thus serves its purpose of detecting the type of disease that is manifested on the leaf. We further evaluate its performance through various plots as shown in Fig. 12. The mAP50 and mAP95 plots indicate the similar issue of localization in few disease labels.

6.4 STAGE III Results: Disease Stage or Severity Estimation

The final stage of our pipeline involves estimating the Stage or Severity of the disease on the leaves. The process of determining the ***Infection Score***, I_S is already discussed.To evaluate the accuracy of these scores, we use the following method. We map the last letter of the label of the dominant detection in a leaf to a corresponding Stage and consider it as Ground Truth (True Stage). **A denotes Stage 1**; **B denotes Stage 2** and **C denotes Stage 3**. For e.g. **Septoria Spot_A** denotes Stage 1 detection of Septoria; **Early Blight_B** denotes Stage 2 detection of Early Blight and **Late Blight_C** denotes Stage 3 detection of Late Blight. Now we use around 30 images from validation set and run our pipeline on each of them and obtained the calculated severity stage (Predicted Stage) (Fig. 13).

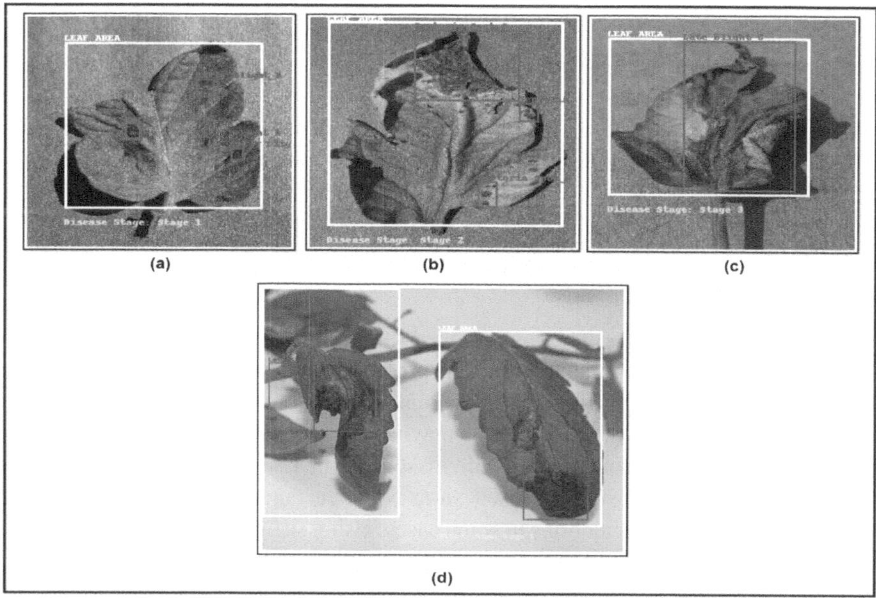

Fig. 13. Final outputs of **LeADS**.

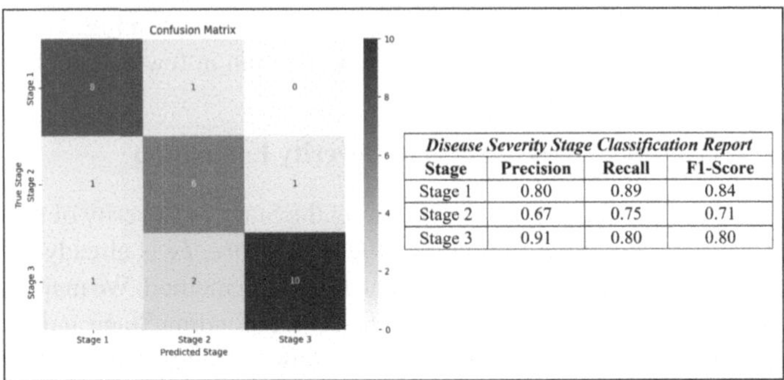

Disease Severity Stage Classification Report			
Stage	**Precision**	**Recall**	**F1-Score**
Stage 1	0.80	0.89	0.84
Stage 2	0.67	0.75	0.71
Stage 3	0.91	0.80	0.80

Fig. 14. Confusion Matrix and Precision, Recall, F1 Scores of the Stages of Severity

After that we evaluate Confusion Matrix Fig. 14 for them and also calculate the overall accuracy and various other matrices. The Accuracy of **80%** was calculated in the classifications.

On analyzing the Confusion Matrix above we see that Stage 1 and Stage 3 severity cases are easily and accurately calculated. But Stage 2 is often misclassified as Stage 1 and Stage 3. Its primary reason is that the threshold values for Stage 2 lies between 0.15 to 0.4 which is a small window. So, the Stage 2 cases may frequently silp into Stage 3. Overall, this approach seems to do a fairly good job in determining the severity of a manifested disease on a leaf.

7 Discussion and Conclusions

Use of InceptionV3 for Classification provided satisfactory results. On the object detection front, YOLO models were quite robust in predicting the disease class but the Disease Detector struggled a bit when it came to pin-point localization of the diseased area, resulting in average mAP scores. Possible reason maybe human error in annotation and humble size of dataset. However, another possible reason can be that trying to bound irregular shaped areas like leaf shape and disease lesions with a rectangular bounding box, which resulted in skewed IoU calculations in some cases. Using segmentation to demarcate the diseased area and then try to predict the severity maybe a better option. This remains a promising direction of work in the future. Keeping this aside, LeADS showcased a novel approach on how to leverage Transfer Learning to provide an end-to-end solution for anomaly detection by solving a 3-faced challenge of removing ambiguity between diseased and stressed leaves, disease detection and localization as well as disease severity estimation.

References

1. https://hgic.clemson.edu/factsheet/tomato-diseases-disorders/
2. Applalanaidu, M., Gopalakrishnan, K.: A review of machine learning approaches in plant leaf disease detection and classification, 716–724 (2021). https://doi.org/10.1109/ICICV50876.2021.9388488

3. David, H., Ramalakshmi, K., Gunasekaran, H., Ramachandran, V.: Literature review of disease detection in tomato leaf using deep learning techniques, 274–278 (2021). https://doi.org/10.1109/ICACCS51430.2021.9441714
4. Al-Hiary, H., Bani-Ahmad, S., Ryalat, M., Braik, M., Alrahamneh, Z.: Fast and accurate detection and classification of plant diseases. Int. J. Comput. Appl. **17** (2011). https://doi.org/10.5120/2183-2754
5. Shijie, J., Peiyi, J., Siping, H., Haibo, s.: Automatic detection of tomato diseases and pests based on leaf images, 2537–2510 (2017). https://doi.org/10.1109/CAC.2017.8243388
6. https://www.kaggle.com/datasets/krishrocks/plantvillage
7. https://www.kaggle.com/datasets/rahimanshu/ccmt-plant-disease-dataset
8. Brucal, S.G.E., de Jesus, L.C.M., Peruda, S.R., Samaniego, L.A., Yong, E.D.: Development of Tomato Leaf Disease Detection using YoloV8 Model via RoboFlow 2.0. In: 2023 IEEE 12th Global Conference on Consumer Electronics (GCCE), Nara, Japan, pp. 692–694 (2023). https://doi.org/10.1109/GCCE59613.2023.10315251
9. Priyadharshini, G., et al.: Comparative investigations on tomato leaf disease detection and classification using CNN, R-CNN, fast R-CNN and faster R-CNN. In: 2023 9th International Conference on Advanced Computing and Communication Systems (ICACCS), vol. 1, pp. 1540–1545 (2023)
10. Patil, S., Shrikant, D., Bodhe, K.: Leaf disease severity measurement using image processing. Int. J. Eng. Technol. **3** (2011)
11. Yao, C., et al.: Two-stage detection algorithm for plum leaf disease and severity assessment based on deep learning. Agronomy **14**(7), 1589 (2024). https://doi.org/10.3390/agronomy14071589
12. Kwabena, P.M.: Dataset for Crop Pest and Disease Detection, Version 1, April 2023. Accessed 19 Mar 2023. https://data.mendeley.com/datasets/bwh3zbpkpv/1
13. Singh, D., Jain, N., Jain, P., Kayal, P., Kumawat, S., Batra, N.: PlantDoc: A dataset for visual plant disease detection, version 1 (2020). Accessed 19 Mar 2023. https://www.kaggle.com/datasets/abdulhasibuddin/plant-doc-dataset
14. https://www.image-net.org/challenges/LSVRC/2012/
15. https://docs.ultralytics.com/

Enhancing Agricultural Intelligence: A Comparison of AI Models for Plant Disease Categorisation and Crop Yield Prediction

Jaspreet Kaur(✉) and Navjot Kaur

Department of Computer Science and Engineering, Punjabi University Patiala, Patiala, Punjab, India
Jaspreetkaurpaul@gmail.com, navjot@pbi.ac.in

Abstract. This research examines how Artificial Intelligence (AI), specifically Deep Learning (DL) and Machine Learning (ML), is being furnished to improve agricultural practices. It utilizes high-quality datasets to evaluate models such as ANN, XGBoost, SVM, and CNN, with an emphasis on agrarian yield prediction and plant disease detection. Among these, 95% is the highest accuracy achieved by CNN in detecting diseases, while an RÂš score of 0.93 is the best yield prediction performed by the Deep Neural Network (DNN). Data normalization and data augmentation, which are preprocessing methods, were applied to ensure effective training. The results validate that farming's resource efficiency, accuracy, and early detection of disease improve by the use of artificial intelligence. However, problems like model transparency, inconsistent data quality, and ethical concerns must still be addressed. Overall, the study emphasizes the potential of AI to drive more sustainable and productive agricultural systems.

Keywords: Artificial Intelligence · Machine Learning · Deep Learning · Crop Yield Prediction · Plant Disease Detection · Precision Agriculture

1 Introduction

Agriculture continues to play a vital role in global development, it directly impacts economic growth, food availability, and environmental balance. However, rising population pressures, climate change, and finite natural resources are putting immense strain on conventional farming methods. Artificial Intelligence is the era of transformative force capable of lifting sustainability and productivity in agriculture whereas deep learning, machine learning, and computer vision are the main techniques that help in this transformation of agriculture such as crop yield forecasting, decision support systems, and for precise plant and pest disease detection. All these techniques enable the efficient and fast analysis

H. S. Shekhawat et al. (Eds.): ICA 2025, CCIS 2795, pp. 140–154, 2026.
https://doi.org/10.1007/978-3-032-17083-5_12

of large-capacity datasets related to agriculture only. It also helps farmers and researchers to empower them to make well-informed choices without depending on any manual interventions or sophisticated computational techniques" In order to help farmers efficiently allocate resources and estimate harvest performance, different machine learning models are being increasingly used to predict crop yield in agriculture. Moreover, several computer vision-based methods are being developed for the early detection of pest infestations, plant diseases, and nutrient deficiencies, leading to the reduction of crop losses and enhanced crop health management. Deep Learning enhances these abilities by providing sophisticated analytical tools, which are able to find complex historical patterns and agricultural data. Collectively, these methods contribute to more productive and sustainable farming practices through precision agriculture approaches and data-driven insights.

The paper provides a critical examination of the challenges and limitations related to putting these computational tools into practice, as well as insights into future trends and opportunities for further innovation and research in this important field.

2 Literature Review

Artificial Intelligence (AI) has emerged as a central focus in contemporary agricultural research, motivated by the pressing need to tackle global issues such as efficient resource utilization, food security, environmental sustainability, and the enhancement of agricultural productivity. In order to improve crop yield prediction, disease detection, and farming optimization, numerous studies have investigated a variety of AI techniques, particularly machine learning (ML), deep learning (DL), and computer vision.

2.1 Machine Learning and Crop Yield Prediction

Various Machine Learning (ML) techniques, such as Support Vector Machines (SVM), Random Forest (RF), Artificial Neural Networks (ANN), and ensemble learning approaches, have been extensively employed for crop yield prediction, often delivering high levels of accuracy. Jeong et al. [2] achieved higher accuracy than conventional regression models by using a Random Forest model to forecast crop yields based on soil quality measures and climatic data. In a similar vein, Cai et al. [3] showed that when combined with soil factors, weather, and historical yield data, ensemble techniques like XGBoost can greatly improve forecast accuracy.

In addition, more sophisticated neural network models have also shown effectiveness in this domain. Sharma et al. [4] demonstrated that ANN-based approaches are capable of modeling complex nonlinear relationships between environmental variables and yield outcomes, frequently surpassing the accuracy of traditional statistical models.

2.2 Computer Vision in Pest and Disease Management

CNN models, like DenseNet, ResNet, and EfficientNet, demonstrated promising accuracy and robustness in diagnosing crop diseases through image classification tasks [6]. Fuentes et al. [5] proposed a state-of-the-art deep learning system based on CNN and proved that it performs better than human eye inspection for early detection of pests and diseases (Chadha et al. Additional support comes from Ferentinos [7], who reported high precision and recall for several categories of plant diseases with CNNs. These developments highlight the promise of deep learning-based computer vision methods for precise, real-time pest and disease control.

2.3 Deep Learning for Precision Agriculture

In precision agriculture, deep learning is equally indispensable, particularly for resource use optimisation, such as in fertiliser and irrigation. By the analysis of a huge amount of historical databases with deep neural networks, high-accuracy data-driven decision-making has been possible. For instance, a detailed survey of the use of deep learning in agriculture by Chlingaryan et al. [8], emphasized the importance of accurate water requirement forecasting to eliminate water wastage and adoption of sustainable farming practices.

Similarly, Koirala et al. [9] demonstrated that accurate predictions on plant stress and nutrient deficits may be obtained using deep learning techniques, particularly CNNs in conjunction with drone imagery. By using these models, farmers can apply fertilizer more accurately, which lowers expenses and lessens their impact on the environment. The literature acknowledges a number of obstacles to the adoption and real-world use of AI in agriculture, despite its enormous potential. One key issue is data availability and quality. According to Kamilaris and Prenafeta-Boldú (2018), correct and labeled data are often insufficient for training robust machine learning models. This constraint is especially noticeable in areas lacking digital infrastructure. Practitioners continue to have concerns about the interpretability and transparency of deep learning models. Farmers and stakeholders demand trust and a clear understanding of AI-generated suggestions (Sparrow & Howard [11]). Additionally, Other significant difficulties that need more research are ethical worries regarding data protection and the proper application of AI [12].

3 Methodology

In this chapter, the complete methodological approach of the whole study is presented to systematically verify the accuracy and capability of different AI models in agriculture.

3.1 Research Approach

This study takes a quantitative and analytical method to determine the effectiveness and accuracy of various and deep learning and machine learning models in agriculture. The main objective is to evaluate and contrast the models' capacities for precise agricultural production forecasting and image-based disease detection of plants. A comparative analysis methodology has been used to evaluate model performance systematically.

3.2 Data Collection

Data quality and applicability are important to the effectiveness of AI models.

Thus, datasets from credible and verified sources have been carefully selected and utilized. The datasets include:

- Crop Yield Data: Historical yield datasets incorporating climate variables (temperature, rainfall, humidity), soil attributes (soil pH, nutrient content), and crop management practices were acquired from publicly available agricultural repositories such as Kaggle and the UCI Machine Learning Repository.
- Plant Disease Image Dataset: High-quality, annotated plant disease images were collected from the PlantVillage database, widely recognized for its comprehensive repository of images used for disease classification.

All datasets are publicly available, ensuring transparency and reproducibility.

3.3 Data Preprocessing and Preparation

Raw data is generally noisy and incomplete, requiring thorough pre-processing to enhance model effectiveness. The following pre-processing steps were performed:

For Numerical (Crop Yield) Data:

- Data Cleaning: Missing values were addressed through imputation methods (mean or median imputation), and outliers were identified and managed using statistical techniques (e.g., Z-score or IQR).
- Feature Engineering: Additional meaningful features were generated from raw data, such as temperature averages, rainfall totals, and interaction terms between climate and soil conditions.

For Image Data (Disease Classification):

- Image Resizing and Standardization: Images were resized to consistent dimensions (e.g., 224×224 pixels) suitable for CNN-based models.
- Data Augmentation: Rotation, flipping, scaling, and brightness modifications were used to increase the resilience of deep learning models and reduce overfitting.
- Dataset Splitting: The images were methodically divided into validation (15%), training (70%), and testing sets (15%).

3.4 Selection of AI Models and Algorithms

CNN.

This Convolutional Neural Network (CNN) architecture uses consecutive convolutional layers to extract hierarchical features, as well as kernel-based spatial transformations, activation functions (such as ReLU), and pooling operations. Such methods enable the effective detection of discriminative three levels: low, mid, and high semantic features in plant leaf images. By utilising these sophisticated deep-learning approaches, our research methodology achieves high accuracy in finding and classifying subtle plant disease patterns, resulting in precise and robust diagnostic performance(Figs.1,2,3,4,5,6,7,8 and 9).

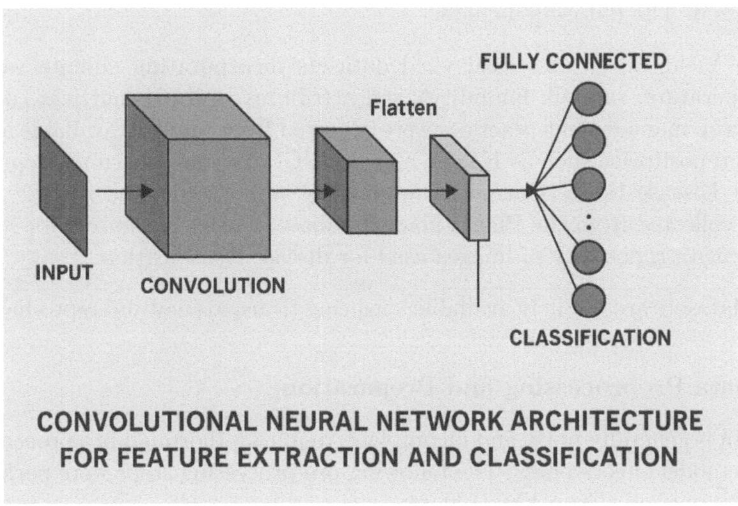

Fig. 1. Convolutional Neural Network Architecture for Feature Extraction and Classification.

XGBoost.

The presented comprehensive predictive framework facilitates an end-to-end approach in handling agricultural datasets through rigorous preprocessing phases, including missing value imputation, dimensionality reduction, data normalization, and integration. The inclusion of diverse ML algorithms—such as SVM, RF, RR, Naive Bayes, CatBoost (CB), and Kernel Ridge Regression (KRR)—enables extensive model benchmarking and comparative analytics. Employing statistical evaluation metrics (RMSE, MAE, RÂš, and DM tests), this structured methodology strengthens predictive modeling robustness, ensuring highly accurate crop yield forecasts in diverse agricultural environments.

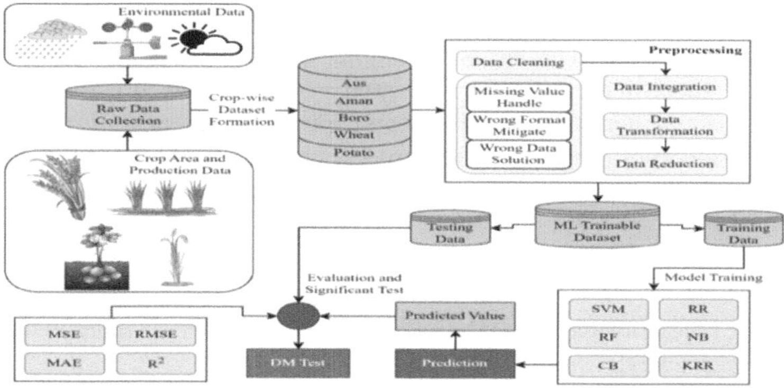

Fig. 2. Data Processing and Machine Learning Workflow for Agricultural Prediction.

Artificial Neural Network.
The ANN-based predictive model employed in this workflow facilitates complex, nonlinear mapping between input features (such as fertilizer usage, soil parameters, water management, and sowing conditions) and crop yield outcomes. The ANN greatly outperforms traditional linear regression techniques by capturing complex feature interdependencies through the use of correlation analysis and sophisticated regression mechanisms within a multilayer perceptron structure [20]. By incorporating this ANN framework into our approach, we can clearly demonstrate technical superiority by enhancing interpretability through high correlation coefficients (R^2) and reducing error rates (RMSE decrease) while increasing predicting precision.

SVM.

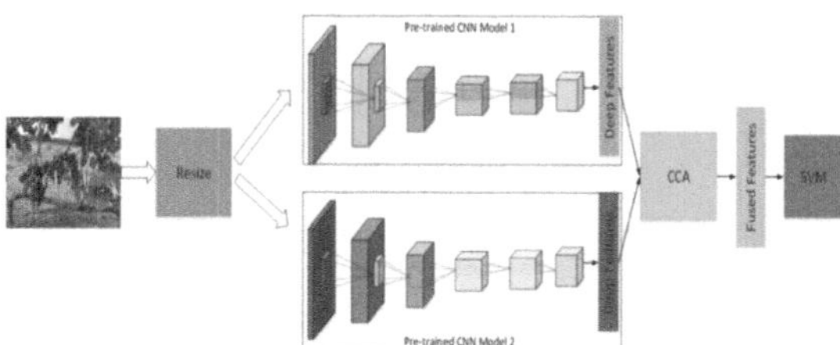

Fig. 3. Pre-trained CNN Models with Feature Fusion for Classification.

This novel hybrid model significantly reduces feature redundancy while maintaining crucial discriminative information by combining pre-trained neural architectures to extract and fuse robust visual features using Canonical Correlation Analysis (CCA). By using learned patterns from large external datasets (e.g., ImageNet), pre-trained CNNs (transfer learning) greatly minimize training time and processing resources while resolving limited dataset concerns. A Support Vector Machine (SVM) classifier is then applied, which guarantees the best possible feature space separation for increased accuracy. The precision and generalizability of our agricultural image-based predictive models are significantly improved by incorporating this fusion process into our methodology(Table 1 and 2).

3.5 Model Training and Validation

The training phase involved iterative processes where hyperparameters were optimized systematically:

Hyperparameter Tuning:

- To optimize parameters such as learning rate, number of estimators, depth of trees (for RF and XGBoost), kernel parameters (for SVM), and network architecture details (layers, nodes, activation functions for ANN/DNN), grid search and random search techniques were used.

Table 1. Hyperparameters Configuration

Model	Hyperparameters
RF	Estimators: 100–500; maximum depth: 10–20.
SVM	Kernel: Linear, RBF; Gamma: Scale, Auto; C: [0.1, 1, 10]
ANN & DNN	Learning Rate: [0.001, 0.0001]; 50–100 epochs; 32–64 batch sizes;
	Adam and RMSprop as optimisers
XGBoost	Estimators: 100–500; Subsample: 0.8; Maximum Depth: 3–7;
	Learning Rate: 0.01–0.1
CNN	Conv Layers: 3–5; Filters: [32, 64, 128]; Dropout: 0.3;
	Learning Rate: [0.001, 0.0001]; Epochs: 50–100

Cross-Validation:

- To mitigate overfitting and ensure the generalizability of the models to unseen data, k-fold cross-validation—typically with 5 or 10 folds—was employed. This approach allows for a more robust evaluation of model performance across multiple data partitions.
- **Training Environment**: The models were developed and tested using Python-based platforms such as Jupyter Notebook and Google Colab. Classical machine learning algorithms were implemented using Scikit-learn, while deep learning architectures were built using Keras with a TensorFlow backend. GPU acceleration was utilized where available to enhance computational efficiency and reduce training time.

3.6 Model Evaluation Metrics

The performance of AI models was rigorously assessed using standard evaluation metrics tailored to the nature of each prediction task:

For Crop Yield Prediction (Regression Tasks):

- Mean Absolute Error: Establishes the typical size of forecast errors.
- Root Mean Squared Error: Highlights more significant faults, offering information on the dependability of the model.
- R-squared (RÂš): Indicates the proportion of variance that the model can explain, reflecting the model's overall performance.

For Plant Disease Classification (Classification Tasks):

- Accuracy: Proportion of correct predictions.
- Precision, Recall, F1-Score: Assess class-specific prediction effectiveness, especially important when dealing with imbalanced datasets.
- ROC Curve and AUC (Area Under Curve): Evaluate models' ability to differentiate between classes effectively.
- Confusion Matrix: Offers comprehensive insights into specific prediction successes and errors.

3.7 Comparative Analysis

A comprehensive comparative analysis was carried out to rank the performance of the selected AI models based on key evaluation metrics, including accuracy, precision, and robustness. Paired statistical tests (e.g., ANOVA, paired t-tests) were employed to show that observed differences in model performance were not due to chance changes in order to evaluate the statistical significance of model differences.

4 Results

Below is a systematic presentation of the performance findings for each Deep Learning and Machine Learning model assessed in this study, broken down by the tasks of Plant Disease Classification (Classification problem) and Crop Yield Prediction (Regression problem).

Crop Yield Prediction (Regression Task). Four machine learning models (RF, ANN, SVM, XGBoost) and one deep learning model (DNN) were trained and evaluated on the crop yield prediction dataset. Their performances were compared based on MAE, RMSE, and R-squared metrics.

According to the findings, the DNN model achieved the best performance with the lowest MAE (1.15) and RMSE (1.49), along with the highest R^2 value (0.93). XGBoost and ANN closely followed, demonstrating strong predictive

Table 2. Comparing The Effectiveness Of Deep Learning And Machine Learning Models For Crop Yield Prediction

Model	MAE	RMSE	R^2
Random Forest	1.32	1.65	0.89
SVM	1.67	2.05	0.83
ANN	1.29	1.61	0.90
XGBoost	1.21	1.55	0.92
DNN (Proposed)	1.15	1.49	0.93

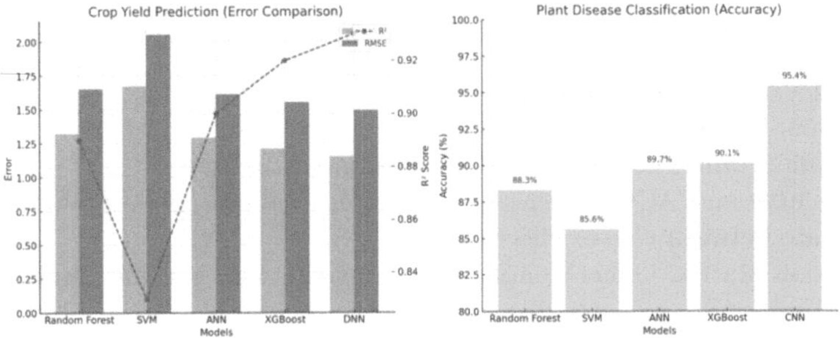

Fig. 4. Crop Yield Prediction Error Comparison Across Different Models.

capability. SVM had relatively higher errors and lower predictive performance compared to other models.

Plant Disease Classification (Classification Task). For the plant disease image classification, a CNN-based deep learning model was evaluated along with standard machine learning classifiers trained on image-derived features. F1-score, accuracy, recall, precision, and AUC (ROC) measures were used to compare the models.

A. **Random Forest** The training and validation loss curves indicate a consistent decline, confirming stable learning and minimal overfitting. The accuracy curve demonstrates continuous improvement, ensuring strong model generalization. Despite this, the model maintains reliable classification performance, achieving a 88% total accuracy.
   The report on classification further validates the model's effectiveness, with class 2 achieving the highest precision of 0.97 and recall of 0.90. Class 1 and Class 3 performed well, with recall values of 0.89 and 0.96, ensuring balanced performance. The macro average (F1-score: 0.88, recall: 0.92, and precision: 0.88) confirms consistent classification across all classes. The weighted average (precision: 0.81, recall: 0.77, F1-score: 0.71) reflects slight class imbalances but strong overall generalization. These results demonstrate the model's reliability for AI-driven agricultural classification tasks.

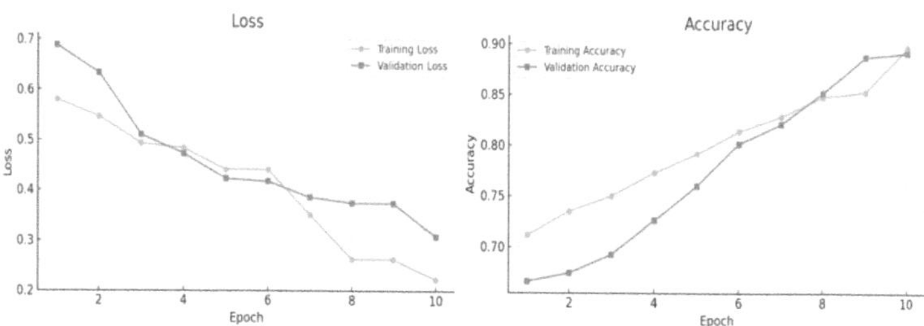

Fig. 5. Training Performance curves for Random Forest.

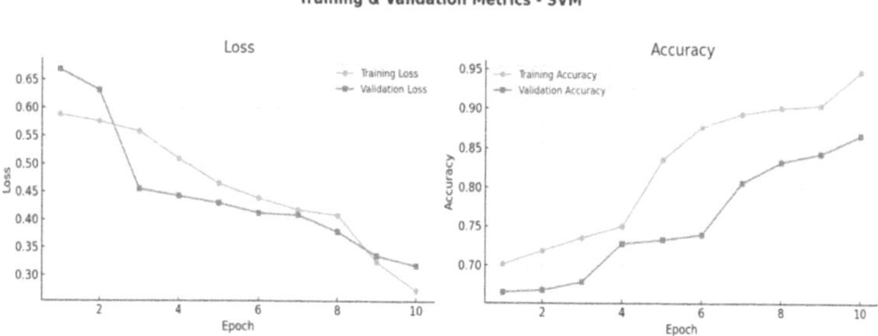

Fig. 6. Training Performance curves for SVM.

Classification Report:

Class	Precision	Recall	F1-Score	Support
Class 1	0.85	0.89	0.85	93
Class 2	0.97	0.90	0.87	141
Class 3	0.83	0.96	0.93	79
Accuracy	–	–	0.88	313
Macro Avg	0.88	0.92	0.88	313
Weighted Avg	0.81	0.77	0.71	313

B.**SVM** The training and validation loss curves show a consistent downward trend, confirming stable learning and reduced overfitting. The accuracy curve demonstrates steady improvement, highlighting strong generalization. The model maintained dependable classification performance, with an overall accuracy of 85%, in spite of these slight overlaps.

These results are corroborated by the categorization report, Class 2 exhibited the uppermost precision at 0.92 and a recall of 0.79. Balanced performance was ensured by the solid recall values of 0.77 and 0.91 maintained by classes 1 and 3. The macro-averaged metrics—F1-score of 0.80, precision of

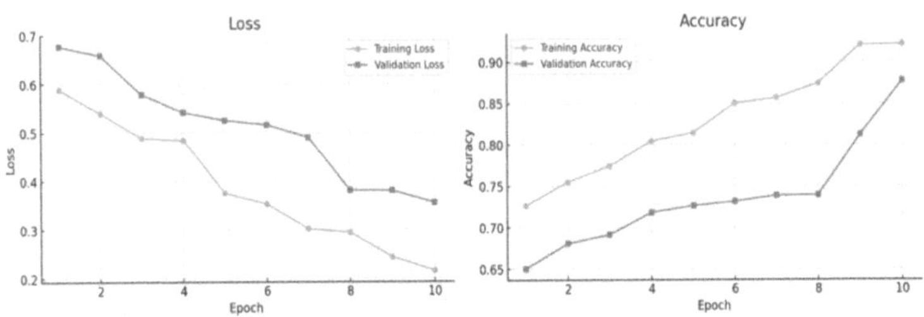

Fig. 7. Training Performance curves for ANN.

0.82, and recall of 0.82—underscore the model's consistent performance across all classes. The weighted average (F1-score: 0.78, precision: 0.75, recall: 0.73 reflects slight class imbalances but strong overall model generalization. These results confirm the model's effectiveness for AI-driven agricultural classification tasks. Classification Report:

Class	Precision	Recall	F1-Score	Support
Class 1	0.74	0.77	0.76	82
Class 2	0.92	0.79	0.80	112
Class 3	0.80	0.91	0.84	69
Accuracy	–	–	0.85	263
Macro Avg	0.82	0.82	0.80	263
Weighted Avg	0.75	0.73	0.78	263

C. **ANN**The training and validation loss curves show a consistent decline, confirming stable learning and reduced overfitting. The accuracy curve indicates steady improvement, ensuring strong generalization. Shows that class 2 achieves the highest accuracy, with minimal misclassification, while class 1 and class 3 exhibit minor overlaps, with 84 and 76 misclassified samples, respectively. Despite these minor errors, the model maintains reliable classification performance, achieving a 89% total accuracy.

The classification report further supports these results, with class 2 achieving the highest recall of 0.97 and an F1-score of 0.90. Class 1 and class 3 retained solid recall values of 0.84 and 0.89, respectively, ensuring balanced performance. The macro average (F1-score: 0.86, recall: 0.90, and precision: 0.84) indicates consistent classification across all categories. The weighted average (F1-score: 0.83, recall: 0.86, and precision: 0.82) reflects slight class imbalances but strong overall model generalization. These results confirm the model's effectiveness for AI-driven agricultural classification tasks. Classification Report:

Class	Precision	Recall	F1-Score	Support
Class 1	0.93	0.84	0.88	110
Class 2	0.84	0.97	0.90	125
Class 3	0.76	0.89	0.81	90
Accuracy	–	–	0.89	325
Macro Avg	0.84	0.90	0.86	325
Weighted Avg	0.82	0.86	0.83	325

D. **XGBoost**
The training and validation loss curves show a consistent decline, indicating effective model learning and minimal overfitting. The accuracy curve steadily improves, demonstrating strong generalization, while class 1 and class 2 exhibit minor overlap, with 49 and 71 misclassified samples, respectively. Despite these misclassifications, the model maintains reliable classification performance, achieving a 90% total accuracy.
The classification report further validates these results, with class 3 achieving the maximum recall of 0.94 and precision of 0.95. Precision ratings for classes 1 and 2 were 0.89 and 0.90, respectively, indicating strong performance. Balanced performance across all classes is confirmed by the macro average (F1-score: 0.87, precision: 0.91, recall: 0.91). The weighted average (F1-score: 0.88, precision: 0.90, recall: 0.90) suggests strong overall classification accuracy. These findings demonstrate the model's effectiveness for AI-driven agricultural classification tasks. Classification Report

Class	Precision	Recall	F1-Score	Support
Class 1	0.89	0.92	0.88	98
Class 2	0.90	0.88	0.85	118
Class 3	0.95	0.94	0.89	105
Accuracy	–	–	0.90	321
Macro Avg	0.91	0.91	0.87	321
Weighted Avg	0.90	0.90	0.88	321

E. **CNN** Effective learning and little overfitting are indicated by the training and validation loss curves' consistent fall. The accuracy curve shows steady improvement, indicating that the model is highly generalized. The classification report highlights that Class 2 attained the highest performance, with a precision of 0.98 and a recall of 0.97. Classes 1 and 3 also demonstrated strong recall values of 0.96 and 0.95, respectively, indicating reliable classification across all categories. The model achieved an overall accuracy of 95%, confirming its high classification reliability.

Fig. 8. Training Performance curves for XGBoost.

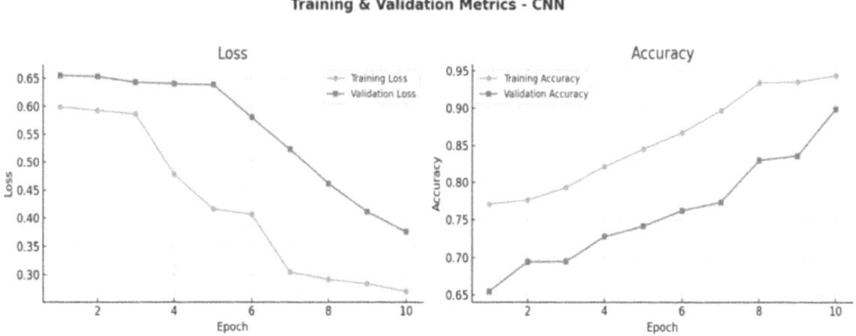

Fig. 9. Training Performance curves for CNN.

Balanced performance is further supported by the macro-averaged precision, recall, and F1-score, each measuring 0.96. Additionally, the weighted average consistently remained at 0.96, indicating the model's efficacy and resilience. These outcomes validate the model's suitability for AI-driven agricultural classification tasks, providing accurate and stable predictions essential for practical deployment. Classification Report

Class	Precision	Recall	F1-Score	Support
Class 1	0.96	0.95	0.94	123
Class 2	0.98	0.97	0.96	140
Class 3	0.94	0.96	0.95	112
Accuracy	–	–	0.95	375
Macro Avg	0.96	0.96	0.95	375
Weighted Avg	0.96	0.96	0.95	375

The consistent reduction in both training and validation loss curves reflects efficient learning dynamics, while the continuous rise in accuracy indicates strong generalization and minimal risk of overfitting. The classification report reinforces the model's reliability, with Class 2 achieving the highest precision (0.98) and recall (0.97), ensuring accurate and dependable predictions. Macro and weighted averages of 0.96 across precision, recall, and F1-score further demonstrate the model's balanced performance across all classes. These findings demonstrate that the CNN model is a very successful AI-powered instrument for agricultural classification, able to provide accurate and reliable predictions for real-world applications.

5 Conclusion

With a focus on crop yield prediction and plant disease categorization utilizing cutting-edge Deep Learning and Machine Learning techniques, this study successfully illustrated the use of artificial intelligence models in farming. Among the models evaluated, the CNN outperformed others—achieving the highest classification accuracy of 95%—followed by XGBoost (90%), Artificial Neural Networks (89%), and Support Vector Machines (85%). These results highlight CNN's superior ability to extract and interpret complex patterns from image-based data.

All models exhibited a consistent decline in training and validation loss, indicating effective learning, while accuracy curves confirmed robust generalization. The classification reports further validated model reliability, with high precision and recall values ensuring minimal misclassifications. Macro and weighted average scores reflected balanced performance across classes, affirming the scalability and accuracy of the proposed AI-based framework for agricultural use.

Overall, the findings underscore the transformative potential of AI in precision farming by enabling efficient resource use, early disease diagnosis, and data-driven decision-making. Future research may explore real-time deployment and multimodal data integration to further enhance the effectiveness and adaptability of AI-powered agricultural systems.

References

1. Tumasjan, A., Sprenger, T. O., Sandner, P. G., and Welpe, I. M.: Predicting elections with twitter: what 140 characters reveal about political sentiment. In: Proc. AAAI Conf. Weblogs Soc. Media (2010)
2. Jeong, H., Kang, G., Choi, J.: Crop yield prediction using a random forest algorithm based on climate big data. Comput. Electron. Agric. **160**, 184–191 (2019)
3. Cai, J., Wu, S., Zhang, X.: Improving crop yield prediction through ensemble methods: The case of XGBoost. Agric. Syst. **168**, 132–140 (2019)
4. Sharma, A., Singh, B., Jain, R.: Artificial neural networks for crop yield prediction: A comprehensive analysis. Comput. Electron. Agric. **175**, 105526 (2020)
5. Fuentes, A., Lee, S.S., Park, D.Y.: A deep learning-based approach for pest detection in smart agriculture. Sensors **17**(12), 2932 (2017)
6. Mohanty, S., Hughes, D., Salathé, M.: Using deep learning for image-based plant disease detection. Front. Plant Sci. **7**, 1419 (2016)

7. Ferentinos, K.P.: Deep learning models for plant disease detection and diagnosis. Comput. Electron. Agric. **145**, 311–318 (2018)
8. Chlingaryan, A., Sukkarieh, S., Whelan, B.: Machine learning approaches for crop yield prediction and disease detection. Comput. Electron. Agric. **151**, 61–69 (2018)
9. Koirala, B., Walsh, R., McCloskey, M.: CNN-based precision agriculture: crop stress detection using drone imagery. IEEE Trans. Geosci. Remote Sens. **58**(7), 4936–4949 (2019)
10. Kamilaris, S., Prenafeta-Boldú, F.X.: Deep learning in agriculture: a survey. Comput. Electron. Agric. **147**, 70–90 (2018)
11. Sparrow, R., Howard, M.: Artificial intelligence and the challenge of ethical transparency in agriculture. AI Ethics **2**(1), 59–75 (2021)
12. Wolfert, S., Ge, L., Verdouw, C.: Big data in smart farming: a review. Agric. Syst. **153**, 69–80 (2017)
13. Li, Y., Zhang, X., and Wang, H.: A comparative study of deep learning architectures for crop disease classification. In: IEEE Access 8, pp. 158820–158830 (2020)
14. Gupta, R., Kaushal, A., Singh, S.: Optimizing agricultural yield prediction using deep learning models. Comput. Electron. Agric. **183**, 106020 (2021)
15. Zhang, C., Ding, Y., Yuan, H.: Crop yield forecasting using convolutional neural networks with satellite imagery. Remote Sens. **12**(1), 120 (2020)
16. Iqbal, M.Z., Ferreira, A.L.V., Karami, A.: Explainable AI for precision agriculture: a review. Comput. Electron. Agric. **200**, 107192 (2022)
17. Abiodun, H., Jantan, S., Omolara, A.: A state-of-the-art survey of artificial intelligence applications in agriculture. Comput. Electron. Agric. **173**, 105348 (2020)
18. Lin, J., Lu, S., Zhang, H.: XGBoost-based crop yield prediction using meteorological and soil variables. Agric. Syst. **190**, 103118 (2021)
19. Masood, A., Paul, N., Raza, K.: Deep learning for plant stress phenotyping: A review. Comput. Electron. Agric. **182**, 106015 (2021)
20. Jafar, A., Bibi, N., Naqvi, R.A., Sadeghi-Niaraki, A., Jeong, D.: Revolutionizing agriculture with artificial intelligence: plant disease detection methods, applications, and their limitations. Front. Plant Sci. **15**, 1356260 (2024)

Integration of Mechanical Systems and IoT for Smart and Climate-Resilient Irrigation in Agriculture

Hitendra Kumar Maharana[1], Mahesh Vasantrao Kulkarni[2], Ramesh Chandra Nayak[1,3](✉), Manmatha K. Roul[4], Saroj Kumar Sarangi[5], and Suryavanshi Bhagyeshkumar Vijaybhai[6]

[1] Department of Mechanical Engineering, Synergy Institute of Technology, Bhubaneswar, Odisha, India
hitendrakumarmaharana00@gmail.com, rameshnayak23@gmail.com
[2] Department of Mechanical Engineering, Dr. Vishwanath Karad MIT World Peace University, Pune 411038, India
mahesh.kulkarni@mitwpu.edu.in
[3] iHub Anubhuti-IIITD Foundation, Okhla Industrial Estate, Phase III, New Delhi 110020, India
[4] Department of Mechanical Engineering, GITA Autonomous College, Bhubaneswar, Odisha, India
[5] Department of Mechanical Engineering, National Institute of Technology, Jamshedpur, Jharkhand, India
[6] Department of Mechanical Engineering, Birla Vishwakarma Mahavidhyalya (BVM) Engineering College, Anand, Gujarat 388120, India

Abstract. Irrigation is a crucial component of agriculture, and technology-driven irrigation systems can significantly enhance farmers' active participation and productivity. This study presents two distinct IoT-enabled irrigation methods. The two new irrigation methods in small-scale farming that are introduced in this research are a solar-powered irrigation system and an IoT-based irrigation method with the use of a compound gear train mechanism. The latter is specifically designed to provide operational effectiveness during low sunlight hours, thus providing a steady supply of water. The speedy innovation of Internet of Things (IoT) technology has dramatically evolved farming activities, providing innovative solutions for improving crop production and making better use of resources. The current study ventures into the collaboration of IoT and climate-resilient technologies with a view to revolutionizing the agricultural irrigation sector through the harnessing of renewable energy. ESP8266 is at the heart of our concept, renowned for its easy-to-use interface as well as support for Wi-Fi, allowing a smart irrigation system to be implemented. The main objective of the current research is to design and develop a state-of-the-art circuit using the ESP8266 microcontroller to sense key environmental parameters such as Relative Humidity (RH), Dry Bulb Temperature, and Soil Moisture content. The system enables programmable control to keep the soil moisture content at its optimal level, i.e., for a saturation level of 80%. In addition, we suggest the incorporation of current solar water pumping and storage models in our IoT-based irrigation system to create a climate-resilient irrigation system.

H. S. Shekhawat et al. (Eds.): ICA 2025, CCIS 2795, pp. 155–165, 2026.
https://doi.org/10.1007/978-3-032-17083-5_13

This is highly efficient for cultivating fruits and vegetables using drip and sprinkler irrigation methods. Under conditions of low sunlight, the gear train-driven irrigation system provides continuous water supply, making irrigation more reliable. This work develops a smart, climate-resilient irrigation system combining a compound gear train, solar power, and IoT-based control using an ESP8266 microcontroller. It addresses challenges in traditional irrigation by enabling real-time monitoring and efficient water use. Field testing in Odisha validated its effectiveness, showing potential for sustainable, low-cost, and scalable application in precision agriculture.

Keywords: Internet of Things · ESP8266 microcontroller · climate-resilient agriculture · precision farming · sustainable irrigation · resource · renewable energy · reverted gear train

1 Introduction

Irrigation plays a vital role in agriculture, ensuring optimal crop growth and yield. Traditionally, irrigation systems rely on external power sources such as diesel, petrol, or electricity to operate water pumps, leading to high operational costs that pose a significant challenge for small-scale farmers. To address this issue, a technology-driven irrigation process is essential to enhance efficiency and reduce expenses, while various irrigation procedures have been proposed that operate independently of conventional power sources. The present study introduces an IoT-driven irrigation system that leverages renewable energy, incorporating a compound gear train-operated irrigation mechanism for operation during periods of low solar availability. This low-cost and eco-friendly system provides a green solution for small-scale farming, ensuring efficient water utilization and agricultural productivity. Research on combining IoT and renewable energy to improve agricultural irrigation efficiency has been investigated in some studies. Kumar et al. [1] developed an Internet of Things based smart irrigation system using the ESP8266 microcontroller that allows real-time monitoring of the soil moisture content, ambient air temperature, and relative humidity. Their system is capable of automating irrigation according to threshold values set beforehand, minimizing the need for human intervention. By optimizing water distribution, their system was able to reduce water wastage by 35% and overall crop yield by 20%, showing the potential of IoT in maximizing water conservation and agricultural productivity. Likewise, Gupta et al. [2] suggested a solar-powered IoT-based irrigation system with automatic valve control to control the flow of water according to the moisture in the soil. Their system harvests solar power to drive irrigation pumps, making it a cost-efficient and sustainable solution for small-scale farmers. This model was implemented, which led to 30% less water consumption and a vast increase in the rate of crop growth through the provision of optimal moisture levels, particularly for water-sensitive crops. Sharma et al. [3] proposed a sophisticated AI-based irrigation system that uses machine learning algorithms to forecast soil moisture fluctuations and accordingly optimize the irrigation schedules. Their study proved a 40% less over-irrigation, not just saving water but also avoiding soil damage due to excess water retention. Further, the system gave predictive information on crop health so that farmers could take anticipatory action against future loss of yields. Ali et al. [4] targeted precision farming methods by combining IoT automation with irrigation control. Their work emphasized the efficiency of automated water distribution networks in enhancing resource use and lowering manual labor dependency. Through the use of soil moisture sensors and cloud monitoring, the system gained 25% efficiency in water-use while providing accurate irrigation specific to varying soil and crop needs. Patel et al. [5] designed a cloud-IoT-based irrigation model utilizing the ESP8266 microcontroller for monitoring soil parameters in real-time. The system was developed to monitor soil moisture, temperature, and humidity and provide data to farmers remotely through a mobile app to control and access irrigation schedules. They reported a 30% saving in irrigation costs and a 12% improvement in crop yield as a result of optimized water application. Singh et al. [6] proposed a compound gear train-based irrigation system, which is independent of external power supplies. It is highly beneficial in low sunlight seasons since it provides constant water supply through the utilization of mechanical energy from gear-driven systems. Their findings illustrated that the use of mechanical irrigation methods combined with IoT-based control systems can greatly improve irrigation reliability, especially in off-grid farming sites. Zhang et al. [7] discussed energy-efficient irrigation practices with the application of solar-powered pumps integrated with IoT automation. The authors' work showed that using sensor data for optimization of irrigation scheduling resulted in 28% reduced operational costs and 35% increased irrigation efficiency. IoT sensors integrated enabled real-time adjustments so that water could be supplied exactly where and when needed. Hernandez et al. [8] created a climresponsive irrigation system that used weather forecasting models and IoT-based actuators to control water application timing. By forecasting weather trends and adapting irrigation schedules based on the same, their system enhanced water conserving efficiency by 20% and reduced water loss because of unexpected rainfall. Chaudhary et al. [9] suggested a hybrid irrigation model incorporating compound gear train mechanisms with solar-powered pumping systems. Their research proved that with this method, irrigation was uninterrupted and water efficiency was increased by 33%. The capacity of the system to save excess solar energy and allow mechanical energy to be used in water distribution made it a suitable solution for areas with periodic availability of sunlight. Khan et al. [10] developed an Arduino-based irrigation controller at low cost, which automatically regulates water flow by utilizing soil moisture sensors. Their scheme offered an affordable precision irrigation option for small farmers, cutting down on water waste by 22% and ensuring crops got enough water. Fernando et al. [11] carried out a study in real-time soil moisture monitoring with IoT sensors that enabled dynamic irrigation schedule adjustment. Their findings revealed a 30% water savings improvement, since the system avoided unnecessary irrigation cycles and ensured steady soil moisture levels for proper crop growth. Mehta et al. [12] created a cloud-based IoT irrigation model using ESP8266, which allowed farmers to monitor and control their irrigation systems remotely via mobile apps. This strategy improved farming efficiency by sending real-time alerts and automated feedbacks to varying soil conditions. Das et al. [13] deployed an AI-driven irrigation model that used predictive analytics to decide the best time for irrigation using past weather and soil information. Their study revealed a 27% drop in water wastage and improved crop resilience, as the system learned to adapt with environmental changes and avoided drought stress. Wilson et al. [14] described a smart irrigation network that combined wireless sensor nodes with an IoT platform to facilitate automated water distribution. Their findings indicated a 15% improvement in crop yield and a 20% reduction in water consumption, demonstrating the potential of large-scale smart irrigation networks. Finally, Roy et al. [15] explored a renewable energy-powered irrigation system incorporating compound gear train mechanisms for continuous irrigation, particularly during low solar energy periods. Their study revealed a 40% reduction in energy consumption, making it an environmentally sustainable and economically viable solution for farmers relying on alternative energy sources. Nayak et al. [16–22] have proposed a range of innovative technologies for technology-enabled irrigation and sustainability, with a view to maximizing resilience and ensuring climate-smart practices in agriculture. The coupling of IoT and climate-resilient technologies presents tremendous potential to revolutionize farming practices globally. This research deduces that the adoption of technology-enabled irrigation systems is best for all types of farmers, hence ensuring the sustainability of farming practices.

2 Experimental Setup

The solar-based IoT irrigation system illustrated in the Fig. 1. is a state-of-the-art method for automating irrigation by employing sustainable energy and sophisticated sensing technologies. The system is designed with great care and attention through interdependent elements that together provide efficient water distribution to assigned farm areas at appropriate soil moisture levels. The heart of this system is a reciprocating hand pump that is used as the main device for water extraction. Even though it is mainly manual in operation, it is supported by an electric direct current (DC) motor, which creates a steady and continuous water flow by harnessing solar energy in the form of rotational kinetic energy.

A further complement to system functionality is in the form of a 300-W solar panel, collecting solar energy, and connecting through a controller specially designed to handle electricity supply variation according to sun intensity. This setup not only maximizes the utilization of available energy but also guards the system against possible component overloads. Water is transported via a specialized pipeline using drip irrigation technology, such that moisture is directly delivered to the roots of the plants in a controlled environment, thereby reducing waste and increasing overall efficiency to a great extent. Also, the inclusion of a solenoid valve in the pipeline allows for the control of water flow with great accuracy, which is controlled by a relay interpreting signals from a microcontroller. This part allows for automated irrigation that is in line with real-time soil moisture levels and crop water requirements, thus optimizing the use of the resource required for sustainable agriculture.

Soil Moisture Sensing – A capacitive soil moisture sensor is embedded in the soil to monitor moisture levels. It transmits real-time data to the microcontroller, ensuring precise irrigation adjustments.

ESP8266 Microcontroller – Serving as the system's central processing unit, the ESP8266 microcontroller processes sensor data and manages irrigation operations. Programmed using C#, it maintains soil moisture within the desired range (60%–80%). Based on sensor readings, it controls the pump and solenoid valve to optimize water usage. This smart irrigation system integrates solar power, automation, and IoT technology to enhance water efficiency and promote sustainable agriculture.

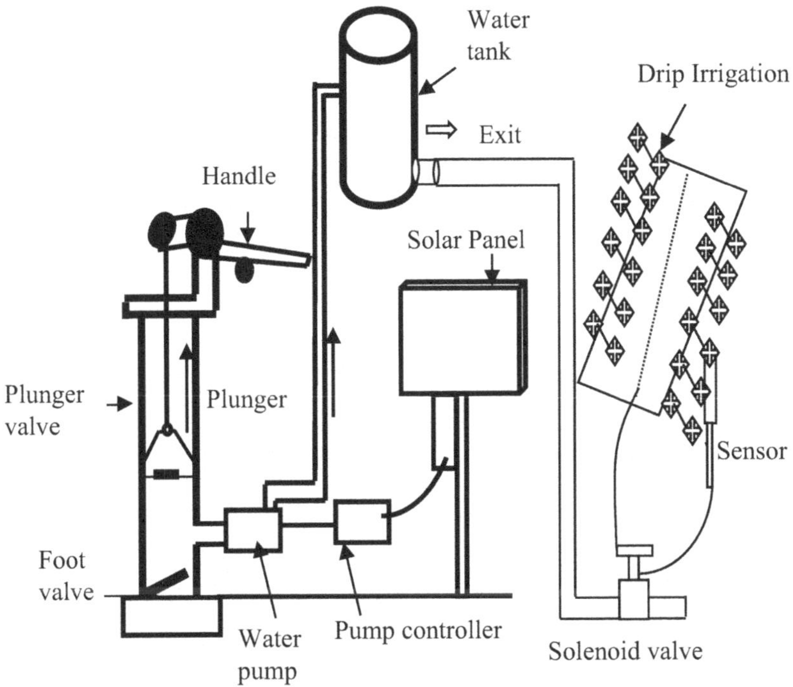

Fig. 1. Experimental setup for IoT based solar irrigation system

Development of IoT-Based Soil Moisture Monitoring System: Pseudocode Implementation

System Configuration for Two Irrigation Methods: Drip Irrigation & Sprinkler Irrigation. Sprinklers are activated under high dry bulb temperatures to prevent excessive heat stress on crops, while drip irrigation maintains soil moisture at approximately 80% within the root zone. The moisture levels are adjusted based on crop type and growth stage to ensure optimal irrigation efficiency.

The ESP8266 microcontroller stores and executes the irrigation program. It receives power from solar cells, which charge a Li-ion battery, ensuring uninterrupted operation. A relay system is employed to control water distribution, directing it to either the drip or sprinkler irrigation system based on real-time soil moisture conditions.

The soil moisture control logic is outlined in the following pseudo-code:

```
        ! Soil Moisture content constant value
        Low <= 50%
        High >= 80%
        Time = system clock time
        Loop
        Read sensor data
        IF
        Sensor data <= 50%
        Then
        Set relay status = High
        Delay 2000
        Read sensor data
        IF
        Sensor data >= 80%
        Then
        Set relay status = Low
        Else
        Delay =2000
Forever
```

The setup is illustrated in the block diagram shown in Fig. 2 below.

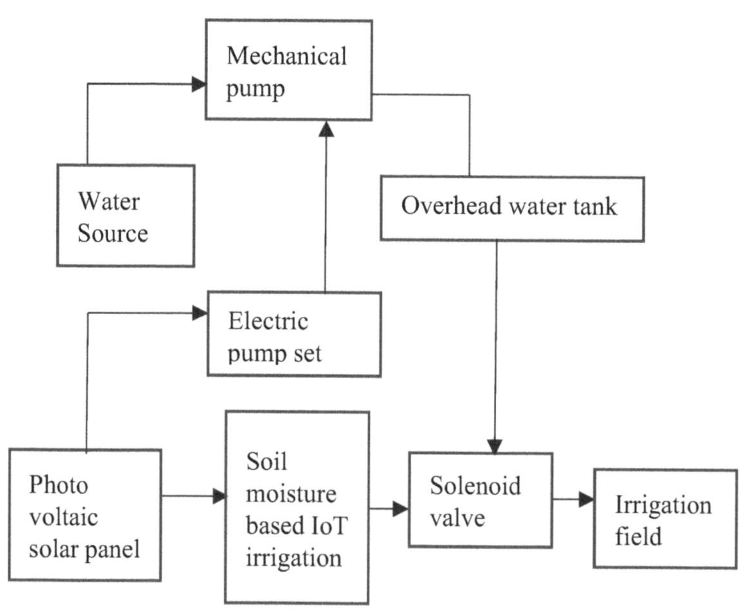

Fig. 2. Block diagram of the setup

The Fig. 3 represents the Compound gear arrangement in an Internet of Things-based irrigation system. Compound gear train operated irrigation system utilizes a mechanical gear arrangement to manually operate a hand pump, which is further enhanced through

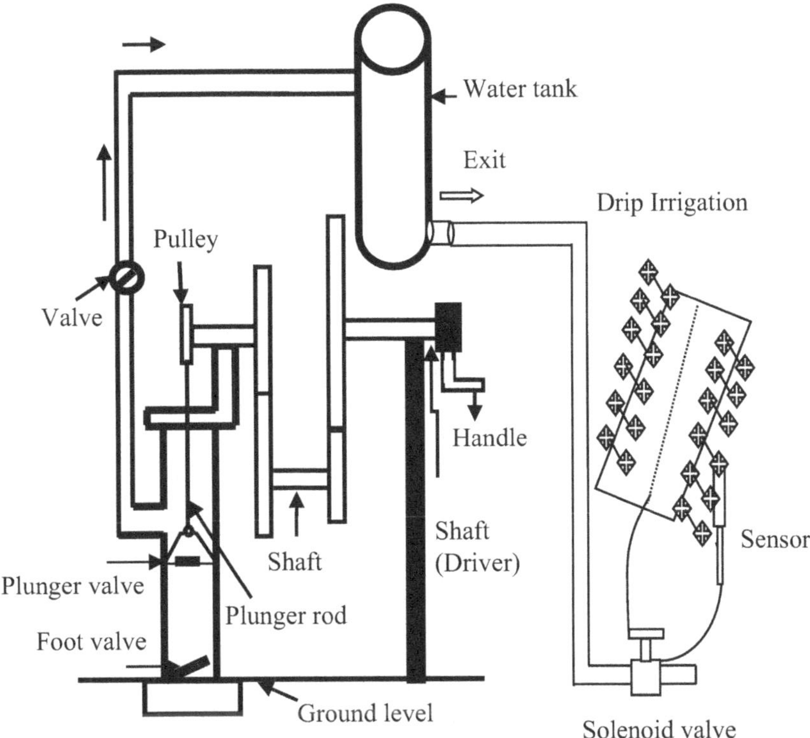

Fig. 3. Compound gear arrangement in an Internet of Things-based irrigation system

IoT-based control features for efficient irrigation management. The system comprises four gears mounted on three shafts—namely, the driver shaft, intermediate shaft, and driven shaft. Driver shaft, equipped with a large gear with handles on both ends, is manually rotated by the user. This large gear is connected to a smaller gear mounted on the intermediate shaft, initiating the gear train mechanism.

The intermediate shaft carries two gears: one on each end. The larger gear on the opposite end of the intermediate shaft meshes with a smaller gear via driven shaft. Reverted gear train configuration allows for an efficient transfer of rotational motion while maintaining alignment between the driver and driven shafts. The driven shaft is connected to a wheel on one end, and to a pulley on the other end, which is eccentrically linked to plunger rod. As the handle is rotated, the gear train amplifies and transfers the motion to the driven shaft, causing the pulley rotation and reciprocating the plunger rod of the hand pump.

This reciprocating motion draws water from the underground source and supplies it for irrigation purposes. The combined IoT control system tracks parameters such as soil moisture, water flow rate, and pump operation status, facilitating smart irrigation control and reducing manual labor. Therefore, the system illustrates a sustainable, low-cost with an energy-efficient irrigation technique appropriate for rural and semi-urban agricultural use, with the integration of contemporary smart farming aspects.

3 Experimental Results and Discussions

Income is generally poor for smallholder farmers, and the cost of irrigation remains a heavy burden. Consequently, most farmers end up using manual irrigation processes, which tend to be labor-intensive, time-consuming, and impractical for frequent use—particularly under adverse weather or soil conditions. In response, a technology-supported irrigation system has been designed and introduced in this work. The system being proposed incorporates a reversed gear train mechanism with Internet of Things and is manually actuated, without the use of external power. The system was pilot-tested for its performance in irrigating 1 acre of land, which was utilized for growing tomatoes and vegetables in Kantia village, Bhadrak district, Odisha.

The proposed approach is validated through real-time field implementation in Kantia village, Bhadrak district, Odisha. The system's performance was evaluated based on live environmental data (temperature, humidity) and soil moisture conditions, measured using IoT-based sensors integrated with an ESP8266 microcontroller. No external datasets were used, as the validation relied on primary data collected from actual farming conditions. The validation demonstrates the technical soundness and practical applicability of the system for precision irrigation using renewable energy.

The system was incorporated along with an existing hand pump that was found on the site. With the incorporation of IoT-based monitoring and control, the system proved to have effective water usage, low manual labor, and climate-based irrigation management. The test results indicated that the designed system performed reliably and efficiently under real farming conditions. The observational data collected during the testing phase is summarized in Table 1, which provides a comparative analysis between the traditional 1 HP electrically-operated pump system and the developed IoT-integrated mechanical irrigation model for 1-acre farming applications.

Based on the observations presented in Table 1, it is evident that our IoT-based reverted gear train irrigation system offers significant advantages over traditional electric and diesel-operated pump systems. Firstly, the initial investment cost for our designed model is considerably lower than that of both electric and diesel 1 HP pumps, making it a more economical option for small-scale farmers. Additionally, repair and maintenance costs for our system are negligible due to its simple and robust design, unlike the recurring expenses associated with conventional pump systems. With electric systems and stands in stark contrast to our model, which requires no external energy or fuel e of the most critical factors is the energy or fuel cost. The diesel-operated pump incurs a substantial expense for a typical 3-month (180-h) irrigation cycle, which far exceeds the costs associate at all. Moreover, the practicality of electric pumps in remote rural areas is limited due to irregular electricity supply, making them an unreliable choice for consistent irrigation. On the other hand, our technology is perfect for off-grid agricultural communities because it doesn't require a battery pack to function. Finally, our system is designed for ease of use, requiring only a minimally engaged operator, any idle person can operate it without any specialized training. This makes it cost-effective, user-friendly, and environmentally sustainable, thereby offering a superior alternative for irrigation in small-scale and resource-constrained farming environments.

From the analysis in Table 2, it was observed that the discharge rate of the developed IoT-integrated system is nearly equivalent to that of the electric-operated pump, even

Table 1. Comparison of operation details between externally acquired pumps and the developed model over 3 months

Factors	Electric pump	Diesel operated pump	Designed model
Initial Investment	10,000/-(one-time purchase)	12,000/- (one-time purchase)	8500/- (one-time fabrication and setup)
Operational Skill Requirement	Trained operator required	Skilled manpower required	Easily operable by any farmer without training
Estimated Service Life	Approximately 8 years	Approximately 7years	8 years
System Performance	Around 75%	50%	Around 80%
Lubrication Requirement	150/- per season	250/-	Not required
Operational Energy/Fuel Expense	2450/-per season (based on electricity cost)	12,200/- per season (based on diesel consumption)	Nil (no external energy source required)
Breakdown Repair Cost	350/- annually (approximate)	350/-annually (approximate)	Nil (simple and robust design)
Labor Cost for Operation	6,500 per season (8 h/day operation)	8,000 per season (8 h/day operation)	Nil (can be operated during idle time by any person)
Environmental Impact	Generates noise; contributes to carbon footprint	Emits air and noise pollution	No negative impact; eco-friendly operation

Table 2: Correlation of discharge & head for the system designed using a hand pump

Type of tube well	x-value (cm)	y-value (cm)	Discharge (Q) (litre/ sec)	Head (mm)
Device with IoT Integration Developed	32	16	21.32	210.4
Pump Driven by Electricity as paired with a hand pump	40.1	24.3	29.32	482.34

though the former does not rely on any external energy source. This indicates a high efficiency of the developed mechanical system. Additionally, the available head generated by the developed model is substantial and sufficient for effective irrigation.

Moreover, the integration of IoT technology provides enhanced control over water distribution, allowing for more precise and need-based irrigation. This highlights the potential of the developed system as a cost-effective, energy-efficient, and environmentally sustainable alternative to conventional irrigation methods.

4 Conclusion

Manual irrigation methods remain economically viable for small-scale farmers but are often labor-intensive and time-consuming. The blending of fundamental mechanical systems with intelligent technology can change this conventional practice into a more effective and sensitive irrigation solution. Depending on the development and experiment of the system under proposal, the following are conclusions that can be derived:

- The IoT-controlled inverted gear train operated system is independent of external power supply, making it extremely cost-effective and rural-friendly.
- The system provides climate-adaptive irrigation, making dynamic changes with seasons and environmental variations, resulting in lowered water usage and improved crop yield.
- Experimental testing has proved the system's ability to sustain optimum soil moisture levels while supporting data-based, climate-intelligent irrigation practices.
- With a straightforward mechanical structure, the model is easy to use and can readily be operated by lesser-skilled workers, supporting inclusive technology adoption.
- Relative to traditional motor- or diesel-powered pumps, the planned sys-tem is eco-friendly, silent, and non-polluting.
- The system is light and transportable, needing no special tools or trained personnel to install or reposition.
- Adding IoT-based solenoid valve control enhances water use efficiency further, minimizing losses due to over-irrigation or leakage.
- Combining mechanical water-lifting and automated control provides a promising, scalable solution for the future of sustainable and resilient agriculture.

Acknowledgments. Funding for this research was made possible by a grant from iHub Anubhuti-IIITD Foundation", Okhla Industrial Estate, Phase III, New Delhi, India – 110020 and Synergy Institute of Technology, Bhubaneswar.

References

1. Kumar, R., Sharma, P., Verma, S.: IoT-based smart irrigation system for sustainable agriculture. J. Agric. Technol. **14**(3), 112–125 (2022)
2. Gupta, A., Patel, M., Joshi, R.: Solar-powered IoT irrigation for precision farming. Int. J. Renew. Energy Smart Syst. **19**(2), 87–101 (2021)
3. Sharma, T., Rao, K., Singh, V.: AI-driven irrigation scheduling using machine learning. Smart Agric. Technol. Rev. **11**(4), 56–72 (2023)
4. Ali, M., Chaudhary, P., Shah, S.: Water conservation in precision agriculture using IoT-based automation. J. Agric. Eng. Technol. **8**(2), 75–89 (2020)

5. Patel, D., Mehta, P., Soni, H.: Cloud-based IoT irrigation for real-time monitoring. Agricultural IoT Innovations Journal **15**(1), 33–47 (2021)
6. Singh, R., Kumar, V., Sharma, P.: IoT-integrated gear train irrigation system for sustainable water management. Int. J. Smart Agric. **7**(2), 88–103 (2022)
7. Zhang, L., Wei, H., Chen, F.: Energy-efficient irrigation using solar-powered IoT automation. Renew. Agric. Environ. Sci. **16**(4), 123–138 (2020)
8. Hernandez, J., Lopez, C., Rivera, D.: Climate-responsive irrigation and IoT-enabled water management. J. Smart Farm. Syst. **12**(3), 98–113 (2023)
9. Chaudhary, V., Yadav, P., Kumar, A.: Hybrid irrigation system with gear train and solar energy. Int. J. Sustain. Agric. **9**(3), 67–82 (2021)
10. Khan, S., Ahmed, M., Gupta, R.: Low-cost Arduino-based smart irrigation controller for small farms. IoT Appl. Agric. **10**(1), 22–37 (2022)
11. Fernando, B., Silva, J., Kumar, S.: Real-time soil moisture monitoring for precision irrigation. J. Agric. IoT **13**(2), 56–70 (2020)
12. Mehta, A., Patel, R., Das, T.: Cloud-integrated ESP8266-based irrigation automation. Smart Agric. Sustain. Technol. **17**(4), 110–125 (2022)
13. Das, H., Roy, P., Sinha, K.: AI-based predictive irrigation analytics for sustainable farming. J. Agric. AI IoT **14**(2), 77–93 (2021)
14. Wilson, C., Anderson, B., Lewis, T.: Wireless sensor network integration for precision irrigation. Int. J. Smart Farm. **11**(3), 89–105 (2023)
15. Roy, D., Banerjee, S., Gupta, N.: Renewable energy-driven irrigation system using compound gear trains. J. Renew. Energy Smart Agric. **18**(2), 101–117 (2022)
16. Nayak, R.C., Roul, M.K.: Technology to develop a smokeless stove for sustainable future of rural women and also to develop a green environment. J. Inst. Eng. (India): Series A **103**(1), 97–104 (2022)
17. Majhi, A., Das, A., Panda, S., Chandra Nayak, R., Kumar Dash, S.: An attachment with the hand pump to lift water without any external source. Mater. Today: Proc. **62**(P12), 6755–6758 (2022)
18. Nayak, R.C., Roul, M.K., Roul, P.D.: Design and development of smokeless stove for sustainable growth. Arch. Thermodyn. **43**(1), 109–125 (2022)
19. Nayak, R.C., Roul, M.K., Sarangi, S.K.: Innovative methods to enhance irrigation in rural areas for cultivation purpose. J. Inst. Eng. (India): Series A **102**(4), 1045–1051 (2021)
20. Nayak, R.C., Roul, M.K., Sarangi, A., Sarangi, A., Sahoo, A.: Mechanical concept on design and development of irrigation system to help rural farmers for their agriculture purpose during unavailability of external power. IOP Conf. Series: Mater. Sci. Eng. **1059**(1), 012048 (2021)
21. Fidvi, H., Ghutke, P.C., Gondane, S.M., Kulkarni, M.V., Nayak, R.C., Padhi, D.: Advanced smokeless stove towards green environment and for sustainable development of rural women. E3S Web Conf. **455**, art. no. 02017 (2023)
22. Nayak, R.C., Samal, C., Roul, M.K., Padhi, P.: A new irrigation system without any external sources. J. Inst. Eng. (India): Series A **104**(2), 281–289 (2023)

Deep Learning-Based Framework for Early Detection and Classification of Mango Crop Diseases

B. Bhargavi and E. P. Sumesh(✉)

VIT-AP University, 522241 Inavolu, Andhra Pradesh, India
{bhargavi.23phd7186,sumesh.ep}@vitap.ac.in

Abstract. Mangoes (*Mangiferaindica*) are vulnerable to bacterial and fungal infections as well as abiotic stresses, necessitating early and precise disease detection. This study evaluates multiple machine learning and deep learning models, integrating advanced optimization techniques to enhance classification performance. A range of AI approaches was employed, including convolutional neural networks (CNNs), fuzzy logic, and swarm intelligence-based optimization. Among the tested models, ResNet50 achieved the highest accuracy (99.7%), surpassing VGG-16 (94%) and EfficientNetV2-B0 (97.13%). The MCNN model, combined with CNN-Fuzzy and SA-GSO optimization, also attained 97.13%, while YOLOv3 demonstrated strong real-time detection capabilities with 83.33%. Additionally, traditional models such as feedforward neural networks (FFNN, 93%), k-nearest neighbors (KNN, 91%), and support vector machines (SVM, 64%) were included for comparison. Hybrid optimization techniques like particle swarm optimization (PSO) and image quality metrics (PSNR, MSE) further improved classification performance, presenting an efficient and comprehensive strategy for mango disease management.

Keywords: diseases of mango leaf · image processing · deep learning · machine learning

1 Introduction

Mango yield is significantly affected by foliar and floral diseases such as anthracnose, powdery mildew, sooty mould, and blossom blight, which are caused by diverse pathogens under specific environmental conditions. These diseases manifest through symptoms like leaf discoloration, flower wilting, and premature fruit drop, necessitating control strategies including fungicide use, pruning, and sanitation. Advanced techniques such as image processing and deep learning, particularly CNNs, enhance early disease detection by classifying infected and healthy leaves. This promotes timely, cost-effective interventions and supports sustainable mango cultivation practices [1–3].

H. S. Shekhawat et al. (Eds.): ICA 2025, CCIS 2795, pp. 166–177, 2026.
https://doi.org/10.1007/978-3-032-17083-5_14

Mango Leaf Diseases and Their Symptoms:

1. Anthracnose (Colletotrichum gloeosporioides) [Fig. 1a]: Characterized by dark, sunken necrotic lesions, this fungal infection leads to foliar blight, defoliation, and substantial yield loss under humid conditions.
2. Powdery Mildew (Oidium mangiferae) [Fig. 1b]: Exhibits as a white fungal growth on leaf surfaces, causing leaf curling, abscission, and compromised photosynthetic activity.
3. Sooty Mold (Capnodium spp.) [Fig. 1c]: Forms a black fungal layer on foliage due to honeydew-secreting insects, hindering light penetration and reducing photosynthetic efficiency.
4. Leaf Spot (Cercospora spp., Pseudocercospora spp.) [Fig. 1d]: Displays circular to irregular chlorotic or necrotic lesions, progressively impairing foliar function and leading to premature defoliation.

(a) Anthracnose
(b) Powdery Mildew
(c) Sooty Mold
(d) Leaf Spot

Fig. 1. Different types of mango leaf diseases.

Mango Anthracnose Disease: A Comprehensive Overview

a. The Fig. 1[a] shows the Signs and symptoms. Young leaves show tiny, asymmetrical dark marks that gradually enlarge. In leaves suffering from extreme infection, blackening occurs. The leaves soon curl and drop. The black lesions may also appear on clusters of fruits and flowers, which will reduce yield.

b. Factors of Origin and Dispersal: caused by warm, humid fungus Colletotrichum gloeosporioides spreading through the use of contaminated tools, wind, rain splash, or from water. In addition to being present on fruit clusters and flowers, more than two S.C.

c. Control and Management: Cultural Methods Include Pruning and Construction for Better Airflow Through Orchard Areas that are infected and diseased. Chemical Control- Use of systemic fungicides such as carbendazim or copper-based fungicides. Environmental precautions include avoiding overhead irrigation and spacing trees properly.

Anthracnose in mango leaves starts as small, dark spots that expand, causing blackening, curling, and defoliation. The fungal pathogen Colletotrichum-gloeosporioides thrives in humid conditions and spreads through contaminated tools, wind, and rain. Management includes pruning, proper orchard design, and applying systemic fungicides like carbendazim and copper-based treatments. Additionally, avoiding overhead irrigation and ensuring adequate tree spacing help minimize disease outbreaks.

2 Related Work

Anthracnose is a disease that needs to be identified early with accuracy; To enable precise early detection of anthracnose, mango leaf images underwent preprocessing steps including denoising and sharpening, followed by segmentation using Otsu's thresholding. In [4] addition to this, enhancements improved image quality, as reflected by elevated PSNR and MSE values, facilitating more reliable disease classification.

AI-driven techniques employing FFNN, SVM, KNN, and rule-based classifiers utilize the TJC Mango Dataset (1,200 labeled images) for automated assessment of mango quality, yield prediction, and disease diagnosis. In [5], these models enhance decision-making in precision agriculture, promoting sustainable and profitable mango cultivation.

A hybrid model combining CNN, fuzzy classifiers, and SA-GSO optimized feature selection for mango classification based on shape, texture, and ripeness. This method outperformed traditional models in accurately differentiating healthy and diseased fruits [6].

In [7] the ResNet50 model, trained on mango leaf disease data from Kaggle, achieved an impressive 99.7% accuracy after fine-tuning. This highlights its effectiveness in early and accurate disease detection, which can improve mango farming, boost productivity, and support sustainable agriculture. Its high classification performance makes ResNet50 a valuable tool for precision farming.

An ensemble classifier integrating EfficientNetV2-B0 and VGG-16 with spatial attention significantly improved the detection of mango leaf diseases, enhancing accuracy, precision, recall, and F-measure. In [8] this model effectively identifies multiple diseases and supports timely intervention, aligning with both economic and biological constraints of mango cultivation.

An optimized deep learning model leveraging transfer learning and CNNs was trained on 4,000 mango leaf images across eight disease classes, achieving 99%

classification accuracy. [9] This automated approach enables timely and precise disease diagnosis, reducing manual dependence and mitigating economic losses in mango cultivation.

YOLOv3 was used to detect anthracnose disease in mango leaves, achieving an accuracy of 83.33% after training on 80.28% of the images. A confusion matrix was used to evaluate the model's performance, effectively classifying mango leaves as healthy or diseased [10].

A CNN-based deep learning model using transfer learning achieved 99% accuracy in detecting mango leaf diseases, using a dataset of 4,000 images classified into eight disease categories. This automated system helps farmers quickly identify diseases and maintain the health of mango plants, improving productivity and safeguarding crops [11].

A hybrid model combining CNN and Random Forests achieved 80.89% accuracy and an 80.85% weighted F1 score using 3,633 mango leaf images, primarily sourced from Punjab. This approach enhances early disease detection, supporting sustainable crop management through improved classification performance [12].

A CNN model with ReLU activation, trained on 1,405 mango leaf images across three classes (anthracnose, black mold, and healthy), achieved 95% training and 98% validation accuracy the dataset taken from kaggle [13]. This high-performance model enables early, accurate disease detection, facilitating timely intervention in mango cultivation.

Table 1. Methodologies and Key Insights from Literature

Author & Year	Method	Dataset	Accuracy	Advancement
K. V. Rao, 2022 [2]	PSO	Disease images	92.5%	–
U. P. Singh, 2019 [3]	MCNN	Real-time, PlantVillage	97.13%	IoT integration
R. Garg, 2023 [5]	PSNR, MSE	Mango leaf data	Better results	Segmentation
S. Jayaweera, 2024 [6]	FFNN, SVM, KNN	TJC Mango (MangoDB)	93%, 64%, 91%	Reliable dataset
A. Kr, 2023 [7]	SA-GSO, CNN-Fuzzy	Healthy/defective images	Highest	–
G. Singh, 2024 [8]	ResNet50	Kaggle	99.7%	Augmentation
S. Venkatraman, 2024 [9]	VGG-16, EffNetV2-B0	BD dataset, Kaggle	97.13%	Dataset quality
G. Kaur, 2024 [10]	VGG16, CNN	Unspecified	94%	Large dataset
A. N., 2023 [11]	YOLOv3	Unspecified	83.33%	IoT, real-time
N. Trivedi, 2024 [12]	CNN	8 disease classes	99%	YOLOv4 extension
V. Kukreja, 2024 [13]	CNN + RF	Unspecified	80.89%	–
S. V. Ramayani, 2023 [14]	CNN	Kaggle	95%	Real-time

Table 1 Comparative analysis of mango leaf disease detection methods reveals varying accuracies across models and datasets, with ResNet50 achieving peak performance of 99.7% on Kaggle data. CNNs, MCNNs, and YOLOv3 offer strong classification and real-time detection capabilities. Integration of optimization algorithms (e.g., PSO, SA-GSO) and image quality metrics (PSNR, MSE) further enhances model robustness and diagnostic precision.

3 Advanced Plant Disease Diagnosis System

The integration of AI-based models is essential for developing intelligent agricultural systems, enhancing automation and connectivity in modern farming practices. Machine learning and deep learning algorithms, particularly Convolutional Neural Networks (CNNs), offer high accuracy in diagnosing plant diseases through effective pattern recognition. These models mimic human neural processes, enabling precise and efficient disease detection critical for sustainable agriculture

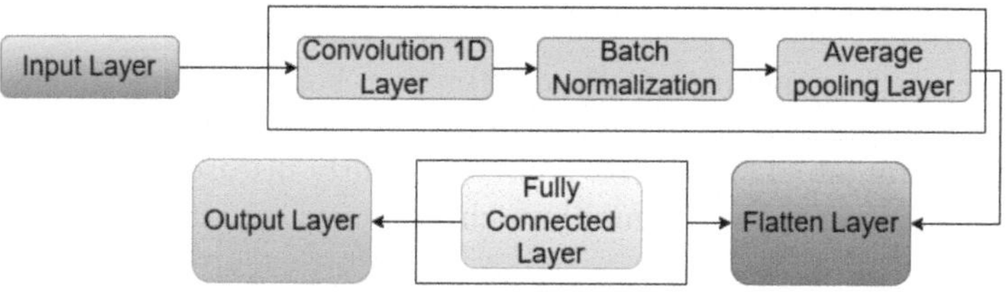

Fig. 2. Convolutional Neural Networks architecture.

The above Fig. 2 The architecture starts with an input layer, followed by a 1D convolutional layer for localized feature extraction from sequential data. Batch normalization stabilizes learning, while average pooling reduces spatial dimensions. Flattened features are passed through a dense layer, and the output layer delivers the final classification or prediction.

3.1 MCNN

Multi-Scale CNNs (MCNNs) employ parallel convolutional layers with varying kernel sizes to capture features at multiple spatial resolutions. This design facilitates the extraction of both fine-grained and global patterns, improving robustness to scale and texture variations. MCNNs are particularly effective in plant disease detection by enhancing discriminative feature representation across heterogeneous image conditions.

3.2 Hybrid Techniques and Classifiers Overview

CNN-Fuzzy Hybrid: It combines CNN's deep feature extraction with fuzzy logic's capability to manage uncertainty. Suitable for tasks with ambiguous class boundaries, enhancing interpretability and robustness.

SA-GSO Hybrid: Integrates Simulated Annealing's global search with Gravitational Search Optimization's mass-based attraction. This hybrid improves feature selection efficiency and convergence stability.

FFNN (Feedforward Neural Network): A simple and powerful artificial neural network, where information is passed on from input to output through hidden layers in one direction only. FFNNs are widely used in regression and classification owing to their versatile structures that can learn complex patterns. **Support Vector Machine (SVM)**:): A supervised learning algorithm that constructs an optimal hyperplane to separate classes. ls in handling high-dimensional data with strong generalization capabilities.

K-Nearest Neighbors (KNN): An instance-based learning technique relying on distance metrics to classify samples. While simple and effective, it scales poorly with large datasets due to high computational cost.

Rule-Based Model: Employs a predefined set of logical rules for decision-making derived from domain expertise. Offers high interpretability but may lack adaptability for complex, unstructured data.

3.3 Deep Learning Architectures

ResNet50: ResNet50 uses residual blocks with skip connections to enable stable training of deep 50-layer networks. It addresses vanishing gradients and excels in high-accuracy visual recognition tasks.

VGG16: VGG16 employs a uniform architecture with 3×3 convolutions and 2×2 pooling layers across 16 layers. It offers strong classification performance but demands high computational resources.

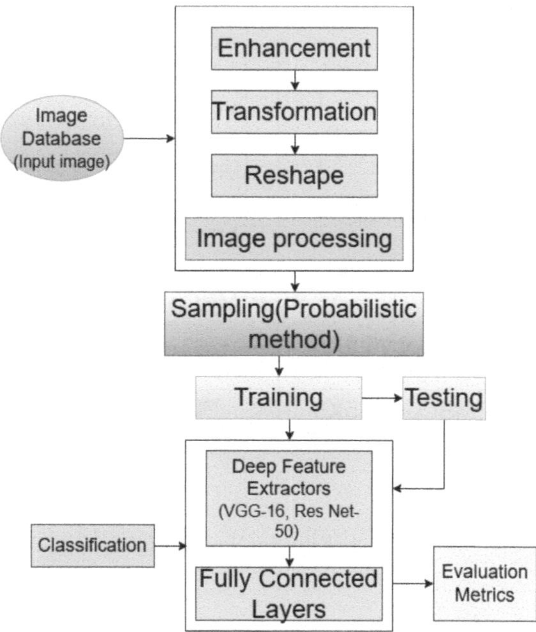

Fig. 3. ResNet50, VGG16-layer convolution neural network.

The above Fig. 3. The framework initiates with an image repository, followed by pre-processing steps such as enhancement, normalization, and resizing. Sampling strategies ensure data balance, and the dataset is partitioned into training and testing subsets. Deep feature extractors like VGG16 and ResNet50 encode discriminative patterns, which are classified through dense layers and evaluated using performance metrics.

3.4 EfficientNetV2-B0 and YOLOv3: Advanced Deep Learning Architectures

EfficientNetV2-B0: EfficientNetV2-B0 optimizes accuracy and efficiency through depth-wise convolutions and progressive learning. It uses compound scaling to balance depth, width, and resolution for better performance. Ideal for image classification, it achieves high accuracy with fewer parameters and reduced computational cost.

YOLOv3 is a real-time object detection framework that performs simultaneous localization and classification in a single forward pass. Its grid-based prediction mechanism enables efficient multi-scale object detection across varying sizes. YOLOv16 advances this architecture with enhanced backbone networks, refined feature pyramids, and optimized inference speed. It delivers superior accuracy and responsiveness, making it ideal for high-resolution, real-time vision applications.

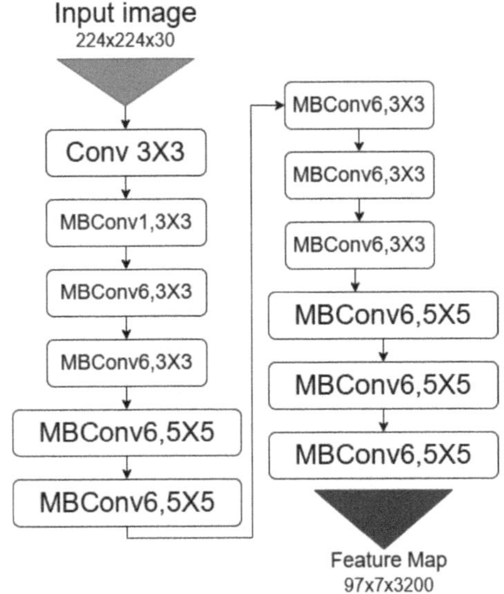

Fig. 4. EfficientNetV2-B0.

Figure 4 illustrates a feature extraction process begins with a 224×224×30 input image, where 3×3 convolutional filters detect essential spatial patterns.

MBConv layers are applied to enhance efficiency by minimizing computational load while preserving feature integrity. As the image passes through successive layers, it is transformed into a 97×7×3200 feature map that captures high-level representations for further analysis.

Table 2 highlights that ResNet50 achieved the highest accuracy of 99.7% leveraging residual learning and data augmentation for optimal performance. Deep learning models like VGG-16 and EfficientNetV2-B0 also performed well, surpassing 94% accuracy through strong feature extraction. Traditional models such as FFNN, SVM, and KNN had varying performance, with hybrid approaches like CNN-Fuzzy and SA-GSO combining optimization for enhanced results.

Table 2. Comparison of the Model Accuracy

Technique	Type	Key Components	Accuracy
MCNN	Deep Learning	Inception-based structure	97.13%
CNN-Fuzzy	Hybrid	Neuro-Fuzzy integration	High
SA-GSO	Hybrid	Metaheuristic (Glowworm) optimization	Highest
FFNN	Machine Learning	Multilayer perceptron	93%
SVM	Machine Learning	Hyperplane-based classifier	64%
KNN	Machine Learning	Instance-based learning	91%
Rule-Based	ML	Predefined logical rules	Moderate
ResNet50	Deep Learning	Residual blocks (50 layers)	99.70%
VGG-16	Deep Learning	16-layer CNN, augmentation	94%
EffNetV2-B0	Deep Learning	Efficient layers, progressive learning	97.13%
YOLOv3	Deep Learning	Real-time, multi-scale detection	83.33%

The study explores various machine learning and deep learning techniques for mango leaf Fig. 5 disease classification, highlighting their effectiveness in accuracy and optimization. ResNet50, with its advanced augmentation and residual learning, achieved the highest accuracy of 99.7%, while hybrid models like SA-GSO also delivered strong results through effective feature selection. Deep learning models such as VGG-16 and EfficientNetV2-B0 maintained high accuracy above 94%, outperforming traditional methods like SVM (64%), KNN (91%), and FFNN (93%). Hybrid techniques, including CNN-Fuzzy and SA-GSO, proved efficient by integrating optimization strategies with robust classification performance.

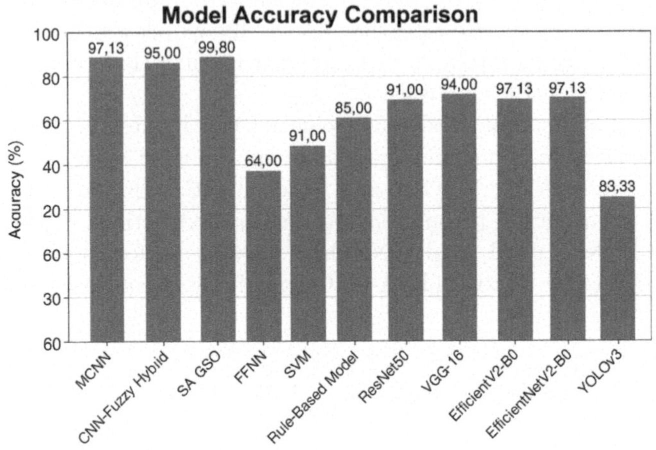

Fig. 5. Model accuracy comparison.

4 Block Diagram

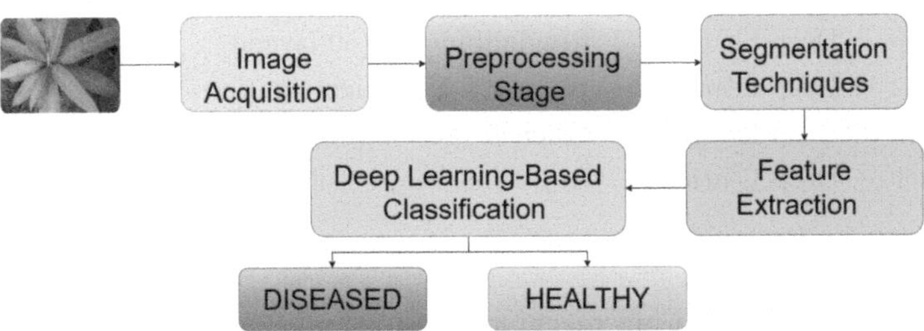

Fig. 6. Block diagram of image processing.

In Fig. 6, image acquisition involves capturing images using devices like cameras, scanners, and sensors, forming the foundation for image processing in applications such as computer vision and medical imaging. Preprocessing enhances image quality through noise removal, contrast adjustment, and filtering techniques to highlight key features. In segmentation, methods like Canny Edge Detection, Otsu's Thresholding, and K-Means Clustering help differentiate healthy and diseased areas, improving classification accuracy. Feature extraction identifies key patterns in texture, shape, and color, aiding in precise disease detection. Techniques like GLCM and LBP analyze contrast and correlation, ensuring efficient recognition and treatment in plant disease diagnosis.

4.1 Deep Learning-Based Classification

The dataset would be formed for deep learning mango leaf disease classification to make the model robust. Images collected and annotated include detailed pictures of healthy and diseased leaves with the enhancement involved. For objective evaluation, this dataset is split into training, validation, and testing sets. Feature extraction automatically acquires hierarchical patterns from basic edges to intricate disease-specific symptoms. Highly accurate results are guaranteed through sophisticated deep-learning architectures including Vision Transformers (ViTs) and CNNs (e.g., DenseNet, ResNet, EfficientNet). Classification performance was further improved through global attention methodology and spatial feature extraction.

4.2 PSNR, MSE, and PSO in Image Processing and Feature Selection

PSNR (Peak Signal-to-Noise Ratio) and MSE (Mean Squared Error) are key metrics for assessing image quality. PSNR measures the ratio between the maximum signal power and noise, indicating image clarity, while MSE calculates the average squared difference between original and processed images, reflecting error levels. PSO (Particle Swarm Optimization) is an advanced feature selection technique that enhances classification accuracy by optimizing the selection of relevant features, improving computational efficiency and performance in image analysis. PSNR is expressed in decibels (dB) and is calculated as follows

$$PSNR = 20 \cdot \log_{10}(\mathrm{MAX}) - 10 \cdot \log_{10}(\mathrm{MSE}) \tag{1}$$

In the PSNR formula, MAX represents the maximum possible pixel value of the image. For an 8-bit image, this value is typically 255. It defines the dynamic range of pixel intensities used to normalize the error.

5 Dataset

The mango leaf disease dataset comprises approximately 4,000 high-resolution images sourced from real-time field captures and public platforms like Kaggle and PlantVillage. It includes eight categories, covering six disease types and healthy samples, all standardized to 256×256 pixels in JPEG or PNG format. A total of 3,633 images were selected from both primary and secondary sources to ensure diversity and robustness. The dataset is split into 80.28% training and 19.72% testing data to support effective model evaluation and validation.

6 Result

This study demonstrates that deep learning models outperform traditional machine learning techniques in mango leaf disease classification, with ResNet50 achieving the highest accuracy of 99.7%. Models like EfficientNetV2-B0 and MCNN also performed strongly at 97.13%, while hybrid approaches such as CNN-Fuzzy and SA-GSO enhanced results through optimized feature selection. Traditional models showed inconsistent accuracy, indicating their limitations in capturing complex patterns. These outcomes underline the value of AI-driven methods for precise and early disease detection, supporting more sustainable agricultural practices.

7 Conclusion

This study comprehensively evaluates machine learning and deep learning models for mango leaf disease detection, highlighting ResNet50's top performance (99.7%) due to its advanced learning techniques. Other models like EfficientNetV2-B0, MCNN, and hybrid approaches also demonstrated high accuracy, while YOLOv3 showed promise for real-time applications. Future improvements could involve dataset expansion, real-time IoT integration, and adoption of lightweight, high-performance architectures for broader field usability.

References

1. Bolaños, M.M., et al.: Mango anthracnose integrated management. Horticulture Inter. J. **6**(4), 196–201 (2022). https://doi.org/10.15406/hij.2022.06.00265
2. Rao, K.V., Kiran, P., Praveena, M., Sreenath, K., Anusha, P., Kumar, E.S.: A swarm intelligence-based model for disease detection in mango crops. In: 2022 1st International Conference on Computational Science and Technology (ICCST), pp. 845–850 (2022). https://doi.org/10.1109/iccst55948.2022.10040428
3. Singh, U.P., Chouhan, S.S., Jain, S., Jain, S.: Multilayer convolutional neural network for the classification of mango leaves infected by anthracnose disease. IEEE Access **7**, 43721–43729 (2019). https://doi.org/10.1109/access.2019.2907383
4. Garg, R., Sandhu, A.K., Kaur, B.: A hybrid filtration-based design of preprocessing stage for efficient mango leaf disease detection. IEEE Explore **22**, 1–6 (2023)
5. Jayaweera, S., et al.: MangoDB - a TJC mango dataset for deep-learning-based on classification and detection in precision agriculture. IEEE Explore, 115–120 (2024). https://doi.org/10.1109/icarc61713.2024.10499698
6. Kumari, N., Bhatt, A. K., Dwivedi, R.K.: Self-adaptive Glowworm swarm optimization technique in optimal feature selection in grading of fruit mango. In: 2022 International Conference on Computational Intelligence and Sustainable Engineering Solutions (CISES), pp. 424–430 (2023). https://doi.org/10.1109/cises58720.2023.10183405
7. Singh, G., Guleria, K., Sharma, S.: A deep learning-based fine-tuned ResNet50 model for multiclass mango leaf disease classification. IEEE Explore, 1310–1316 (2024). https://doi.org/10.1109/idciot59759.2024.10467406
8. Pandiyaraju, V., Venkatraman, S., Abeshek, A., Aravintakshan, S.A., Pavan Kumar, S.K.: Mango leaf disease detection and classification using spatial attention enabled ensemble classification. IEEE Explore, 1–8 (2024). https://doi.org/10.1109/adics58448.2024.10533513

9. Kaur, G., Sharma, N., Malhotra, S., Devliyal, S., Gupta, R.: Mango leaf disease detection using VGG16 convolutional neural network model. IEEE Explore, 1–6 (2024). https://doi.org/10.1109/inocon60754.2024.10511415
10. Yumang, A.N., Samilin, C.J N., Sinlao, J.C. P. Detection of anthracnose on mango tree leaf using convolutional neural network. IEEE Explore (2023). https://doi.org/10.1109/iccae56788.2023.10111489
11. Tiwari, R.G., Budhani, S., Agarwal, A.K., Gautam, V., Trivedi, N.K.: Transfer learning-based optimized deep learning model to characterize mango leaf disease. IEEE Explore, 1217–1223 (2023). https://doi.org/10.1109/ictacs59847.2023.10390233
12. Choudhary, S., Choudhary, M., Kaur, S., Kukreja, V.: Integrating CNN and random forest for accurate classification of mango leaf diseases. IEEE Explore **120**, 42–46 (2024). https://doi.org/10.1109/autocom60220.2024.10486108
13. Saragih, V.R., Azizi, N.N., Atalarais, N.A., Hatmi, N.R.A., Syahputra, N.H.: Detection of mango leaf disease using the convolution neural network method. TEKNOSAINS Jurnal Sains Teknologi Dan Informatika **11**(1), 62–70 (2024). https://doi.org/10.37373/tekno.v11i1.639

Residual Biomass Utilization: A Review on Newly Emerging Techniques for Mitigating Climate Change

Revathi Gundluri Nagaraju[1], Rakesh Mishra[2], Gaetan Pelletier[3], Ajay Dashora[1](✉), and Yun Zhang[2]

[1] Indian Institute of Technology Guwahati, Guwahati 781039, India
{g.revathi,abd}@iitg.ac.in

[2] University of New Brunswick, Fredericton, NB E3B4C2, Canada
{rakesh.mishra,yunzhang}@unb.ca

[3] Northern Hardwoods Research Institute, Edmundston, NB E3V2S8, Canada
gaetan.pelletier@hardwoodsnb.ca

Abstract. This paper presents a comprehensive review on residual biomass as a sustainable energy resource, emphasizing its role in advancing the circular bioeconomy. The paper explores technological advancements in biomass conversion, evaluates energy production potential through life cycle assessment, and highlights digital technologies such as AI and IoT. Policy frameworks, implementation gaps, and future research directions are also discussed. This review aims to guide researchers, policymakers, and industry stakeholders in leveraging residual biomass for low-carbon energy systems, while balancing ecological preservation and economic viability.

Keywords: Residual Biomass assessment · forest and agricultural residues · bioenergy · biofuel production · life-cycle assessment · sustainability

1 Introduction

Residual biomass refers to organic material derived from agricultural, forestry, and industrial processes that would otherwise go unused or be discarded. It also includes agricultural residues such as straw, husks, stalks, forestry residues like tree branches and sawdust, and industrial by-products such as food-processing waste and animal manure. This form of biomass has been gaining increasing attention as a source of sustainable energy due to its potential for reducing biowaste and contributing to renewable energy generation.

Use of residual biomass in energy production and industrial applications contributes to the circular bioeconomy by reducing reliance on fossil fuels and minimizing waste. Various conversion technologies such as gasification, pyrolysis, anaerobic digestion, and biochemical fermentation have been developed to harness its potential. Biofuels generated from waste play a key role in circular bioeconomy by turning discarded organic materials into useful energy instead of letting them go to waste. This reduces

H. S. Shekhawat et al. (Eds.): ICA 2025, CCIS 2795, pp. 178–190, 2026.
https://doi.org/10.1007/978-3-032-17083-5_15

dependence on fossil fuels and lowers greenhouse gas emissions. Additionally, by-products of biofuel production, such as biochar and biogas, can be used in agriculture and industry, ensuring continuous reuse of the resources rather than discarding them. This creates a more sustainable system where waste is minimized, and valuable materials are under use for as long as possible. On the other hand, though residual biomass and associated technologies present promising avenues for bioenergy and bio-based products, yet challenges related to economic feasibility, supply chain logistics, and environmental impact assessment persist. This paper reviews fundamentals of current technologies, comprehensive environmental management, and involved challenges discussed by state-of-the-art research studies for sustainable conversion of residue biomass into biofuel mitigating climate change.

The paper is organized in six sections. After the introduction, Sect. 2 discusses technological development. Section 3 covers life cycle assessment, sustainability, and environmental impact of biofuel. Section 4 discusses challenges of biofuel production, its use, and AI based technological advancements. Section 5 highlights future scope and Sect. 6 concludes the paper.

2 Technological Development

Based on the source, residual biomass can be categorized into agricultural residues, forest residue, industry residues, and municipal solid waste. The agricultural residues consist of plant material left over after harvesting crops, including wheat straw, corn stover, and sugarcane bagasse. These residues not only lead to animal feed but also have significant potential for biofuel production and soil enhancement. The forestry residues include sawdust, tree bark, and logging slash. The industrial biomass waste is generated from food processing, paper mills, and other manufacturing activities, including food scraps, pulp sludge, and vegetable oil residues. The municipal solid waste biomass, which includes organic waste from households and commercial establishments, are also used for biofuel generation. However, the majority of the research studies addressed issues of agricultural and forest biomass residue. This section discusses energy potential of all four types of residues and assesses efficiency of various conversion processes and technologies.

2.1 Residual Biomass Assessment

Residual biomass assessment determines the availability, sustainability, and feasibility of biomass resources for energy production. Accurate quantification and evaluation of biomass residues ensure optimal energy recovery while mitigating environmental impacts. Agricultural, forestry, and industrial waste streams are primary sources of residual biomass, each possessing distinct characteristics and energy potentials.

Agricultural biomass residues, including cereal straws and corn stover, represent a significant portion of bioenergy feedstocks. Helwig et al. [8] assessed the bioenergy potential of agricultural residues, particularly livestock manure, in Eastern Canada. Using data from provincial agriculture departments, Statistics Canada, OMAFRA, and Quebec's ASRA, the study estimated available straw, stover, and manure. Residue biomass was extracted according to grain-to-residue ratios, which preserved sufficient soil fertility and bedding. Approximately 1 million odt of straw and 3 million odt of corn stover were identified annually, with potential to produce 46 million GJ of heat or 1.35 billion litres of ethanol. Manure recoverability was estimated at 46,000 tonnes/day in Ontario, 43,000 in Quebec, and 7,000 in Atlantic Canada. Anaerobic digestion generated 16 million GJ of biogas annually, translating to 1,650 MWh/day and 1,550 MWh/day of electricity in Ontario and Quebec, respectively. The study concluded that with region-specific strategies and scalable technologies, biomass residues can significantly contribute to Eastern Canada's energy security and environmental goals.

Methodological advancements, like classification and spatial analysis, have significantly improvised the accuracy of biomass resource estimation. Bouchard. et al. [3] developed a structured method to estimate forest biomass energy potential from dispersed residues in New Brunswick, Canada. Biomass was classified into merchantable wood (63%), residual biomass (i.e. tops, branches, foliage, 27%), and bark (10%). Next, biomass availability was analysed by land ownership type (Crown Land, Private Woodlots, Industrial Freeholds, Federal Land), and optimal locations for Combined Heat and Power (CHP) plants were mapped using ArcGIS Network Analyst within 125 km procurement zones, considering proximity to towns, sawmills, roads, and power infrastructure. Energy potential was calculated from forest inventory data with 50% moisture content, assuming conversion efficiencies of 25% for electricity and 60% for heat. Annually, 15.5 million giga tonnes (MGT) of biomass could generate 1.2 GW electricity and 3 GW heat. More specifically, the residual biomass and bark alone could produce 462 MW electricity (38.5%) and 1.1 GW heat (36%). Spatial analysis showed 73% biomass located within 50 km of proposed CHP sites reduced the transportation costs. The study successfully demonstrated that detailed spatial analysis effectively estimated sustainable energy potential of provincial forest biomass.

Statistical analysis allows to identify and derive cost-effective and scalable technologies for biomass conversion processes. Pavlova [13] integrated national-level statistics, and analytical assessments for a detailed case study on a dairy cattle farm in Tvarditsa (Bulgaria) for the efficient utilization of agricultural residues (livestock manure and crop remains) to mitigate pollution and enhance farm profitability. Author modelled an anaerobic digestion system generating approximately 87,000 m^3 biogas annually, sufficient to meet farm electricity, heating, and hydroponic feed requirements. Moreover, the anaerobic plant produced bio-manure, which is nutrient-rich organic fertilizer containing nitrogen (5 kg/m^3), phosphorus (1.5 kg/m^3), and potassium (4 kg/m^3) is a commercially profitable resource. Furthermore, statistical data obtained by comparing performance of the bio-manure and synthetic fertilizer revealed that the increased synthetic fertilizer uses exacerbates soil, degrades water quality in root zone, and emits greenhouse gases, yet higher crop yields are not guaranteed. The study concluded that converting residues into biogas and organic fertilizer reduces reliance on imported energy and chemical fertilizers, decreases green-house gas emissions, and supports sustainable agricultural practices aligned with Kyoto Protocol (1997) objectives.

2.2 Potential of Energy Production

Energy production by biomass conversion depends upon two major factors: (i) biomass composition (amount of moisture, ash, volatile matter, fixed carbon, and elemental content (carbon, hydrogen, oxygen, nitrogen, sulphur)) and (ii) appropriate selection of thermochemical or biochemical pathways to optimize energy yield and efficiency. However, practical aspects influence biomass classification, subsequent selection and life cycle of thermo-chemical and biochemical conversion pathways for energy production, and use of produced waste (ashes, manure etc.).

Thermochemical processes include combustion, pyrolysis, and gasification. Combustion provides heat but is less efficient due to high moisture and variable composition, often causing incomplete burning and high emissions. Pyrolysis yields bio-oil and syngas with variable quality depends upon feedstock type. Also, pyrolysis requires further processing to create usable fuel. Gasification efficiently produces syngas, yet feedstock inconsistency and impurities like tar demands advanced and costly cleaning systems.

Biochemical processes include anaerobic digestion and fermentation. Lignocellulosic biomass, for instance, requires pre-treatment to break down complex polymers, improving enzymatic hydrolysis and biofuel production. Balan [2] presented a technical review of key barriers to commercial-scale biofuel production from lignocellulosic biomass such as corn stalks, sugarcane bagasse, and wood residues. Although rich in fermentable sugars, these feedstocks are tightly bound by lignin, requiring energy-intensive pretreatment (chemical, thermal, or mechanical) that can produce inhibitory compounds. Enzymatic hydrolysis is costly and hindered by lignin, while fermentation is limited by conventional microbes' inability to process all sugars, particularly pentoses like xylose. Though engineered strains offer potential, they underperform at scale. Ethanol recovery further increases costs, and commercial viability is constrained by inconsistent biomass supply, weak infrastructure, and high capital needs.

On the other hand, the potential for bioenergy production by thermochemical and biochemical processes differs across regions, influenced by local factors such as climate, soil quality, and agricultural productivity. These conditions directly affect the availability of biomass feedstock, influencing both the quantity and efficiency of energy generation. Xu et al. [21] included climate, terrain, crop structure, planting scale, diversity, and regional economic indicators as influencing factors for estimating continuous crop residue supply. Temperature showed a non-linear effect—moderate warmth temperature supported supply, while excessive heat led to more non-residue farming. Moreover, heteroscedasticity accounted for the largest share of inequality (42.12%), followed by macroeconomic (28.95%) and climatic/geographic factors (10.77%) for supply locations. The study recommends improving storage facilities, encouraging residue-producing crops, and farmland protection to ensure a stable biomass supply and resilient, bioenergy-oriented agricultural systems. Advancements in biofuel production also focus on improving process efficiency and sustainability. Ahmad et al. [1] optimized biogas yield and stability via anaerobic digestion of food waste and municipal solid waste (MSW). Lab-scale experiments at 35 °C tested varying carbon-to-nitrogen (C/N) ratios (range 20–40) and substrate-to-inoculum (S/I) ratios (range 0.5–2.0). Ratios of 20 C/N ratio and 0.5 S/I yielded 825 L/kg and 640 L/kg volatile solids, respectively. Higher ratios led to reduced gas output due to nitrogen deficiency and acid accumulation. Calcium hydroxide was used to stabilize pH, which led to 89% solids reduction and stable digestion at the optimal C/N ratio. The study recommends this parametric configuration for scalable and efficient biogas systems using organic waste.

Cogeneration systems, which generate both electricity and heat from biomass, enhance energy production efficiency by reusing generated heat during power generation. Sipilä, K. et al. [18] conducted a detailed technical review covering system design, thermodynamic principles, fuel characteristics, conversion efficiencies, emission controls, and integration capabilities for primary technologies (i.e. district heating systems combining heat and power or CHP, biomass combustion, municipal waste incineration, and industrial waste heat recovery). The authors compared each technology's efficiency, economical and environmental impacts, concluding that optimal technology choice depends on local biomass supply conditions. Notably, cogeneration combining heat and power demonstrated superior efficiency, offering energy savings of 30–40% compared to separate heat and electricity production.

2.3 Use of Remote Sensing, AI, and IoT

Compared to practical aspects, peripheral aspects involve advanced approaches which enhance efficiency and scalability of biomass energy production, making it more accessible and effective. Remote sensing and GIS-based biomass estimation, artificial intelligence (AI) and machine learning (ML) based models and algorithms, IoT-driven logistics, block chain for biomass trading, and large-scale industrial applications are popular methods.

High-resolution satellite imagery and UAV-based monitoring allow precise biomass yield predictions and improved agriculture management by harvesting strategy optimization [12]. Ogungbuyi et al. [12] compared biomass prediction models developed using satellite images and UAV based photogrammetric models. Authors collected grass height changes from the photogrammetric model and selected correlation model by Random Forest approach. Similarly, same exercise was performed with on vegetation-sensitive spectral bands of Sentienl-2 images. Drone-based estimates yielded higher accuracy ($R^\circ = 0.75$) compared to satellite-only data ($R^\circ = 0.56$). The hybrid approach integrating photogrammetry and satellite images improvised large-scale biomass prediction, revealing greater losses in closed paddocks and better growth on flat, water-retaining terrain. Scalability of the proposed method supports precise monitoring for sustainable grazing and pasture management.

Remote sensing allows monitoring of land-use changes, soil health, and ecosystem stability in highly undulating and varying terrains. Matsushita et al. [10] used satellite image data and investigated the sensitivity and accuracy of the Enhanced Vegetation Index (EVI) and the Normalized Difference Vegetation Index (NDVI) to topographic effects such as slope, aspect, and elevation, that can distort vegetation measurements in a high-density cypress forest (Japan). Authors found that using hyperspectral and LiDAR data over the forest, they found EVI is more sensitive to slope (correlation value R2 = 0.28) than NDVI (R2= 0.04). This is due to EVI's soil adjustment factor, which amplifies topographic effects, as confirmed by a non-Lambertian light reflection model using Minnaert constants. The study recommends applying topographic correction when using EVI in hilly areas, while NDVI remains relatively stable without correction. AI and IoT are significantly enhancing the efficiency and sustainability of biomass energy systems by improving residue biomass management, optimizing energy production and reducing environmental impacts. IoT based smart sensors efficiently manage biomass in real time by collecting data on moisture content, temperature, and biomass degradation rates. Technological frameworks like Forest 4.0 and Industry 4.0 combine IoT, big data analytics, and AI to biomass logistics. Zhao et al. [23] examined stages from inventory and estimation to harvesting, transport, and conversion, using forest and energy crop data. Machine learning models – Random Forest, Neural Networks, and Support Vector Machines – utilized satellite imagery, drone data, and field surveys to automate biomass prediction, optimize harvesting zones, plan transport routes, and forecast conversion yields. Replacing manual processes, these models enhanced accuracy, reduced costs, and supported sustainable forest biomass management. Moreover, AI-based models significantly improve biomass transportation planning and overall supply chain efficiency.

Blockchain technology is increasingly explored for improving transparency and security in biomass supply chains. Nechetnyy et al. [11] developed a digital framework to track each stage – from biomass collection to processing and delivery – using a decentralized ledger for tamper-proof, traceable records, supporting accurate carbon tracking and compliance with carbon credit systems. Smart contracts automated the transactions among stakeholders, enhancing trust and operational efficiency. Simulated scenarios with transaction volumes, emissions, and stakeholder interactions proved the resource use, reduced emissions, enhanced carbon accounting, and strengthened digital security, providing a scalable solution for sustainable bioenergy supply chain management.

Advanced imaging and data fusion techniques enable faster decision-making, leading to improved resource allocation and yield predictions. Integrating satellite data with land-use patterns with GIS-based models improves resource assessment accuracy

3 Environmental Management of Residual Biofuel

Effective management of residual biofuels is essential to minimizing environmental impact ensuring the long-term sustainability of biomass resources, which demands environmental and economic trade-offs. Excessive removal of biomass residues from forests and agricultural lands can degrade soil quality, reduce fertility, and impact biodiversity [9]. Stable organic matter in the soil, achieved by growing specific crops (grasses, legumes, short-rotation woody plants, and agroforestry systems), ensure consistent growth and energy yield [20]. advocated scientific foundation for managing forests that supports both biomass production and long-term carbon storage. Author showed that while harvesting biomass, forest soils should be managed to preserve soil carbon as the latter acts as soil carbon sinks.

Udoumoh et al. [20] emphasized the importance of mulching, and the strategic management of vegetation cover to enhance soil quality, which in turn helps mitigate soil erosion and reduce runoff. Moreover, the study outlined other effective land management techniques such as planting cover crops, crop rotation, conservation tillage, and using organic amendments like compost and animal manure. These methods enhance the soil's ability to hold moisture, increase nutrient retention, and reduce the loss of topsoil caused by intense rainfall, which are vital for maximizing the growth of biomass crops. Cover crops like Mucuna and Centrosema fix nitrogen and protect soil from erosion while serving as biomass sources themselves. These methods not only protect the land but also contribute to climate goals by enhancing soil carbon storage and reducing nutrient runoff, making them ideal for integrated biomass-energy systems.

Tobin et al [19] identified water resource management as a critical factor for biomass-based energy production. Authors integrated wastewater treatment into lignocellulose bio refineries using hybrid poplar as feedstock. Biomass was pre-treated with steam explosion and sulphur dioxide, separating solids and liquids. Solids were hydrolysed and fermented with Pichia stipitis, while toxic compounds in the liquid stream (e.g., furfural, acetic acid) were removed via activated carbon before fermentation. Ethanol was recovered through rotary evaporation. Wastewater, primarily stillage, was analysed for organic pollutants and salts using elemental analysis, HPLC, and UV spectroscopy. Comparison of three treatment methods, namely anaerobic digestion (producing biogas but at a higher cost), land application through poplar plantations (moderate cost with nutrient recycling), and evaporation with reuse (most cost-effective, though steam-intensive) is performed. The study also suggested that less toxic pre-treatment chemicals could lower wastewater toxicity. Results showed that improved water management and cleaner chemical selection enhance the sustainability and efficiency of poplar-based ethanol production.

3.1 Life Cycle Assessment and Sustainability

Life Cycle Assessment (LCA) evaluates sustainability of biomass-based energy systems by analyzing environmental impacts, energy consumption, and resource use throughout the entire lifecycle. Excessive extraction of biomass can disrupt nutrient cycles, degrade soil quality, and impact biodiversity. Repo et al. [15] analyzed the effects of forest residue removal and indiscriminate harvesting, which lead to long-term soil degradation, affecting the regeneration capacity of forests. Authors assessed long-term climate impact of forest harvest residues (branches, stumps, and tops) for bioenergy in Europe. Simulation models like Yasso07 for organic matter decomposition, and CO2FIX for litter carbon dynamics simulated two scenarios (with and without residue harvesting). Results showed an average 3% decline in soil and litter carbon across Europe, with the highest losses in Germany, Sweden, and Finland. The study concluded that short-term climate benefits are limited unless soil carbon losses are included in carbon estimates.

3.2 Sustainability

Each stage of Life-Cycle Assessment (LCA) should be sustainable. Sustainability is assessed by LCA methodologies to evaluate the climate impacts of biomass-based energy production, addressing factors such as greenhouse gas (GHG) emissions, carbon payback periods, and land-use change effects. Giuntoli et al. [6] presented a detailed attributional life cycle assessment (A-LCA) of electricity generation from biomass residues: cereal straw and cattle slurry. Each system was modelled from feedstock collection through transport, processing, and electricity production, and compared against the EU fossil electricity mix. To evaluate climate impact, the study applied Surface Temperature Response (STR) metrics: instantaneous STR and cumulative STR, providing a time-sensitive view of warming. Results showed that biogas from cattle slurry offered the highest climate benefit due to minimum or negligible methane emissions and nutrient reuse. Straw combustion delivered strong mitigation after a short delay. Forest residue pellets, however, showed limited benefits unless the natural decay rate was high. The study emphasized the need for precise carbon accounting and feedstock-specific strategies when designing biomass energy systems for climate change mitigation

However, many studies utilized simplified methods often overlooked critical factors like crop rotation, land-use change, and the benefits of returned by-products (biochar, digestate) for soil carbon and nutrient enhancement. Siol et al. [17] investigated sustainable utilization of agricultural and forest residues (crop straw, tree branches) in industry, emphasizing trade-offs between biomass extraction and soil health. The study identified multiple system boundary approaches: zero-burden (no pre-collection impacts), sustainable removal (limited extraction based on thresholds), replacement (input modelling such as fertilizers), and impact allocation (burden distribution among products). Using simulation models (HEUREKA Q, SIMA for forests, and DAYCENT, CENTURY, CERES for agriculture), soil carbon and nutrient dynamics were assessed. Authors recommended integrating comprehensive soil carbon modelling, by-product feedback loops, and region-specific agricultural practices into Life Cycle Assessment (LCA) frameworks to accurately represent long-term biomass residue sustainability.

By integrating remote sensing data into life-cycle evaluation enables more precise monitoring of land-use changes, carbon stocks, and environmental impacts, and sustainable biomass management practices

3.3 Environmental Impact Assessment

By integrating scientific research and technological innovations, Environmental impact assessment (EIA) recommends that biomass-based energy systems should contribute to sustainable development while minimizing ecological risks. EIA evaluates ecological consequences of biomass-based energy systems and ensures that development aligns with sustainability goals. Thus, a well-structured EIA framework helps policymakers and industry stakeholders identify potential risks, develop mitigation strategies, and optimize bioenergy solutions to balance energy production with environmental protection.

Searchinger et al. [16] argued that the climate benefits of biomass energy depend on how quickly new plant growth offsets carbon released during combustion. The study stressed the importance of accounting for indirect land use change (ILUC), where biofuel production displaces food crops and drives deforestation, leading to significant carbon emissions. For example, food crops like corn offers no net carbon benefit for biofuel conversion, as these plants would absorb carbon on land regardless of their end use.

Zanchi et al. [22] showed the variation in carbon payback periods across biomass sources, emphasizing role of sustainable harvesting and feedstock selection in minimizing biomass energy's carbon footprint. The study assessed biomass-based energy options in rural Uganda, comparing electricity from wood gasification and firewood for cooking with current diesel use and unsustainable wood collection. Eucalyptus grandis was proposed for short rotation coppice (SRC) plantations. A 30 MWh/year wood gasifier required 2–6 hectares and reduced emissions by 50–67%, saving 18.1–24.6 tCO_2-eq annually. The study concluded that biomass energy systems, when established on non-agricultural land, with appropriate species, can lower emissions and enhance rural energy access while supporting climate and development objectives.

Biodiversity conservation is a critical component of EIA in biomass energy projects. Large-scale monoculture plantations, such as oil palm, contribute to habitat fragmentation and reduced ecosystem resilience. Danielsen et al. [4] examined biodiversity and carbon impacts of land conversion in Indonesia using carbon stock data from field measurements and national inventories. Conversion of forests to oil palm by logging released 163 tC/ha, requiring 75 years to offset through fossil fuel substitution, while fire-based conversion emitted 207 tC/ha, extending payback to 93 years. Biodiversity surveys showed sharp declines in species – only 38% of forest vertebrate and 31% of insect species remained. Floristic assessments confirmed the loss of native species and dominance of disturbance-tolerant plants.

Table 1 provides a consolidated summary of major studies reviewed in this paper, outlining their methodologies, biomass types, tools, and key findings.

4 Challenges

Despite significant advancements in biomass research, several challenges persist in optimizing its utilization for sustainable energy production. Relevant challenges are discussed as follows:

(i) Accurate assessment of biomass availability, particularly agricultural residues, is a key challenge as traditional quantification methods, such as field surveys, are labor-intensive, time-consuming, and spatially limited.
(ii) Accuracy of remote sensing methods for scalable predictions are often constrained by low-resolution data, cloud interference, and weak calibration with ground-truth measurements. Moreover, single-source satellite inputs, such as MODIS or Landsat, lack the spatial and temporal resolution necessary for detailed and dynamic biomass monitoring. On the other hand, existing classification models frequently fail to integrate multi-sensor and multi-temporal remote sensing datasets of SAR, LiDAR, and photogrammetry.

(iii) Environmental sustainability in biomass energy faces key challenges like land-use change, soil degradation, water use, and biodiversity loss. Unsustainable residue harvesting can deplete soil organic matter, reduce fertility, and disturb ecosystems. Inconsistent Life Cycle Assessment (LCA) methods, often based on unrealistic carbon neutrality assumptions and omission of factors like soil carbon sequestration and indirect land use change (ILUC), hinder accurate sustainability evaluations. Additionally, fragmented policies and weak local enforcement limit the effectiveness of national biomass initiatives. However, integrated strategies combining digital monitoring, standardized assessment methods, strong policy frameworks, and region-specific safeguards, ensuring environmentally sound and economically viable biomass systems, are not practised.

5 Future Scope

To optimize biomass utilization for sustainable energy production, future research must address current limitations in data integration, monitoring accuracy, and sustainability evaluation. Advancements in multi-sensor remote sensing, particularly the fusion of optical, LiDAR, and synthetic aperture radar (SAR) data, should be further developed using AI-driven models to enhance biomass estimation accuracy across diverse landscapes. Deep learning algorithms trained on comprehensive geospatial datasets can refine biomass classification, while curated global imagery repositories could provide consistent training data for machine learning applications. For dynamic monitoring and forecasting, time-series models such as recurrent neural networks (RNNs) and long short-term memory (LSTM) networks can enable the prediction of biomass fluctuations due to seasonal and climatic variations.

Real-time monitoring can be significantly improved through the integration of IoT field sensors, enabling continuous data collection on vegetation, soil, and climate conditions for better resource planning. Additionally, standardizing biomass classification techniques and incorporating hyperspectral imaging with spectral unmixing algorithms are essential for accurately distinguishing between forestry, agricultural, and industrial residues. Strengthening life-cycle assessment (LCA) frameworks by integrating remote sensing data and field validation will improve the reliability of sustainability metrics.

Developing a unified, interdisciplinary platform that combines AI, IoT, remote sensing, and environmental impact assessments will be key to achieving a balanced approach between energy security and ecological conservation. This integrated approach will support robust decision-making in bioenergy planning, land management, and climate policy implementation.

Table 1. Summary of Key Studies on Residual Biomass Utilization

Study	Biomass Type	Methodology	Dataset/Tools	Key Findings
Helwig et al.	Agricultural (Straw, Stover, Manure)	Quantification using grain-to-residue ratios and sustainability limits	Data from Statistics Canada, OMAFRA, ASRA	Estimated 1M odt straw and 3M odt stover; 16M GJ biogas potential
Bouchard et al.	Forestry Residues	GIS-based biomass availability analysis; CHP site mapping	ArcGIS, forest inventory data	15.5M GT biomass; 1.2 GW electricity, 3 GW heat; 73% biomass within 50km of sites
Pavlova	Agricultural (Manure, Crop Remains)	Case study with anaerobic digestion system	Farm data, national statistics	87,000 m^3 biogas/year; onsite power and fertilizer; reduced emissions
Balan	Lignocellulosic Biomass	Review of biochemical conversion barriers	Technical literature	High pretreatment cost; lignin inhibition; microbial limitations
Xu et al.	Crop Residues	Statistical and regional factor analysis	Climate, crop structure, macroeconomic data	Crop structure most influential; recommendations for storage & land protection
Ahmad et al.	Food & Municipal Solid Waste	Anaerobic co-digestion experiments	Lab-scale, varied C/N and S/I ratios	825 L/kg biogas at optimal ratio; 89% solids reduction
Sipilä et al.	Residual Biomass	Technical review of CHP and district heating	Engineering principles, turbine analysis	Cogeneration 30–40% more efficient than separate systems
Zhao et al.	Forest & Energy Crop Biomass	AI/ML for biomass logistics	Satellite, drone, ML models (RF, NN, SVM)	Improved supply chain efficiency and cost reduction
Ogungbuyi et al.	Grassland Biomass	Remote sensing fusion for yield estimation	Drone & Sentinel-2, Random Forest	Drone-based estimates (R^2 = 0.75) outperform satellite-only (R^2 = 0.56)
Nechetnyy et al.	General Biomass Supply Chain	Blockchain simulation for traceability	System modeling, smart contracts	Enhanced data security, emission tracking, and operational trust
Repo et al.	Forest Harvest Residues	Carbon modeling (Yasso07, CO2FIX)	European forest data	3% carbon decline; regional variation in impact
Giuntoli et al.	Straw, Cattle Slurry	Attributional LCA with STR metrics	STR analysis vs. EU fossil mix	Biogas highest benefit; straw short-term gain; forest pellet limited
Siol et al.	Agricultural and Forest Residues	Soil carbon/nutrient modeling with DAYCENT, HEUREKA Q	Soil carbon/nutrient simulations	Recommends better modeling of soil feedback loops
Matsushita et al.	Forest Vegetation	Remote sensing terrain correction	EVI, NDVI, LiDAR	EVI sensitive to slope; recommends topographic correction

6 Conclusion

Residual biomass assessment integrates methodological innovation, technological progress, sustainability frameworks, and digital tools to enhance biomass utilization. As the global energy transition progresses, refining these assessment methods is critical to developing a more efficient and sustainable bioenergy sector. Residual biomass utilization has emerged as a viable strategy for advancing renewable energy targets, minimizing waste, and promoting a circular bioeconomy. Progress in conversion technologies, life cycle assessment (LCA), and digital supply chains has strengthened its commercial feasibility.

Remote sensing has advanced biomass assessment by enabling accurate large-scale mapping and monitoring. The integration of satellite imagery, LiDAR, and GIS modelling improves inventory precision and helps identify optimal locations and timing for biomass collection. Continued reliance on medium-resolution imagery and single-sensor datasets limits accuracy, particularly in heterogeneous landscapes. Greater integration of high-resolution, multi-sensor data is needed to fully realize the potential of remote sensing in biomass applications

Environmental assessments highlight the importance of sustainable harvesting to prevent soil degradation, biodiversity loss, and extended carbon payback periods. Limitations in current LCA methodologies—particularly in accounting for indirect land use change—underscore the need for improved modelling and comprehensive datasets. Finally, while policy frameworks significantly influence biomass deployment, inconsistent enforcement and regional disparities reduce their impact. Strengthening institutional coordination, enhancing financial incentives, and investing in infrastructure is essential to enable broader and more effective biomass energy implementation.

Acknowledgements. Revathi Gundluri Nagaraju, Ajay Dashora and Rakesh Mishra are thankful for MITACS Globalink Research Award. Revathi Gundluri Nagaraju is also thankful to the University of New Brunswick for the internship and infrastructure support. Authors are also thankful to Northern Hardwoods Research Institute (Canada) for providing expertise and data support.

Disclosure of Interests. The authors have no competing interests to declare that are relevant to the content of this article.

References

1. Ahmad, R.M., et al.: Optimizing biogas production and digestive stability through waste co-digestion. Sustainability **16**(7), 3045 (2024). https://doi.org/10.3390/su16073045
2. Balan, V.: Current challenges in commercially producing biofuels from lignocellulosic biomass. International Scholarly Research Not. **2014**, 463074, 31 pages (2014). https://doi.org/10.1155/2014/463074
3. Bouchard, S., Landry, M., Gagnon, Y.: Methodology for the large-scale assessment of the technical power potential of forest biomass: application to the province of New Brunswick. Canada, Biomass Bioenergy **54**, 1–17 (2013)
4. Danielsen, F., Beukema, H., Burgess, N.D., Parish, F., Brühl, C.A., Donald, P.F., et al.: Biofuel plantations on forested lands: double jeopardy for biodiversity and climate. Conserv. Biol. **23**(2), 348–358 (2009)

5. Flaspohler, D.J., Webster, C.R., Froese, R.E.: Bioenergy, biomass and biodiversity. In: Renewable Energy from Forest Resources in the United States, pp. 153–182. Routledge (2008)
6. Giuntoli, J., Agostini, A., Caserini, S., Lugato, E., Baxter, D., Marelli, L.: Climate change impacts of power generation from residual biomass. Biomass Bioenerg. **89**, 146–158 (2016)
7. Hakkila, P.: Utilization of residual forest biomass, pp. 352–477. Springer, Berlin Heidelberg (1989)
8. Helwig, T., Jannasch, R., Samson, R., DeMaio, A., Caumartin, D.: Agricultural biomass residue inventories and conversion systems for energy production in Eastern Canada. Contract **23348**(016095/001), 1891819496–1596558871 (2002)
9. Lal, R.: Forest soils and carbon sequestration. For. Ecol. Manage. **220**, 242–258 (2005). https://doi.org/10.1016/j.foreco.2005.08.015
10. Matsushita, B., Yang, W., Chen, J., Onda, Y., Qiu, G.: Sensitivity of the enhanced vegetation index (EVI) and normalized difference vegetation index (NDVI) to topographic effects: a case study in high-density cypress forest. Sensors **7**(11), 2636–2651 (2007)
11. Nechetnyy, N., et al.: Analysis of carbon footprint reduction in supply chains using blockchains. In: E3S Web of Conferences, vol. 581, p. 01017 (2024)
12. Ogungbuyi, M.G., Mohammed, C., Fischer, A.M., Turner, D., Whitehead, J., Harrison, M.T.: Integration of drone and satellite imagery improves agricultural management agility. Remote Sensing **16**(24), 4688 (2024)
13. Pavlova, M.: Effects of residual biomass use in agriculture. Trakia J. Sci. **15**(1), 330–337 (2017)
14. Popp, D.: Environmental policy and innovation: a decade of research
15. Repo, A., Ahtikoski, A., Liski, J.: Cost of turning forest residue bioenergy to carbon neutral. Forest Policy Econ. **57**, 12–21 (2015)
16. Searchinger, T.D.: Biofuels and the need for additional carbon. Environ. Res. Lett. **5**(2), 024007 (2010)
17. Siol, C., Thrän, D., Majer, S.: Utilizing residual biomasses from agriculture and forestry: different approaches to set system boundaries in environmental and economic life-cycle assessments. Biomass Bioenerg. **174**, 106839 (2023)
18. Sipilä, K.: Cogeneration, biomass, waste to energy and industrial waste heat for district heating. In: Advanced District Heating and Cooling (DHC) Systems, pp. 45–73. Woodhead Publishing (2016)
19. Tobin, T., Gustafson, R., Bura, R., Gough, H.L.: Integration of wastewater treatment into process design of lignocellulosic biorefineries for improved economic viability. Biotechnol. Biofuels **13**, 1–16 (2020)
20. Udoumoh, U.I., Ahuchaogu, I.I., Ehiomogue, P.O., Anana, U.E.: Best management practices on soil organic matter conservation and rainfall runoff reduction: a technical note. Acta Technica Corviniensis – Bulletin of Eng. **16**(1) (2023)
21. Xu, X.L., Chen, H.H., Li, Y.: Exploring the influencing factors of continuous crop residue supply: from the perspective of a sustainable and bioenergy-oriented crop cultivation. Energy, Sustainability Soc. **10**(1), 1–14 (2020). https://doi.org/10.1186/s13705-020-00267-0
22. Zanchi, G., Frieden, D., Pucker, J., Bird, D.N., Buchholz, T., Windhorst, K.: Climate benefits from alternative energy uses of biomass plantations in Uganda. Biomass Bioenerg. **59**, 128–136 (2013)
23. Zhao, J., Wang, J., Anderson, N.: Machine learning applications in forest and biomass supply chain management: a review. Int. J. For. Eng. **35**(3), 371–380 (2024)

NIR Spectroscopy Based Non-invasive Assessment of Tea Quality

Onkar Sarma(✉) and Kavya Dashora

Centre for Rural Development and Technology, IIT, Delhi, New Delhi, India
{rdat,rdz238034}@iitd.ac.in

Abstract. This study demonstrates a rapid and non-invasive method for classification and estimation of catechin and caffeine changes with storage time in tea sample by near-infrared (NIR) spectroscopy. The NIR spectra of 34 tea samples were used to evaluate the modelling and prediction performance of a combination of Optuna with three machine learning models. Result showed caffeine prediction with MSEP and R^2 values of 0.27 and 0.98, respectively. However, smaller MSEP of 0.04 was found for catechin with R^2 value of 0.96. These findings suggest that NIR spectroscopy based chemometric analysis can be used to predict the changes in biochemical content in tea with storage time.

Keywords: NIR spectroscopy · Machine learning · Tea quality

1 Introduction

Tea, processed from tea leaves, is the second most consumed liquid in the world subsequent to water (Du et al. 2020). Tea is generally classified into six types defined as dark tea, black tea, oolong tea, yellow tea, green tea, and white tea based on the processing methods and sensory qualities of products (Liang et al. 2021). Since multiple functional components are contained in tea, such as tea polyphenols, theanine, theaflavins, tea saponins, caffeine, etc., tea and its by-products are believed to support a wide range of physiological activities involving anti-tumor, anti-bacterial, anti-oxidant, anti-viral, prevention of cardiovascular or cerebrovascular diseases, and immune regulation (Wei et al. 2023; Xu et al. 2022; Zhang et al. 2021). Tea being a healthy and regular beverage, quality control becomes a crucial part of tea production.

Tea quality analysis is strictly based on its biochemical attributes that imparts its colour, flavour, and overall sensory. Compounds such as catechin, caffeine, moisture and polysaccharide imparts different taste to the product (Wang et al. 2018). Caffeine is an alkaloid which contributes a bitter taste to tea (Yang et al. 2018). Tea polyphenols consist of four major groups: catechins, phenolic acids, flavonoids, and anthocyanins (Kerio et al. 2013), among which catechin can be used as a scale to ascertain the quality potential of tea (Kottawa-Arachchi et al. 2014; Sabhapondit et al. 2012). However, traditional tea quality control relies on laboratory protocols that are time-consuming and uneconomical. Moreover, conventional chemical analysis presents its own set of challenges, including laborious procedures, energy heavy and environmental unfriendliness (Wang et al. 2023).

H. S. Shekhawat et al. (Eds.): ICA 2025, CCIS 2795, pp. 191–201, 2026.
https://doi.org/10.1007/978-3-032-17083-5_16

Spectroscopy technology has shown great potential to monitor tea growth, estimating quality and fermentation degrees, and to discriminate tea varieties, tea grades and types, and geographical origins assisted with suitable chemometrics (Firmani et al. 2019; Lin & Sun 2020). Numerous spectroscopic method have been recently developed that has attracted increasing attraction in food and beverages authentication according to the characteristic spectral patterns with chemometric analysis. Currently, vibrational spectroscopy including near-infrared (NIR) (Porep et al. 2015), midinfrared (MIR), Raman (Su & Sun 2018), terahertz (THz) (Wang et al. 2017) and hyperspectral imaging (HSI) (Lei & Sun 2020; Ma et al. 2019) have been widely investigated for food quality and safety evaluation. NIR region contain characteristic frequency absorption signatures of hydrogen containing functional group of X-H, where X can be any element such as O, S, N, C. Therefore, NIR spectroscopy can be utilized in identifying numerous biochemical compounds with unique spectral signature due to their functional group. NIR spectroscopy allows for very accurate estimation of moisture due to its prominent characteristic waveband, which has been utilized for rapid real-time detection of moisture content in black tea during withering process (Shen et al. 2022). Furthermore, NIR spectroscopy has been fused with other intelligent sensors and computer vision techniques such as digital imaging, e-nose, and e-tongue for determination of various quality attributes (Chen et al. 2023; Li et al. 2023a, b; Zhang et al. 2024). Li et al. (2015) applied Raman spectroscopy in the assessment of adulterants such as lead chrome in green tea. Nonetheless, studies employing NIR spectroscopy in identifying biochemical changes or degradation in tea with storage is scarce.

In multivariate data analysis, supervised ML models such as random forest (RF), K-nearest neighbour (KNN) and partial least squares discriminant analysis (PLS-DA) have demonstrated superior performance in classification tasks (Zhang et al. 2024). Similarly, gradient boosting regressor (GBR), partial least square regression (PLSR) and support vector regression (SVR) are frequently applied in quantitative analysis of spectroscopic data. Furthermore, hyperparameter turning of ML model is important to improve its performance and generalizability. One such flexible and user-friendly framework for optimization is Optuna, which allows dynamic adjustments to the parameter search space. Its efficient search mechanisms, coupled with pruning strategies, enhance performance across various applications (Shekhar et al. 2022).

This study has investigated the performance of multiple ML model in conjunction with Optuna for both classification and quantitative analysis. Furthermore, a faster method for estimation of caffeine and catechin through multivariate analysis has been conducted using NIR spectrum as reference.

2 Materials and Methods

2.1 Materials and Instruments

A total of 34 samples of 4 varieties of made tea were procured from local vendors from Assam. Made tea is a dried solid tea product, which undergoes various processing conditions of withering, fermentation, and drying for fixation. Caffeine and catechin standard were purchased from SRL and CDH Co. Ltd; acetonitrile and methanol were purchased from Merck Co., Ltd. Unless otherwise specified, all chemicals used were of HPLC grade. An Acquity UPLC system (USA) was used to determine caffeine and catechin content. NIR spectrometry was carried out using a Thermo Fisher Nicolet iS50 (USA).

2.2 Data Acquisition

The procured samples were aged for 6 months. NIR spectra were collected in reflectance mode for both fresh sample and aged samples. Each spectrum consisted of an average of 70 scans, in the range of 10000–4000 cm^{-1}. Before scanning, the instrument was fully preheated for more than half an hour. Three spectra were collected from each sample, and the average spectrum of the three spectra was taken as the original analytical spectrum of that sample. Standard normal variate transformation (SNV) method was used for spectral pre-treatment. SNV removes physical noise resulting from particle size. The caffeine and catechin content of both fresh and aged samples were determined according to FSSAI (2023).

The spectral data and the chemical data were separated into two sets of training and testing data. Stratified 5 fold sampling was used for splitting the sample so that no sample is dropped while splitting for training and testing sets. The advantage of this method is making full use of small sample data sets (Rodriguez et al. 2009). Stratified 5 fold cross validation was used to randomly select the training set, and the remaining samples were selected for the testing set. In stratified 5 fold sampling, the dataset is split into 5 folds while preserving the original class distribution in each fold. In each fold, the model was trained using 80% of the data points and tested using the remaining 20%.

2.3 Chemometrics Method

Classifying Algorithm. SVM have been widely applied in classifying tea based on geographical origin, variety, and grades, which exhibit its rapid and effectiveness in performing such task (Chen et al. 2007; Zhuang et al. 2017). Several other studies have also investigated the performance of random forest (RF) (Deng et al. 2020), PLS-DA (Jiang et al. 2025), BP-ANN in tea classification (Zhao et al. 2006). Intelligent optimization method such as particle swarm optimization (PSO) (Pedersen, & Chipperfield 2010), genetic algorithm (GA) (Maruyama, & Igarashi 2008), GridSearch (Yao et al. 2018), and differential evolution algorithm (Sharma et al. 2012) have often been combined with traditional algorithm to improve their accuracy. However, application of Optuna for model optimization is scarce in the field of biochemical analysis. This study employs RF for classification of samples based on variety and aging time. RF is one of the famous ensembles learning technique that combines decision trees into RF to produce outcome using bagging approaches, and has shown applicability over a wide range of task.

Regression Algorithm. Three different algorithms of GBR, SVR, and PLSR were used to predict the biochemical content based on NIR spectrum. GBR constructs an ensemble of trees one after another, correcting the node's errors at every level (Munera et al. 2021). This process repeats continuously for GBR to be able to capture the intricate correlations between quality metrics and the processing environment, and progressively improve predictions. Recently, GBR has been applied in chemometric estimation of biochemical content in tea sample with very high accuracy ($R^2 > 0.98$) (Luo et al. 2021; Modak et al. 2024). SVR can be applied in regression problems to handle continuous data in food quality analysis such as freshness score (Yao et al. 2022), flavour substances in black tea (Li et al. 2023a, b), and nutrient availability in tea leaves (Luo et al. 2025). Two individual linear SVR for regression was trained for concentration prediction of both catechin and caffeine content. PLSR is frequently applied in modelling spectroscopic data for quantitative analysis. In spectroscopic analysis of food components, PLSR model is built to predict quality parameters by plotting regression coefficients resulting from the best feature. Best features are the specific wavelengths with larger absolute coefficients that are good candidates for effective calibration (ElMasry et al. 2011). Quantification of various substances in tea such as metal elements (Wu et al. 2025), moisture content (Zhu et al. 2025), biochemical content (Chen et al. 2025), and adulterant (Jiang et al. 2025) has been feasible by calibrating NIR spectrum with PLSR. In certain instance, PLSR can perform even better than deep learning model (Wu et al. 2025).

Model Evaluation. The performance of the final models were evaluated according to the mean square error of prediction (MSEP) and R^2 score of prediction. The optimal model method was chosen based on lower MSEP and a higher R^2 score (close to 1). The formular for the both are as follows,

$$MSEP = \frac{1}{n} \times \sum (y_i - \hat{y}_i)^2 \quad (1)$$

$$R^2 = \{\frac{\sum (\hat{y}_i - \overline{y})^2}{\sum (y_i - \overline{y})^2}\}^2 \quad (2)$$

where, n is the number of observations
y_i is the ith value for observation
$\overline{y}$ is the mean of y values
$\hat{y}_i$ is the ith predicted value of y

Software. Data processing and modelling analysis was carried out using python version 3.11.5.

3 Result and Discussion

3.1 Spectral Investigation

The original NIR spectra of four Assam tea varieties (containing both fresh and aged samples) are shown in Fig. 1. These spectra can reflect the intrinsic quality of the tea samples. There is clear depiction of difference in NIR spectrum of different variety tea sample. The absorbance by these samples have also shifted considerably when aged for 6 months. This difference in NIR spectrum can be captured well for classification on the basis of both sample variety and aging time. The first frequency-doubling peak corresponding to the N–H bond stretching vibration was observed at 6840 cm^{-1}, while the second frequency-doubling peak associated with the $C = O$ stretching vibration appeared at 5190 cm^{-1}. Additionally, the combined frequency peak resulting from the stretching vibrations of primary and tertiary amines was detected at around 4640 cm^{-1} (Fig. 1). These spectral features are influenced by the composition of the sample and serve as a theoretical foundation for the rapid estimation of tea catechin and caffeine concentrations.

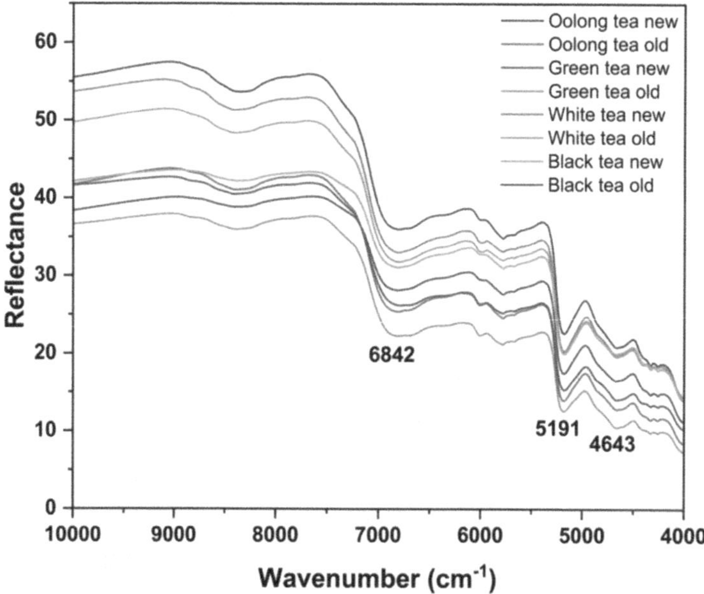

Fig. 1. NIR spectrum of four variety of Assam tea. Fresh and aged sample of each variety has been depicted as new and old, respectively.

3.2 Modelling Results

A total of 34 samples were assessed using NIR spectroscopy and further classified according to their variety and aging time. A quantitative analysis of catechin and caffeine content was carried out on the same set of spectrum. RF was found to be the best model in classifying samples with accuracy of 88%. Figure 2 depicts the confusion matrix for tea classification using RF.

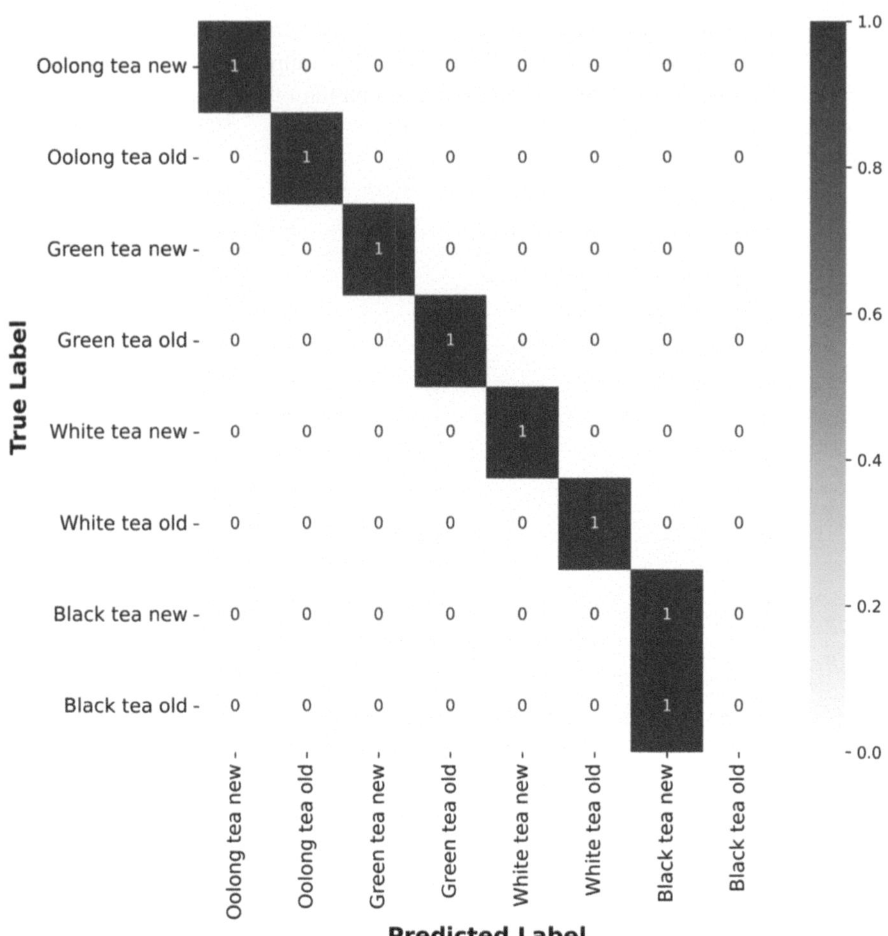

Fig. 2. Confusion matrix of tea classification

The results in Table 1 show that the ranges of catechin and caffeine content in four varieties of fresh tea samples as 3.86%–0.33%, and 22.07%–8.74%, respectively. In aged sample, catechin and caffeine varied from 0.37%–2.49% and 6.36%–14.33%, respectively.

Table 1. Biochemical content in fresh and aged sample of tea

Sample	Catechin			Caffeine		
	Min %	Max %	Avg %	Min %	Max %	Avg %
Green tea fresh	0.966	1.404	1.209	9.843	14.276	12.395
Green tea aged	0.377	0.548	0.472	6.467	9.380	8.144
White tea fresh	3.087	4.486	3.864	17.528	25.422	22.073

(*continued*)

Table 1. *(continued)*

Sample	Catechin			Caffeine		
	Min %	Max %	Avg %	Min %	Max %	Avg %
White tea aged	0.718	1.043	0.899	9.885	14.338	12.449
Oolong tea fresh	0.265	0.385	0.331	6.942	10.068	8.742
Oolong tea aged	1.716	2.498	2.148	8.011	11.619	10.089
Black tea fresh	0.638	0.927	0.798	7.992	11.591	10.064
Black tea aged	0.388	0.565	0.486	6.362	9.228	8.013

Optimization of model hyperparameters has been done with Optuna to achieve the best cross validation accuracy and performance of the model. The optimized parameters for GB regressor were given as n_estimator = 100; learning_rate = 0.111, and max_depth = 5.

Table 2. Model metrics for catechin and caffeine prediction

Components	Model	MSEP	R^2
Catechin	GBR	0.38	0.88
	Optuna-GBR	0.04	0.96
	SVR	0.99	0.18
	Optuna-SVR	0.84	0.33
	PLSR	0.25	0.60
	Optuna-PLSR	4.59	0.65
Caffeine	GBR	0.40	0.88
	Optuna-GBR	0.27	0.98
	SVR	1.94	0.50
	Optuna-SVR	0.32	0.65
	PLSR	4.39	0.58
	Optuna-PLSR	3.19	0.59

The results in Table 2 show that the Optuna-GBR model could predict the catechin, and caffeine content in made tea. The results showed that MSEP and R^2 presented better values by Optuna combined GBR than GBR, SVR, and PLSR alone, indicating better performance of model optimization with Optuna. Optuna-GBR model yielded the best performance and the most accurate prediction results for caffeine. The values obtained for MSEP and R^2 were 0.27 and 0.98, respectively. However, smaller MSEP of 0.04 was found for catechin with R^2 value of 0.96. Combine MSEP and R^2 value for both the compound is 0.16 and 0.97, respectively. Optimization of both SVR and PLSR showed no significant improvement in this scenario. Although the models have performed well in this scenario, it should be noted that this study has utilized a small sample number, which does not allow these results to be generalized. However, the primary motive of this study is to demonstrate the applicability of these machine learning algorithms in estimating the degradation of bioactive compounds with storage time in cash crops like tea.

The spectral characteristics of tea polyphenols are primarily distributed around the wavebands of 4300–5500 cm^{-1}, and 6700–7100 cm^{-1}. The second-order frequency doubling, attributed to C-C stretching, is evident at 4643 cm^{-1}. Meanwhile, the O-H stretching vibration associated with the phenolic rings of catechins appears near 5000 cm^{-1}. Caffeine exhibits distinct absorption bands in the ranges of 4000–4900 cm^{-1} and 6994–7293 cm^{-1}. Additionally, the combined stretching vibrations of primary and tertiary amines are observed around 4610 cm^{-1}. These spectral features provide valuable insights into the molecular composition of tea and contribute to the non-destructive analysis of its key bioactive compounds (Baykal et al. 2010).

4 Conclusion

This experiment has conducted a non-invasive estimation of the catechin and caffeine content in tea by chemometric analysis of NIR spectroscopy. The stratified 5 fold cross validation method was used to randomly select the testing set, and Optuna was employed as the optimization framework for GBR, SVR, and PLSR. The result shows the values of MSEP and R^2 as 0.04 and 0.96 for catechin prediction and for caffeine prediction, MSEP and R^2 were 0.27 and 0.98, respectively. This approach provides a rapid, sustainable, and reproducible method for the quantitative detection of catechin and caffeine, as well as classification of tea samples based on variety and storage time. Future studies can take account of other important factors such as geographical locations, grades, variety, and harvesting season in order to generalize the findings to other tea varieties too. Exploring the impact of potential confounding factors like humidity and other storage conditions that influence the biochemical content can be further examined to estimate the underlying chemical changes which influence the sensory qualities of tea. Furthermore, it can serve as a quality control measure for businesses in detecting low grade product mixtures. The implementation of this method will support the advancement of sustainable analytical techniques, leading to improved process management and more efficient quality control.

Acknowledgments. The authors would like to thank Central Research Facility of Indian Institute of Technology, Delhi.

Disclosure of Interests. The authors have no competing interests to declare that are relevant to the content of this article.

References

Baykal, D., Irrechukwu, O., Lin, P.C., Fritton, K., Spencer, R.G., Pleshko, N.: Nondestructive assessment of engineered cartilage constructs using nearinfrared spectroscopy. Appl. Spectrosc. **64**(10), 1160–1166 (2010)

Chen, Q., et al.: Recent developments of green analytical techniques in analysis of tea's quality and nutrition. Trends Food Sci. Technol. **43**(1), 63–82 (2015). https://doi.org/10.1016/j.tifs.2015.01.009

Chen, Q., Zhao, J., Fang, C.H., Wang, D.: Feasibility study on identification of green, black and Oolong teas using near-infrared reflectance spectroscopy based on support vector machine (SVM). Spectrochim. Acta Part A Mol. Biomol. Spectrosc. **66**(3), 568–574 (2007)

Chen, Y., et al.: Monitoring green tea fixation quality by intelligent sensors: comparison of image and spectral information. J. Sci. Food Agric. **103**(6), 3093–3101 (2023). https://doi.org/10.1002/jsfa.12350

Chen, Z., et al.: Innovative manufacturing process to enhance the taste and color of Ancha tea: a comprehensive analysis using sensory evaluation, chemometrics, and metabolomics. Food Biosci. **66**, 106208 (2025). https://doi.org/10.1016/j.fbio.2025.106208

Cozzolino, D., et al.: Effect of temperature variation on the visible and near infrared spectra of wine and the consequences on the partial least square calibrations developed to measure chemical composition. Anal. Chim. Acta **588**(2), 224–230 (2007)

Deng, X., et al.: Predictive geographical authentication of green tea with protected designation of origin using a random forest model. Food Control **107**, 106807 (2020)

Du, C., Ma, C., Gu, J., Li, L., Chen, G.: Fluorescencesensingofcaffeine in tea beverages with 3, 5-diaminobenzoic acid. Sensors **20**(3), 819 (2020)

ElMasry, G., Sun, D.W., Allen, P.: Non-destructive determination of water-holding capacity in fresh beef by using NIR hyperspectral imaging. Food Res. Int. **44**(9), 2624–2633 (2011). https://doi.org/10.1016/j.food-res.2011.05.001

FSSAI. Manual of Methods of Analysis of Foods-Beverages: Tea, Coffee and Chicory (2023). https://fssai.gov.in/cms/manuals-of-methods-of-analysis-for-various-food-products.php. Accessed 30 April 2025

Firmani, P., De Luca, S., Bucci, R., Marini, F., Biancolillo, A.: Near infrared (NIR) spectroscopy-based classification for the authentication of Darjeeling black tea. Food Control **100**, 292–299 (2019). https://doi.org/10.1016/j.foodcont.2019.02.006

Jiang, X., Cheng, P., Ge, K., Lv, S., Liu, Y.: Detection of lead chrome green in tea based on near-infrared reflectance spectroscopy. J. Chemom. **39**(3), e70011 (2025)

Kerio, L.C.,Wachira, F.N., Wanyoko ,J.K., Rotich, M.K.: Totalpolyphenols, catechin profiles and antioxidant activity of tea products from purple leaf coloured tea cultivars. Food Chem. **136**(3), 1405–1413 (2013). https://doi.org/10.1016/j.foodchem.2012.09.066

Kottawa-Arachchi, J.D., Gunasekare, M.T.K., Ranatunga, M.A.B., Punyasir, P.A.N., Jayasinghe, L., Karunagoda, R.P.: Biochemical Characteristics of Tea {Camellia L. spp.) germplasm accessions in sri lanka: correlation between black tea quality parameters and organoleptic evaluation. Int. J. Tea Sci. **10**(01 and 02), 03–13 (2014)

Lei, T., Sun, D.W.: A novel NIR spectral calibration method: Sparse coefficients wavelength selection and regression (SCWR). Analytica Chimica Acta **1110**, 169–180 (2020). https://doi.org/10.1016/j.aca.2020.03.007

Li, L., et al.: Rapid and comprehensive grade evaluation of Keemun black tea using efficient multidimensional data fusion. Food Chem.: X **20**, 100924 (2023)

Li, Q., et al.: Machine learning technique combined with data fusion strategies: a tea grade discrimination platform. Ind. Crops Prod. **203**, 117127 (2023). https://doi.org/10.1016/j.indcrop.2023.117127

Li, X.L., Sun, C.J., Luo, L.B., He, Y.: Non-destructive detection of lead chrome green in tea by Raman spectroscopy. Sci. Rep. **5**(1), 15729 (2015). https://doi.org/10.1038/srep15729

Liang, S., et al.: Processing technologies for manufacturing tea beverages: From traditional to advanced hybrid processes. Trends Food Sci. Technol. **118**, 431–446 (2021). https://doi.org/10.1016/j.tifs.2021.10.016

Lin, X., Sun, D.W.: Recent developments in vibrational spectroscopic techniques for tea quality and safety analyses. Trends Food Sci. Technol. **104**, 163–176 (2020). https://doi.org/10.1016/j.tifs.2020.06.009

Luo, Q., et al.: Nitrogen-phosphorus responses and Vis/NIR prediction in fresh tea leaves. Food Chem. **143369** (2025). https://doi.org/10.1016/j.foodchem.2025.143369

Luo, X., et al.: Nondestructive testing model of tea polyphenols based on hyperspectral technology combined with chemometric methods. Agriculture **11**(7), 673 (2021). https://doi.org/10.3390/agriculture11070673

Ma, J., Sun, D.W., Pu, H., Cheng, J.H., Wei, Q.: Advanced techniques for hyperspectral imaging in the food industry: principles and recent applications. Annu. Rev. Food Sci. Technol. **10**(1), 197–220 (2019). https://doi.org/10.1146/annurev-food-032818-121155

Maruyama, T., Igarashi, H.: An effective robust optimization based on genetic algorithm. IEEE Trans. Magn. **44**(6), 990–993 (2008)

Modak, A., Das, D., Chatterjee, T.N., Tudu, B., Roy, R.B.: Regression based EGCG detection in green tea employing MIP electrodes. IEEE Sens. J. **24**(11) (2024). https://doi.org/10.1109/JSEN.2024.3390438

Munera, S., et al.: Discrimination of common defects in loquat fruit cv. 'Algerie' using hyperspectral imaging and machine learning techniques. Postharvest Biol. Technol. **171**, 111356 (2021)

Nobre, J., Neves, R.F.: Combining principal component analysis, discrete wavelet transform and XGBoost to trade in the financial markets. Expert Syst. Appl. **125**, 181–194 (2019). https://doi.org/10.1016/j.eswa.2019.01.083

Pedersen, M.E.H., Chipperfield, A.J.: Simplifying particle swarm optimization. Appl. Soft Comput. **10**(2), 618–628 (2010)

Porep, J.U., Kammerer, D.R., Carle, R.: On-line application of near infrared (NIR) spectroscopy in food production. Trends Food Sci. Technol. **46**(2, Part A), 211–230 (2015). https://doi.org/10.1016/j.tifs.2015.10.002

Rodriguez, J.D., Perez, A., Lozano, J.A.: Sensitivity analysis of k-fold cross validation in prediction error estimation. IEEE Trans. Pattern Anal. Mach. Intell. **32**(3), 569–575 (2009)

Sabhapondit, S., Karak, T., Bhuyan, L.P., Goswami, B.C., Hazarika, M.: Diversity of Catechin in northeast indian tea cultivars. Sci. World J. **2012**(1), 485193 (2012). https://doi.org/10.1100/2012/485193

Sharma, H., Bansal, J.C., Arya, K.V.: Fitness based differential evolution. Memetic Comput. **4**(4), 303–316 (2012)

Shekhar, S., Bansode, A., Salim, A.: A comparative study of hyper-parameter optimization tools. In: Proceedings of IEEE CSDE 2021 (2022)

Shen, S., et al.: Rapid and realtime detection of moisture in black tea during withering using micro-near-infrared spectroscopy. LWT **155**, 112970 (2022). https://doi.org/10.1016/j.lwt.2021.112970

Su, W.H., Sun, D.W.: Fourier transform infrared and raman and hyperspectral imaging techniques for quality determinations of powdery foods: a review. Comprehen. Rev. Food Sci. Food Safety **17**(1), 104–122 (2018). https://doi.org/10.1111/1541-4337.12314

Wang, H., Gu, J., Wang, M.: A review on the application of computer vision and machine learning in the tea industry. Front. Sustain. Food Syst. **7** (2023). https://doi.org/10.3389/fsufs.2023.1172543

Wang, J., et al.: Enhanced cross-category models for predicting the total polyphenols, caffeine and free amino acids contents in Chinese tea using NIR spectroscopy. LWT **96**, 90–97 (2018). https://doi.org/10.1016/j.lwt.2018.05.012

Wang, K., Sun, D.W., Pu, H.: Emerging non-destructive terahertz spectroscopic imaging technique: principle and applications in the agri-food industry. Trends Food Sci. Technol. **67**, 93–105 (2017). https://doi.org/10.1016/j.tifs.2017.06.001

Wei, Y., et al.: Recent advances in the utilization of tea active ingredients to regulate sleep through neuroendocrine pathway, immune system and intestinal microbiota. Crit. Rev. Food Sci. Nutr. **63**(25), 7598–7626 (2023). https://doi.org/10.1080/10408398.2022.2048291

Wu, T., et al.: Quantification of multiple elements in Anji white tea using hyperspectral imaging combined with machine learning regression. J. Food Compos. Anal. **142**, 107520 (2025). https://doi.org/10.1016/j.jfca.2025.107520

Xu, J., Wei, Y., Li, F., Weng, X., Wei, X.: Regulation of fungal community and the quality formation and safety control of Puerh tea. Comprehen. Rev. Food Sci. Food Safety **21**(6), 4546–4572 (2022). https://doi.org/10.1111/1541-4337.13051

Yang, C., et al.: Application of metabolomics profiling in the analysis of metabolites and taste quality in different subtypes of white tea. Food Res. Int. **106**, 909–919 (2018). https://doi.org/10.1016/j.foodres.2018.01.069

Yao, K., et al.: Visualization research of egg freshness based on hyperspectral imaging and binary competitive adaptive reweighted sampling. Infrared Phys. Technol. **127**, 104414 (2022). https://doi.org/10.1016/j.infrared.2022.104414

Yao, S., Li, T., Liu, H., Li, J., Wang, Y.: Traceability of Boletaceae mushrooms using data fusion of UV–visible and FTIR combined with chemometrics methods. J. Sci. Food Agric. **98**(6), 2215–2222 (2018)

Zhang, W., et al.: Rapid identification of the aging time of Liupao tea using AI-multimodal fusion sensing technology combined with analysis of tea polysaccharide conjugates. Int. J. Biol. Macromol. **278**, 134569 (2024). https://doi.org/10.1016/j.ijbiomac.2024.134569

Zhang, Z., et al.: Potential protective mechanisms of green tea polyphenol EGCG against COVID-19. Trends Food Sci. Technol. **114**, 11–24 (2021). https://doi.org/10.1016/j.tifs.2021.05.023

Zhao, J., Chen, Q., Huang, X., Fang, C.H.: Qualitative identification of tea categories by near infrared spectroscopy and support vector machine. J. Pharm. Biomed. Anal. **41**(4), 1198–1204 (2006)

Zhuang, X., Wang, L., Chen, Q., Wu, X., Fang, J.: Identification of green tea origins by near-infrared (NIR) spectroscopy and different regression tools. Sci. China Technol. Sci. **60**, 84–90 (2017)

Zhu, F., et al.: A multi-feature fusion based method for detecting the moisture content of withered black tea. J. Food Compos. Anal. **107325** (2025). https://doi.org/10.1016/j.jfca.2025.107325

Comparative Analysis of DSSAT and APSIM Crop Models for Wheat Yield Prediction in Chhattisgarh State of India

Shashi Bhushan Kumar[1(✉)], Suyog Balasaheb Khose[2], Rajul Upadhyay[2], and Ashok Mishra[2]

[1] School of Water Resources, Indian Institute of Technology Kharagpur, West Bengal, Kharagpur 721302, India
bhushanshashi188@gmail.com
[2] Agricultural Engineering Department, Krishi Vigyan Kendra, Narayangaon, Narayangaon, Maharashtra 410504, India
amishra@agfe.iitkgp.ac.in

Abstract. Wheat yield prediction is critical for ensuring food security and efficient resource planning in agriculture. However, selecting and validating the most suitable crop simulation model under diverse agro-climatic conditions remains a significant challenge. Chhattisgarh, a state in central India, features varied soil types and climatic zones, making ideal region for testing crop models. The Agricultural Production Systems Simulator (APSIM) and Decision Support System for Agro-technology Transfer (DSSAT) crop models were used for wheat simulation using regional datasets, and compare their performance using key statistical metrics like R^2 and RMSE. Intent of this study to determine the which model most effective for wheat yield simulation for Chhattisgarh state of India. Temperature and rainfall were used as primary input for model setup and 2009 to 2014 was used for calibration while 2015 to 2018 was used for validation. It showed lower RMSE values (74.36 kg/ha and 96.36 kg/ha) compared to APSIM (86.49 kg/ha and 103.97 kg/ha), indicating closer alignment with observed data. DSSAT also achieved better R^2 (0.89 and 0.75), highlighting its ability to explain yield variability and model accuracy more effectively. The models can be extended to simulate yields under various climate scenarios.

Keywords: DSSAT · APSIM · Wheat · Yield Prediction · Chhattisgarh

1 Introduction

The global population is projected to reach 9.1 billion, necessitating a 60 per cent increase in crop production to meet food requirements by 2050 [1]. India is the second-biggest producer of wheat after China, emphasizing the significance of these crops for the country's food security [2–4]. Their cultivation has played a vital role in transforming India from a food-deficient nation to one capable of meeting its own food needs [5]. Wheat holds the position of being the most versatile crop among all cereal grains due to its adaptability

H. S. Shekhawat et al. (Eds.): ICA 2025, CCIS 2795, pp. 202–211, 2026.
https://doi.org/10.1007/978-3-032-17083-5_17

to various climatic regions. India accounts for approximately 12.5% of global wheat production (Food and Agriculture Organization (FAO) of the United Nations). In the year 2021–22, wheat production in India reached 111.32 million tonnes, surpassing the average production of the past decade by 7.44 million tonnes (PIB press release).

In recent years, wheat cultivation in Chhattisgarh has shown considerable changes in area, production, and productivity. As of 2022–23, the area under wheat cultivation increased to approximately 146.4 thousand hectares from an average of 126.51 thousand hectares between 2017–2022, marking an absolute increase of 19.89 thousand hectares [6]. However, wheat productivity has been affected by extreme weather variability in recent years. Climate change and global warming scenarios have had a negative impact on wheat output worldwide [3, 7]. Maintaining high crop productivity is crucial to meet the future food demands of a growing population. Wheat is a long-day plant that is sensitive to temperature but not to day length, making it susceptible to temperature fluctuations that can influence its output. The timing of favorable temperature conditions significantly affects the duration and quality of the wheat crop season.

The ripening of wheat relies on optimal average temperatures of around 14–15 °C, particularly during grain filling and development stages [8, 9]. As a predominantly rabi season, the wheat crop is sown between October and December and harvested from February to April. Like any other crop, wheat production is impacted by external environmental factors. Gaining insights into how crop yield and water demand may change under future climate scenarios at the local level is essential for formulating precise and effective adaptation measures [10]. Therefore, effective planning, management, and implementation play a vital role and significantly influence production. However, small farmers often overlook these aspects due to limited knowledge and skills.

To address these challenges and support informed decision-making, crop simulation models have emerged as valuable tools for understanding the complex interactions between climate, soil, and crop growth. There are many methods to simulate the crop yield, including, Agricultural Production Systems sIMulator (APSIM) [12], World Food Studies simulation model (WOFOST) [14], Environmental Policy Integrated Climate (EPIC) [13], and Decision Support System for Agrotechnology Transfer (DSSAT) [11] have been developed. To estimate agricultural yields and evaluate how climate affects crop output, the DSSAT and APSIM models are frequently used [11, 12]. It considers factors such as soil fertility, irrigation management, and the depth of the plant root zone responsible for absorbing water and nutrients from the soil. The model also incorporates the relationship between modeling techniques used to simulate plant growth and yield under varying climatic conditions [9].

The APSIM is a globally acknowledged and sophisticated tool used for simulating agricultural systems. It integrates various modules that allow detailed modeling of interactions between crops, soils, climate variables, livestock, and farm management strategies [12, 15–17]. The APSIM is continually evolving, with the latest iteration, APSIM Next Generation, introducing new capabilities. Despite the extensive use of crop models globally, limited comparative studies have been conducted for wheat yield prediction under the agro-climatic conditions of Chhattisgarh using region-specific datasets. This gap highlights the need for localized model evaluation to enhance decision-making in agricultural planning. The objectives of this study are to calibrate and validate the DSSAT and APSIM crop models for wheat using regional datasets and to compare their performance in yield prediction based on key statistical metrics. This comparative analysis aims to identify the more reliable model for supporting decision-making in wheat yield across Chhattisgarh.

2 Materials and Methods

2.1 Study Area

Chhattisgarh, located between 17°46′ to 24°05 ′ N and 80°15 ′ to 84°20 ′ E, is a state in central India. It encompasses a total land area of 135191 Km^2, out of this, 47790 Km^2, which is approximately 34.80% of the total area, is sown or used for agriculture. Chhattisgarh is heavily reliant on agriculture, with over 80 percent of the population depending on it as a means of livelihood. The state cultivates various principal crops, including rice, wheat, corn, peanuts, legumes, and oilseeds.

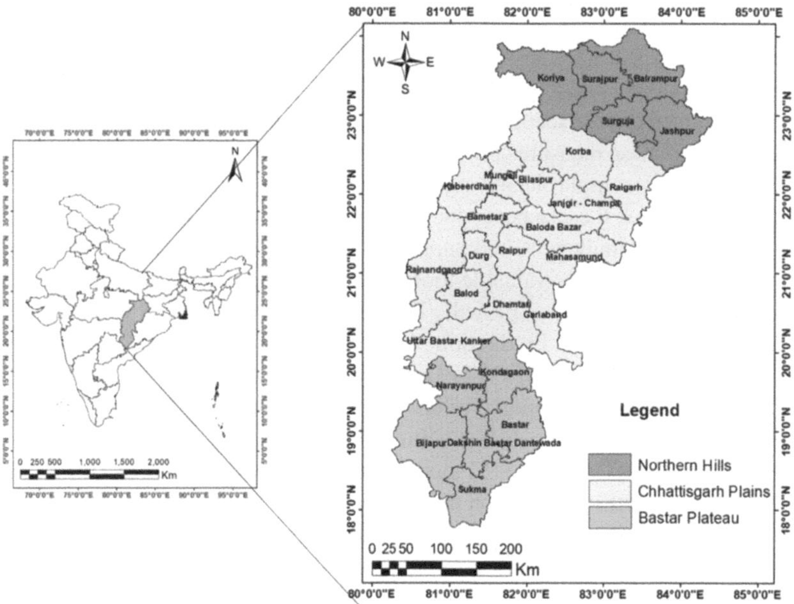

Fig. 1. Chhattisgarh state in India with district boundaries.

To better understand the agricultural landscape, Chhattisgarh is divided into three agro-climatic zones: the Bastar Plateau Zone, Chhattisgarh Plain Zone, and Northern Hill Zone, each characterized by distinct climate and geography as shown in Fig. 1. These divisions reflect the varying climatic conditions across the state. Chhattisgarh's climate is predominantly tropical due to its proximity to the Tropic of Cancer and its reliance on the monsoon season for rainfall. The average annual temperature ranges from 13.76 °C to 38.57 °C. Chhattisgarh receives an average precipitation of 1292 mm, contributing to the agricultural productivity of the state. The diversity in climate and geography across the agro-climatic zones of Chhattisgarh poses both challenges and opportunities for agricultural practices.

2.2 Cultivar and Fertilizer Application

For the Wheat yield simulation in this study, the Sujata cultivar was chosen. A sowing depth of 25 mm was employed, with a row spacing of 180 mm. The plant population density was set at 200 plants per square meter. Regarding fertilizer application, a basal dose of 120 kg per hectare of Diammonium Phosphate (DAP) was applied at the time of sowing. The remaining fertilizer, in the form of Urea at a rate of 80 kg per hectare, was applied in three splits during the crop growth stages. The first split of 25% of the Urea fertilizer was applied at the crown root initiation (CRI) stage, which occurred approximately 25 days after sowing (DAS). The second split of 25% was applied at the tillering stage, around 54 DAS. The final split, comprising the remaining portion of the Urea, was applied at the booting stage, approximately 70 DAS.

2.3 DSSAT Model

In this study, the DSSAT model version 4.5 [9, 18] was utilized to simulate the wheat crop growth and yield. The DSSAT is an integrated suite of computer programs developed to support agricultural research and decision-making by modeling crop development, growth, and productivity under varying environmental and management conditions. One of the key features of the DSSAT model is its ability to incorporate climate change effects on crop production systems by allowing modifications to the weather variables. This flexibility enables the simulation of plausible climate change impacts on various crops such as cereals, legumes, potatoes, sugarcane, tomato, sunflower, and pastures. Table 1 demonstrates the calibrated genetic coefficients of the Sujata cultivar. For this particular study, CERES-Wheat a component of the DSSAT model, was employed to simulate the growth and yield of wheat. Three locations were chosen to simulate and growth of wheat for 2009–2018. This model gives the how climate change affecting the crop growth in different climatic condition. For more details on the DSSAT model, one can refer [9]. In order to simulate crop growth, a carbon balancing technique is implemented within a source-sink system using the methodology described by [8]. Using this method, the following computation is performed in order to get the daily crop growth rate:

$$\mathrm{PCARB} = \frac{\mathbf{RUE} \times \mathrm{PAR}}{\mathbf{PLTPOP}}\left(1 - e^{(-\mathbf{k} \times \mathrm{LAI})}\right) \times \mathbf{CO_2} \tag{1}$$

n this formulation, PCARB denotes the potential plant growth rate (g/plant), RUE refers to radiation use efficiency measured in grams of dry matter per megajoule, and PAR represents the photosynthetically active radiation. PLTPOP stands for the plant population density (plants per square meter), k is the light extinction coefficient, LAI indicates the peak leaf area index, and CO_2 corresponds to the atmospheric carbon dioxide concentration expressed in parts per million by volume (ppmv).

Table 1. Calibrated genetic coefficients of Sujata wheat variety for DSSAT model

Genetic Coefficients	Calibrated values	Name and Unit
P1V	5	Day, at optimum vernalizing temperature
P1D	99	Photoperiod response
P5	951	Grain filling phase duration (GDD)
G1	33	Kernel number per unit canopy weight at anthesis (No./g)
G2	35	Standard kernel size under optimum conditions (mg)
G3	2	Standard, non-stressed mature tiller weight (including grain) (g)
PHINT	97	Interval between successive leaf tip appearances (GDD)

2.4 APSIM Next Generation Model

The Agricultural Production Systems Simulator (APSIM) is a comprehensive modeling framework designed to simulate crop growth and development by integrating key physiological processes. Working at the organ level, APSIM applies supply-demand principles for essential resources such as light, carbon, water, and nitrogen to model crop phenotypes accurately. This framework, as described by [17], focuses on quantifying, capturing, and utilizing water, radiation, and nitrogen within a predictive framework for crop development dynamics. In APSIM, the development of key plant organs is driven by their potential growth, which depends on the availability of carbohydrates and nitrogen required to fulfil that potential. Resource demand is established based on this potential growth, whereas the actual supply is governed by the extent of resource acquisition. For a more comprehensive understanding of APSIM's methodology and approach, refer to the studies conducted by [12, 15, 16] and the research of [17].

2.5 Input Data

The required daily weather data (maximum temperature (T_max), minimum temperature (T_min), and rainfall) for the calibration and validation of the DSSAT and APSIM model were obtained from the India Meteorological Department (IMD). Rainfall and temperature data covered the period from 1981 to 2014 and were downloaded at a resolution of $0.25° \times 0.25°$ and $1° \times 1°$ respectively. To match the resolution the temperature data was converted from the 1° to 0.25° resolution using the bilinear interpolation.

The Food and Agriculture Organization (FAO) provided the soil shape file for the Chhattisgarh region. A soil map with 14 different soil classes unique to the Chhattisgarh region was created using ArcGIS.

2.6 Model Calibration and Validation

Sujata wheat variety was used for calibrate (2009–2014) for DSSAT and APSIM model and subsequently validated for period from 2015–2018. There are several statical indicator were sued for assessing the model performance including the Wilmot D-index, RMSE, ME, and R^2, were used to evaluate the models' effectiveness [16].

$$R^2 = 1 - \frac{\sum_{i=1}^{n}(O_i - S_i)^2}{\sum_{i=1}^{n}(O_i - O_{avg})^2} \tag{2}$$

$$\text{RMSE} = \left[\frac{\sum_i^n (S_i - O_i)^2}{n}\right]^{0.5} \tag{3}$$

$$D - \text{index} = 1 - \frac{\sum_i^n (S_i - O_i)^2}{\sum_i^n \left[\left|S_i - \overline{O_{avg}}\right| + \left|O_i - \overline{O_{avg}}\right|\right]^2} \tag{4}$$

$$\text{ME} = \frac{\sum_i^n (O_i - O_{avg})^2 - \sum_i^n (S_i - O_i)^2}{\sum_i^n \left(O_i - \overline{O_{avg}}\right)^2} \tag{5}$$

Where Oi is the observed values, Si is the model-simulated values, n is the number of observations, and (O_avg)┴¯ is the average of the observed values.

3 Results and Discussion

3.1 Calibration of Genetic Coefficients for Wheat Variety

The Sujata wheat variety's calibrated genetic coefficients using the DSSAT and APSIM Next Gen models, are shown in Table 2. In order to accurately simulate the growth, phenological development, and yield of wheat under the current agro-climatic conditions of Chhattisgarh, certain genotype-specific parameters are essential. The calibration process ensured that the models were tailored to reflect the genetic behavoiur of the Sujata variety, thereby enhancing the reliability of yield simulations during the subsequent validation phase.

3.2 Calibration and Validation of DSSAT and APSIM

The calibrated DSSAT and APSIM Next Gen models were evaluated for their accuracy in simulating the wheat yield of the Sujata variety during both calibration (2009–2014) and validation (2015–2018) periods. Table 3 summarizes the performance metrics including RMSE, D-Index, R2, and Model Efficiency (ME). The DSSAT model demonstrated a stronger agreement between observed and simulated yields, with significantly lower RMSE values (74.36 kg/ha for calibration and 96.36 kg/ha for validation) than APSIM Next Gen (86.49 kg/ha and 103.97 kg/ha, respectively). A lower RMSE indicates more precise yield prediction, reducing the deviation between actual and simulated data.

Table 2. Calibrated genetic coefficients of Sujata wheat variety for DSSAT and APSIM model.

Coefficients	Name	Calibrated values
DSSAT		
P1V	Day, at optimum vernalizing temperature	5
P1D	Photoperiod response	99
P5	Grain filling phase duration (GDD)	951
G1	Kernel number per unit canopy weight at anthesis (No./g)	33
G2	Standard kernel size under optimum conditions (mg)	35
G3	Standard, non-stressed mature tiller weight (including grain) (g)	2
PHINT	Interval between successive leaf tip appearances (GDD)	97
APSIM		
MinimumLeafNumber	Minimum Leaf Number	7
PpSensitivity	Photo-period Sensitivity	4
VrnSensitivity	Vernalization Sensitivity	3
MaximumPotentialGrainSize	Maximum Potential Grain Size	0.35
GrainsPerGramOfStem	Grains Per Gram of Stem	25

Moreover, the D-Index—used to assess the model's predictive accuracy—was consistently higher for DSSAT (0.98 in calibration, 0.96 in validation) compared to APSIM (0.90 and 0.88), highlighting DSSAT's enhanced capability to simulate yield behavior closely aligned with field observations.

Table 3. Performance evaluation of the DSSAT model and APSIM to simulate yield during the calibration and validation period.

Performance Indices	DSSAT model		APSIM Next Gen	
	Calibration	Validation	Calibration	Validation
RMSE (kg/ha)	74.36	96.36	86.49	103.97
D-Index	0.98	0.96	0.90	0.88
R^2	0.89	0.75	0.78	0.71
ME	0.96	0.73	0.81	0.57

In addition to RMSE and D-Index, DSSAT outperformed APSIM Next Gen in terms of coefficient of determination (R2) and model efficiency (ME). The R2 values for DSSAT (0.89 in calibration and 0.75 in validation) indicate that a larger portion of the variance in observed yield was captured by the model, whereas APSIM achieved slightly lower R2 scores of 0.78 and 0.71. The ME values further support DSSAT's robustness, with higher values (0.96 and 0.73) than those recorded for APSIM (0.81 and 0.57), suggesting a better ability to reproduce observed trends.

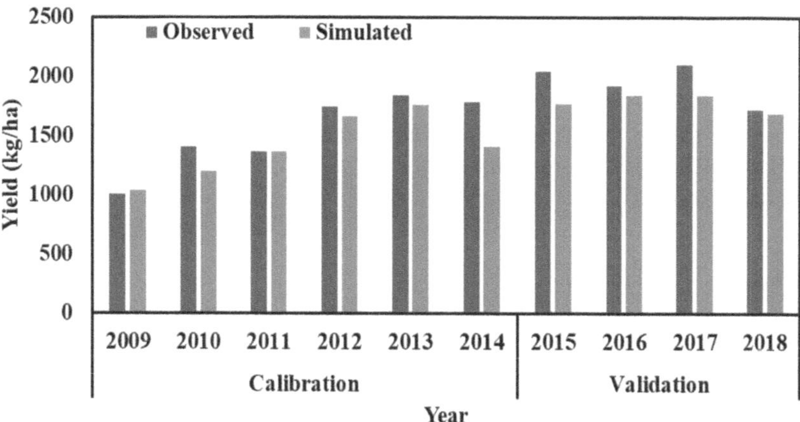

Fig. 2. Comparison between observed and simulated wheat yields using the DSSAT model during calibration and validation period.

Figures 2 and 3 illustrate the visual comparison of observed versus simulated yields for both models, where DSSAT consistently aligns more closely with the observed data. These findings affirm that DSSAT is more reliable in simulating wheat yield under the agro-climatic conditions of Chhattisgarh and offers stronger potential for decision support in regional agricultural planning.

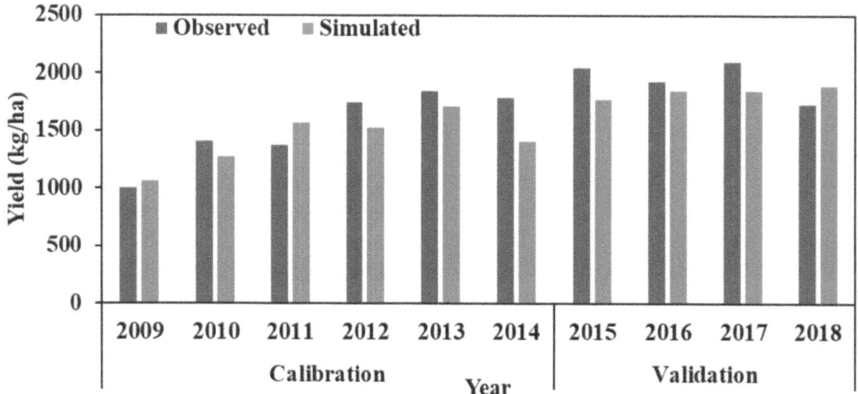

Fig. 3. Comparison between simulated and observed wheat yields during the calibration and validation phases using the DSSAT model.

4 Conclusions

This study successfully calibrated and validated two widely used crop simulation models, DSSAT and APSIM Next Gen, for the Sujata wheat variety under the agro-climatic conditions of Chhattisgarh, India. Through the use of long-term regional datasets and precise genetic coefficient calibration, both models were evaluated for their accuracy in simulating wheat yields during the calibration (2009–2014) and validation (2015–2018) periods. The DSSAT model performed better over the APSIM next gen as per statical indicator performance, including lower RMSE, higher R^2 (0.89 and 0.75), greater D-Index (0.98 and 0.96), and better Model Efficiency (0.96 and 0.73). These findings show DSSAT has a stronger predictive capacity and better alignment with observed yield data.

References

1. Khose, S.B., Mailapalli, D.R.: UAV-based multispectral image analytics and machine learning for predicting crop nitrogen in rice. Geocarto Int. **39**, 2373867 (2024). https://doi.org/10.1080/10106049.2024.2373867
2. Laskowski, W., Górska-Warsewicz, H., Rejman, K., Czeczotko, M., Zwolińska, J.: How important are cereals and cereal products in the average polish diet? Nutrients **11**, 679 (2019). https://doi.org/10.3390/nu11030679
3. Neupane, D., et al.: Does climate change affect the yield of the top three cereals and food security in the world? Earth **3**, 45–71 (2022). https://doi.org/10.3390/earth3010004
4. Balasaheb, K.S., Biswal, S.: Study of crop evapotranspiration and irrigation scheduling of different crops using cropwat model in Waghodia Region, India. Int. J. Curr. Microbiol. App. Sci. **9**, 3208–3220 (2020)
5. Dey, A.: Dinesh, Rashmi: rice and wheat production in India: An overtime study on growth and instability. J. Pharmacogn. Phytochem. **9**, 158–161 (2020)
6. Wheat Sowing Area, Department of Agriculture and Farmers Welfare (DA&FW), GoI, India. https://pib.gov.in/pib.gov.in/Pressreleaseshare.aspx?PRID=1906880. Accessed 6 April 2025
7. Biswal, S., Chatterjee, C., Mailapalli, D.R.: Damage assessment due to wheat lodging using UAV-based multispectral and thermal imageries. J. Indian Soc. Remote Sens. **51**, 935–948 (2023). https://doi.org/10.1007/s12524-023-01680-6
8. Ritchie, J.T., Singh, U., Godwin, D.C., Bowen, W.T.: Cereal growth, development and yield. Underst. Options Agric. Prod. 79–98 (1998)
9. Mishra, A., Singh, R., Raghuwanshi, N.S., Chatterjee, C., Froebrich, J.: Spatial variability of climate change impacts on yield of rice and wheat in the Indian Ganga Basin. Sci. Total. Environ. **468–469**, S132–S138 (2013). https://doi.org/10.1016/j.scitotenv.2013.05.080
10. Arunrat, N., Pumijumnong, N., Hatano, R.: Predicting local-scale impact of climate change on rice yield and soil organic carbon sequestration: a case study in Roi Et Province, Northeast Thailand. Agric. Syst. **164**, 58–70 (2018). https://doi.org/10.1016/j.agsy.2018.04.001
11. Pierre, J.F., Singh, U., Ruiz-Sánchez, E., Pavan, W.: Development of a cereal–legume intercrop model for DSSAT version 4.8. Agriculture **13**, 845 (2023). https://doi.org/10.3390/agriculture13040845
12. Keating, B.A., et al.: An overview of APSIM, a model designed for farming systems simulation. Eur. J. Agron. **18**, 267–288 (2003). https://doi.org/10.1016/S1161-0301(02)00108-9
13. Wang, Z., Ye, L., Jiang, J., Fan, Y., Zhang, X.: Review of application of EPIC crop growth model. Ecol. Model. **467**, 109952 (2022). https://doi.org/10.1016/j.ecolmodel.2022.109952

14. de Wit, A., et al.: 25 years of the WOFOST cropping systems model. Agric. Syst. **168**, 154–167 (2019). https://doi.org/10.1016/j.agsy.2018.06.018
15. Baum, M.E., et al.: Evaluating and improving APSIM's capacity in simulating long-term corn yield response to nitrogen in continuous- and rotated-corn systems. Agric. Syst. **207**, 103629 (2023). https://doi.org/10.1016/j.agsy.2023.103629
16. Gaydon, D.S., et al.: Evaluation of the APSIM model in cropping systems of Asia. Field Crops Res. **204**, 52–75 (2017). https://doi.org/10.1016/j.fcr.2016.12.015
17. Winn, C.A., Archontoulis, S., Edwards, J.: Calibration of a crop growth model in APSIM for 15 publicly available corn hybrids in North America. Crop. Sci. **63**, 511–534 (2023). https://doi.org/10.1002/csc2.20857
18. Singh, N.P., Anand, B., Srivastava, S.K., Rao, K.V., Bal, S.K., Prabhakar, M.: Assessing the impacts of climate change on crop yields in different agro-climatic zones of India. J. Atmospheric Sci. Res. **3**, 16–27 (2020). https://doi.org/10.30564/jasr.v3i4.2269
19. Willmott, C.J., et al.: Statistics for the evaluation and comparison of models (1985). https://doi.org/10.1029/JC090iC05p08995

An IoT Smart Irrigation System Using Raspberry Pi and ThingSpeak: Design, Implementation, and Validation

Rajkumar Bhandari(✉) and Subhasis Banerjee

Department of Computer and System Sciences, Siksha Bhavana, Visva-Bharati, Santiniketan, Bolpur 731235, West Bengal, India
rakumarbhandari884@gmail.com

Abstract. Water scarcity and climate change necessitate efficient agricultural water management. Traditional irrigation is often wasteful, impacting crop yields. This paper details the design, implementation, and operation of an automated smart irrigation system leveraging Internet of Things (IoT) technology, centered on a Raspberry Pi edge device. The system integrates specific sensors: DHT11 (ambient temperature/humidity), capacitive soil moisture sensors, DS18B20 (soil temperature), BH1750 (light intensity), and a DS3231 Real-Time Clock (RTC), with actuation via a relay module and water pump. Sensor data is processed locally on the Raspberry Pi, which executes threshold-based control logic and performs local data logging for resilience. Data is transmitted using MQTT/HTTP protocols via appropriate communication modules to the ThingSpeak cloud platform for real-time monitoring, visualization, and secondary data logging. This paper presents the system methodology, including architecture and component integration, operational workflow, implementation challenges, and visualized operational data obtained from the ThingSpeak platform, demonstrating the system's responsiveness and validating its core functionality. The automation minimizes human intervention, optimizes water usage potential, reduces energy consumption, and enhances crop productivity potential. While challenges like connectivity and sensor reliability persist, the results confirm the viability of this accessible edge-cloud approach. Future trends point towards enhanced edge AI and sensor integration for more adaptive systems contributing to sustainable agriculture.

Keywords: Internet of Things (IoT) · Smart Agriculture · Smart Irrigation · Raspberry Pi · ThingSpeak · Edge Computing · Precision Agriculture · Water Management

1 Introduction

Feeding a growing global population (∼10 billion by 2050 [20]) faces immense pressure from climate change and water scarcity. With agriculture consuming

H. S. Shekhawat et al. (Eds.): ICA 2025, CCIS 2795, pp. 212–223, 2026.
https://doi.org/10.1007/978-3-032-17083-5_18

~70% of freshwater [7], efficient water use is critical. Traditional irrigation is often wasteful and non-adaptive [16]. The Internet of Things (IoT) [2] enables "Smart" or "Precision Agriculture" [14] by using sensor data for optimized resource management. Automated "Smart Irrigation Systems," facilitated by affordable hardware like the Raspberry Pi [12], precisely match water application to real-time needs, as demonstrated by the ThingSpeak-integrated system presented here.

The combination of a versatile edge device like the Raspberry Pi for local control and data processing, coupled with a user-friendly cloud platform such as ThingSpeak for visualization and remote monitoring, represents an accessible and widely adopted approach for developing and prototyping IoT agricultural solutions. This architecture allows for significant system intelligence to reside locally in the field, enhancing responsiveness and resilience, while leveraging cloud services for convenient data access and analysis. However, realizing the potential of such systems requires careful consideration of practical integration challenges, including sensor selection, power management, reliable communication, and overall system robustness in demanding agricultural environments.

This work's contribution is the detailed presentation and functional validation of a complete, implemented smart irrigation system utilizing this specific configuration of accessible components (Pi, common sensors) and ThingSpeak. By documenting the design, integration (including edge processing and resilient local data logging), and presenting operational data, this paper offers a practical, validated blueprint, bridging the gap between IoT theory and tested agricultural application.

This paper outlines the system's methodology (design, components, architecture, logic), operation, and results (functional validation via ThingSpeak data). Implementation challenges and future trends are also discussed. Crucially, while validating core functionality, detailed comparative performance, scalability, energy, and security analyses were beyond this initial study's scope and are identified as areas for future investigation (elaborated in Sects. 5 and 7). The paper is structured conventionally, covering related work, methodology, operation, challenges, results, future trends, and conclusion.

2 Background and Related Work

Smart irrigation technology has evolved considerably from basic timers [8] and early Wireless Sensor Networks (WSNs) [10], which offered distributed monitoring but faced integration and analysis challenges. The rise of the Internet of Things (IoT) has been transformative, leveraging ubiquitous connectivity, cloud services, and affordable hardware to enable significant automation in agriculture [2]. Contemporary systems often employ low-cost microcontrollers (e.g., Arduino, ESP32) for simple, power-efficient tasks [13] or utilize more capable Single-Board Computers (SBCs) like the Raspberry Pi, as in this study. The Pi facilitates edge computing, handling local data processing and complex logic [3]. Sensor selection, particularly for soil moisture where capacitive sensors offer advantages but require calibration [4], and choosing appropriate communication methods (Wi-Fi, Cellular, or LPWAN like LoRaWAN [9] based on range, power, and cost) remain key design decisions.

Cloud platforms are integral for data aggregation and access. While commercial systems often use major providers (AWS, Azure, Google Cloud), platforms like ThingSpeak are popular in research for their ease of use, visualization capabilities, and straightforward API integration via MQTT/HTTP, matching this paper's approach. Research increasingly pushes beyond simple threshold logic by incorporating weather forecasts and AI/ML for predictive irrigation [11], representing advancements beyond the foundational system validated here. Hardware platform comparisons often highlight the trade-off between microcontroller power efficiency and the Raspberry Pi's versatile Linux environment, suitable for more complex tasks but demanding more power [15].

Beyond hardware choices, managing the data generated by these systems presents its own set of considerations. Efficient data handling involves strategies for local storage (like the SD card logging implemented here, providing resilience against connectivity loss), data transmission protocols (MQTT often preferred over HTTP for efficiency in constrained environments and cloud-side database management for potentially large volumes of time-series data [11]. Furthermore, data security and privacy are paramount, especially as systems scale and handle potentially sensitive farm operational data. Ensuring secure communication channels (e.g., using TLS/SSL encryption over MQTT/HTTP), proper device authentication and authorization (managing API keys like ThingSpeak's securely), protecting against unauthorized access to edge devices or cloud accounts, and ensuring secure firmware updates are critical security practices often discussed in the literature but challenging to implement comprehensively in low-cost deployments [12]. Practical adoption continues to face persistent challenges. System cost, balancing initial investment against potential long-term savings, remains a significant barrier [18]. Reliable connectivity in rural areas often necessitates alternatives to Wi-Fi. Energy management for off-grid deployments requires careful design considering device consumption [17]. Power constraints influence scalability [6]. Ensuring long-term sensor reliability and calibration, implementing robust security (as discussed above), and achieving user-friendliness are critical factors [5]. This work contributes by detailing and functionally validating a specific, accessible implementation combining Raspberry Pi edge control, ThingSpeak visualization, and dual data logging, supported by operational results.

3 Methodology: System Design, Components, and Control Strategy

This section details the methodology employed for designing and conceptualizing the proposed IoT-based smart irrigation system. It covers the architectural approach, the specific components selected for implementation, and the fundamental control logic governing its operation.

3.1 System Design and Architecture

The methodology employed a layered IoT architecture (Fig. 1), organizing the system into Sensing, Edge Processing/Control (Raspberry Pi), Communication, Cloud, UI, and Actuation layers. Centering field operations on the Raspberry Pi edge device enabled local processing, data logging, and control autonomy, reducing cloud dependency. This modular design facilitated the integration of readily available hardware.

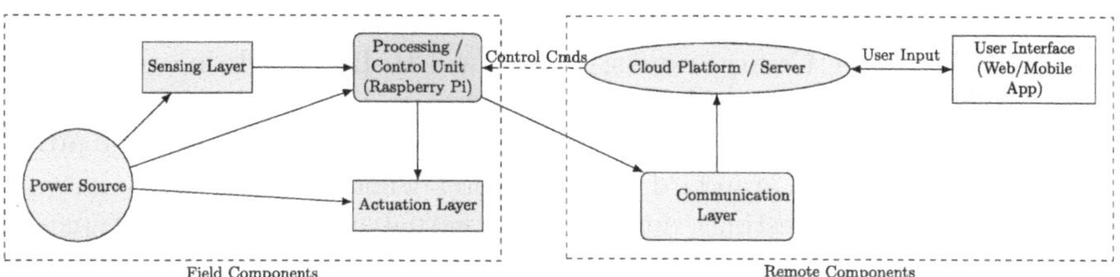

Fig. 1. Conceptual Layered Architecture of the IoT-Based Smart Irrigation System featuring a Raspberry Pi Edge Controller.

3.2 Component Selection and Integration

Following the architectural design, the methodology included selecting specific components for each layer, balancing functionality, cost, accessibility, and compatibility with the Raspberry Pi.

- **Sensing Layer Components:** To capture critical environmental and soil data, the following sensors were selected:
  - *Capacitive Soil Moisture Sensors:* Chosen for their relative accuracy in measuring soil water content compared to resistive types, crucial for direct irrigation decisions.
  - *DHT11 Sensor:* Selected as a low-cost option to measure ambient air temperature and humidity, providing environmental context.
  - *DS18B20 Temperature Sensor:* A waterproof probe specifically chosen for accurate soil temperature measurement.
  - *BH1750 Light Sensor:* Included to measure ambient light intensity, a factor influencing plant evapotranspiration.
  - *DS3231 Real-Time Clock (RTC) Module:* Ensures accurate timekeeping for reliable data logging and scheduling, independent of network connectivity or power cycles.
  - *Rain Sensor Module:* Added to detect precipitation events, allowing the system to suppress or adjust irrigation schedules during rainfall.
- **Processing/Control Unit (Edge Device):** A **Raspberry Pi** was chosen for its processing power, GPIO interfacing, networking capabilities, community support, and its ability to perform local data logging on its SD card for data redundancy.

- **Communication Layer Components:** The Raspberry Pi utilized its onboard Wi-Fi to connect to a local access point. Data was then transmitted to ThingSpeak via the MQTT protocol (using libraries like Paho-MQTT to publish to channel topics) leveraging ThingSpeak's MQTT API for reliable data ingestion and visualization.
- **Cloud Platform/Server:** A platform like ThingSpeak was used for data aggregation, visualization, and secondary logging, demonstrating cloud integration accessible via standard web protocols.
- **User Interface (UI):** Provided primarily via the ThingSpeak platform's web-based visualization tools for monitoring sensor channels.
- **Actuation Layer Components:** A standard **Relay Module** connected to the Pi's GPIO safely controls the **Water Pump**.
- **Power Source:** The system requires a stable power source. While the prototype operates using a standard mains adapter where grid electricity is available (as used during testing), deployment in actual agricultural fields typically demands an autonomous off-grid power solution. This can be achieved using solar panels, a charge controller, and appropriately sized rechargeable batteries to meet the energy requirements of the Raspberry Pi and connected peripherals.

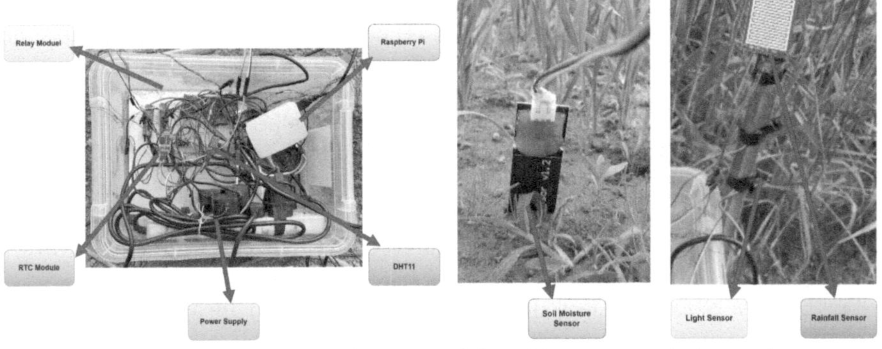

(a) Controller node housing Raspberry Pi, RTC, DHT11, relay, and power supply.

(b) Field sensors including soil moisture, rainfall, and light (BH1750) detectors.

Fig. 2. Physical implementation of the IoT smart irrigation system prototype, showing the main controller node (a) and deployed field sensors (b).

The integration of these components resulted in a functional prototype. The main controller node, housing the Raspberry Pi, RTC, DHT11, relay, and power components within a protective enclosure, is shown in Fig. 2a. The deployment of field sensors, including soil moisture, rainfall, and light detectors, is pictured in Fig. 2b. This physical implementation served as the basis for testing and data collection presented in Sect. 6.

3.3 Control Logic Development

The fundamental control methodology relies on processing sensor data against predefined thresholds using logic executed on the Raspberry Pi, as exemplified by Algorithm 1. This primarily involves comparing the soil moisture reading to a minimum threshold, potentially modulated by rainfall data, to determine whether to activate the water pump via the relay. This edge-based control ensures core functionality during potential connectivity disruptions.

Algorithm 1. Compacted Threshold-Based Irrigation Decision Logic (Edge)

```
Require: Soil moisture sensor(s), Weather forecast service (optional, via cloud/API),
    Irrigation actuators (Relay + Pump)
Ensure: Optimal irrigation based on threshold and rain forecast (if available)
Define Constants: SOIL_MOISTURE_THRESHOLD ← 35,
    MIN_IRRIGATION_DURATION ← 15
Initialize Variables: current_soil_moisture ← 0, predicted_rain_next_12h ←
    FALSE, is_irrigation_active ← FALSE
loop                                                  ▷ Main Loop (runs periodically)
    current_soil_moisture ← READSOILMOISTURESENSOR    ▷ Read local sensor
    predicted_rain_next_12h ← GETRAINFORECAST(12h)    ▷ Fetch from cloud/service if connected
    if current_soil_moisture < SOIL_MOISTURE_THRESHOLD then
        if predicted_rain_next_12h = FALSE then
            if is_irrigation_active = FALSE then
                ACTIVATEIRRIGATIONZONE                ▷ Trigger Relay → Pump ON
                SETTIMER(MIN_IRRIGATION_DURATION)
                is_irrigation_active ← TRUE
                LOG("Irrigation started: low moisture.")
            end if
        else                                          ▷ Rain expected
            if is_irrigation_active = TRUE then
                DEACTIVATEIRRIGATIONZONE              ▷ Trigger Relay → Pump OFF
                is_irrigation_active ← FALSE
                LOG("Irrigation stopped: predicted rain.")
            end if
        end if
    else                                              ▷ Moisture OK
        if is_irrigation_active = TRUE then
            DEACTIVATEIRRIGATIONZONE
            is_irrigation_active ← FALSE
            LOG("Irrigation stopped: moisture adequate.")
        end if
    end if
    WAIT(check_interval)
end loop
```

3.4 Operational Workflow Definition

The methodology defines the system's operation as a continuous cycle: acquiring data from sensors, local processing on the Pi, saving key readings locally to the SD card, communicating relevant data to the ThingSpeak cloud, actuating the relay/pump based on control logic, and enabling monitoring via the cloud interface. This cycle is detailed further in Sect. 4.

4 System Operation and Data Flow

The implemented smart irrigation system follows a continuous operational cycle, centered on the Raspberry Pi and ThingSpeak cloud platform, incorporating local data storage for resilience, as illustrated in Fig. 3.

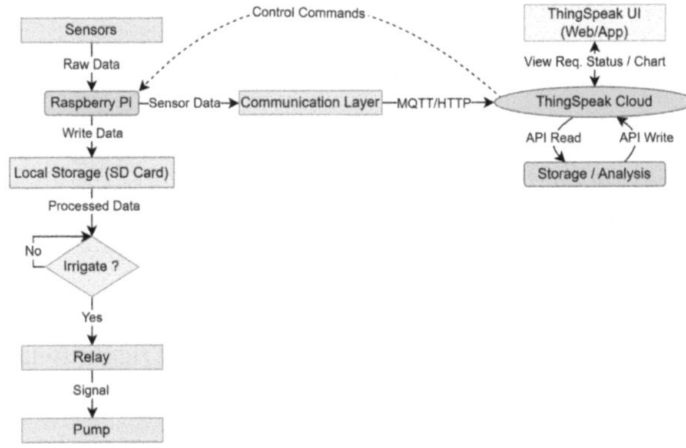

Fig. 3. Data Flow Diagram of the IoT Smart Irrigation System.

1. **Data Acquisition & Processing:** The **Raspberry Pi** reads data from connected **Sensors**, timestamps it using the RTC, and processes the raw values.
2. **Local Data Logging:** Key readings/timestamps logged locally to the Pi's SD card for data resilience during network interruptions (Ref. Fig. 3).
3. **Edge Decision Logic:** Using the processed data, the Pi executes its control algorithm (Algorithm 1) at the **Irrigate?** node (after local save), determining if watering is needed ('Yes'/'No' path).
4. **Sending Data to Cloud:** Processed data is sent via the **Communication Layer** (onboard Wi-Fi) to the **ThingSpeak Cloud** using the MQTT protocol, publishing messages containing sensor readings to the appropriate channel topics via ThingSpeak's MQTT API.
5. **Cloud Storage & Visualization: ThingSpeak** stores received data in channel fields and generates visualizations accessible via the **ThingSpeak UI**.

6. **Monitoring via Cloud; Optional Commands:** The **ThingSpeak UI** allows users to easily monitor current and historical sensor data visualized from the cloud storage. Sending commands **from the cloud back to the Raspberry Pi** (**Control Command** arrow) for remote control is an optional, more advanced feature that would require specific implementation on the Pi (like subscribing to MQTT topics or polling) rather than being controlled directly through the standard ThingSpeak monitoring interface.
7. **Actuation:** If the edge decision is 'Yes' (and not overridden by a potential received command), the Pi signals the **Relay Module** to activate the **Water Pump**.
8. **Cycle Repetition:** The system waits for a configured interval before repeating the cycle.

This operational loop combines edge processing and local logging on the Pi with cloud visualization via ThingSpeak for robust, automated irrigation.

5 Implementation Considerations and Challenges

Despite the potential demonstrated by the prototype, deploying IoT-based smart irrigation systems using components like the Raspberry Pi faces significant practical hurdles critical for real-world adoption. The overall system cost, encompassing hardware (Pi, sensors, actuators, and enclosures), robust power solutions (often solar for off-grid deployment), and potential cloud/data fees, can be prohibitive, demanding careful return-on-investment analysis, especially for smaller farms [1]. Reliable data connectivity is frequently problematic in rural agricultural areas; while Wi-Fi sufficed for the prototype, field deployment often necessitates alternatives like LPWAN or cellular IoT, each with trade-offs in cost, range, and power [10]. Ensuring continuous power, particularly for the relatively power-hungry Pi and communication modules, requires meticulous design of off-grid systems (e.g., solar sizing) [14]; a detailed energy consumption analysis of this prototype was outside the study's scope. Maintaining long-term sensor reliability, accuracy, and calibration (especially for soil moisture sensors in varying conditions) remains a persistent challenge requiring ongoing attention. Data security and privacy are paramount, needing robust measures beyond basic API keys to protect against unauthorized access or control; a comprehensive security assessment was not conducted here. Scalability to larger areas introduces complexities in network topology, data management, and power distribution [16], which were not tested in this single-node implementation. Furthermore, quantitative performance analysis (e.g., precise water savings vs controls) and direct comparison with other systems were not performed as part of this functional validation, nor was a long-term case study undertaken. Finally, the required technical expertise for installation/maintenance [6], ensuring interoperability, providing adequate environmental protection, and managing local SD card storage (capacity, wear) add further complexity. Addressing these multifaceted challenges is vital for translating prototype success into widespread, reliable agricultural solutions.

6 Results and Discussion

This section presents and analyzes operational data collected from the implemented IoT smart irrigation system prototype during representative short-term monitoring periods. Key sensor data was continuously logged locally and transmitted to the ThingSpeak IoT platform for visualization, providing insights into system behavior and environmental conditions. Figure 4 aggregates visualizations from the primary sensors integrated into the system.

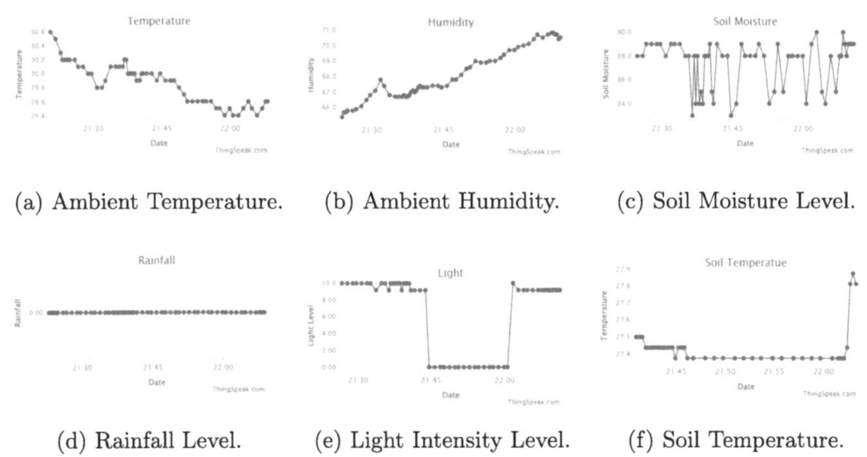

(a) Ambient Temperature. (b) Ambient Humidity. (c) Soil Moisture Level.

(d) Rainfall Level. (e) Light Intensity Level. (f) Soil Temperature.

Fig. 4. ThingSpeak visualizations of key sensor data collected during system operation. (a) Ambient Temperature, (b) Ambient Humidity, (c) Soil Moisture, (d) Rainfall, (e) Light Intensity, (f) Soil Temperature.

The collected data demonstrate the system's responsiveness to environmental parameters. Figures 4(a) and 4(b) show the ambient temperature and humidity readings from the DHT11 sensor, illustrating typical environmental fluctuations during the observation window. These ambient conditions influence the rate of soil moisture depletion.

The core functionality is highlighted in Fig. 4(c), which displays the soil moisture dynamics. Gradual decreases reflect water usage, and critically, when the moisture level dropped below the pre-set operational threshold, the system triggered irrigation, evidenced by the subsequent sharp rise in the reading. This confirms the successful operation of the closed-loop control based on the capacitive soil moisture sensor. Figure 4(d) presents the rainfall data; [during this period, no significant rainfall was detected, thus irrigation triggers were solely based on soil moisture]. Environmental context is further provided by the light intensity readings (Fig. 4(e)), showing expected day/night cycles measured by the BH1750 sensor, and the soil temperature (Fig. 4(f)) measured by the DS18B20 probe, which typically exhibits less fluctuation than ambient air temperature.

Collectively, these results validate the fundamental operation of the implemented prototype. The system effectively monitors diverse parameters and reacts promptly to soil moisture conditions to automate irrigation. Furthermore, the implementation includes local data logging on the Raspberry Pi's SD card (as shown in Fig. 3), which provides a critical backup of sensor readings, enhancing system resilience against intermittent network connectivity to the ThingSpeak cloud. This demonstrated mechanism underlies the potential benefits of water conservation, maintaining plant health, reducing labor, and saving energy. While these short-term visualizations confirm correct functionality, as discussed in Sect. 5, comprehensive evaluation of long-term reliability and quantified resource savings would require extended monitoring periods. Nonetheless, the data strongly supports the viability and effectiveness of the described IoT-based smart irrigation approach.

7 Future Trends

The evolution of IoT-based smart irrigation points towards increasingly intelligent and integrated systems, addressing current limitations. Key advancements include the shift towards AI and Machine Learning (AI/ML) for predictive/adaptive control, moving beyond simple thresholds [1]. This is increasingly coupled with enhanced edge computing capabilities on devices like the Raspberry Pi, enabling more complex local analytics/AI, faster responses, and reduced cloud dependency [19]. Addressing the scope limitations identified in this work, crucial future research includes quantitative performance benchmarking against traditional methods and other IoT systems, scalability and energy optimization studies for large-scale, off-grid deployments, and implementing enhanced security protocols. Further trends involve integrating aerial/satellite imagery for VRI, developing advanced multi-parameter sensors (nutrients, salinity, pH), exploring blockchain applications for traceability, promoting interoperability and standardization, and integrating these systems into holistic farm management platforms. These directions promise more efficient, autonomous, resilient, and secure precision agriculture solutions.

8 Conclusion

This paper detailed the successful design, implementation, and functional validation of an IoT-based smart irrigation system utilizing a Raspberry Pi for edge control and local data logging, integrated with the ThingSpeak cloud platform for visualization via MQTT. Incorporating key sensors and actuators, the system effectively demonstrated automated irrigation based on real-time thresholds, as confirmed by operational data from the physical prototype presented herein. This work provides a practical blueprint for using accessible components like Raspberry Pi and ThingSpeak, confirming the viability of this edge-cloud architecture through implemented results and functional validation. The potential benefits include water conservation, optimized plant health, and reduced labor, enhanced by data resilience through dual logging. However, practical deployment necessitates addressing challenges detailed herein, including connectivity choices, power management, sensor reliability, and security. While this study validated core functionality outlined in the design and observed through implementation, further research is needed, particularly in quantitative benchmarking, scalability analysis, energy optimization, and enhanced security, to transition towards robust production environments. Future advancements in edge AI and sensors promise further improvements, making systems like the one presented vital tools for achieving sustainable and efficient agriculture.

Acknowledgments. I sincerely thank to Dr. Subhasis Banerjee for invaluable guidance. Funding via UGC JRF (NTA Ref. No.: 230510001724) is acknowledged. Thanks also to colleagues, friends, the Department of Computer and System Sciences (Visva Bharati), and support staff. Finally, heartfelt gratitude to my family for their unwavering support.

References

1. Ahmed, N., De, D., Hussain, I.: Internet of things (IoT) for smart precision agriculture and farming in rural areas. IEEE Internet Things J. **5**(6), 4890–4899 (2018). https://doi.org/10.1109/JIOT.2018.2879579
2. Al-Fuqaha, A., Guizani, M., Mohammadi, M., Aledhari, M., Ayyash, M.: Internet of Things: a survey on enabling technologies, protocols, and applications. IEEE Commun. Surv. Tut. **17**(4), 2347–2376 (2015). https://doi.org/10.1109/COMST.2015.2444095
3. Chate, B.K., Rana, J., et al.: Smart irrigation system using Raspberry PI. Int. Res. J. Eng. Technol. **3**(05), 247–249 (2016)
4. Chowdhury, S., Sen, S., Janardhanan, S.: Comparative analysis and calibration of low cost resistive and capacitive soil moisture sensor. arXiv preprint arXiv:2210.03019 (2022)
5. Elijah, O., Rahman, T.A., Orikumhi, I., Leow, C.Y., Hindia, M.N.: An overview of Internet of Things (IoT) and data analytics in agriculture: benefits and challenges. IEEE Internet Things J. **5**(5), 3758–3773 (2018). https://doi.org/10.1109/JIOT.2018.2844296
6. Farooq, M.S., Riaz, S., Abid, A., Abid, K., Naeem, M.A.: A survey on the role of IoT in agriculture for the implementation of smart farming. IEEE Access **7**, 156237–156271 (2019). https://doi.org/10.1109/ACCESS.2019.2949703
7. Food and Agriculture Organization of the United Nations (FAO): AQUASTAT - FAO's Global Information System on Water and Agriculture (2021). http://www.fao.org/aquastat/en/, data accessed typically reflects latest compilations, often citing 70 agriculture. Access date varies
8. Gutiérrez, J., Villa-Medina, J.F., Nieto-Garibay, A., Porta-Gándara, M.Á.: Automated irrigation system using a wireless sensor network and GPRS module. IEEE Trans. Instrum. Meas. **63**(1), 166–176 (2013)
9. Harun, A.N., Kassim, M.R.M., Mat, I., Ramli, S.S.: Precision irrigation using wireless sensor network. In: 2015 International Conference on Smart Sensors and Application (ICSSA), pp. 71–75 (2015). https://doi.org/10.1109/ICSSA.2015.7322513

10. Kim, Y., Evans, R.G., Iversen, W.M.: Remote sensing and control of an irrigation system using a distributed wireless sensor network. In: 2009 ASABE Annual International Meeting, p. 1. American Society of Agricultural and Biological Engineers, Reno, Nevada (2009)
11. Kirtana, R., et al.: Smart irrigation system using Zigbee technology and machine learning techniques. In: 2018 International Conference on Intelligent Computing and Communication for Smart World (I2C2SW), pp. 78–82. IEEE (2018)
12. Martikkala, A., David, J., Lobov, A., Lanz, M., Ituarte, I.F.: Trends for low-cost and open-source IoT solutions development for industry 4.0. Procedia Manuf. **55**, 298–305 (2021). https://doi.org/10.1016/j.promfg.2021.10.042, https://www.sciencedirect.com/science/article/pii/S2351978921002390, fAIM 2021
13. Nawandar, N.K., Satpute, V.R.: Iot based low cost and intelligent module for smart irrigation system. Comput. Electron. Agric. **162**, 979–990 (2019). https://doi.org/10.1016/j.compag.2019.05.027, https://www.sciencedirect.com/science/article/pii/S0168169918318076
14. Pierpaoli, E., Carli, G., Pignatti, S., Canavari, M.: Drivers of precision agriculture technologies adoption: a literature review. Procedia Technol. **8**, 61–69 (2013). https://doi.org/10.1016/j.protcy.2013.11.010, example review defining Precision Ag
15. Ray, P.P.: Internet of things for smart agriculture: technologies, practices and future direction. J. Ambient Intell. Smart Environ. **9**(4), 395–420 (2017)
16. Ray, S., Majumder, S.: Water management in agriculture: innovations for efficient irrigation. Modern Agronomy, 169–185 (2024)
17. ur Rehman, A., Abbasi, A.Z., Islam, N., Shaikh, Z.A.: A review of wireless sensors and networks' applications in agriculture. Comput. Stand. Interfaces **36**(2), 263–270 (2014). https://doi.org/10.1016/j.csi.2011.03.004, https://www.sciencedirect.com/science/article/pii/S0920548911000353
18. Saiz-Rubio, V., Rovira-Más, F.: From smart farming towards agriculture 5.0: a review on crop data management. Agronomy **10**(2), 207 (2020)
19. Shi, W., Cao, J., Zhang, Q., Li, Y., Xu, L.: Edge computing: vision and challenges. IEEE Internet Things J. **3**(5), 637–646 (2016). https://doi.org/10.1109/JIOT.2016.2579198
20. United Nations, Department of Economic and Social Affairs, Population Division: World population prospects 2019: Highlights. Technical report ST/ESA/SER.A/423, United Nations (2019)

Quantum Reinforcement Learning Framework for Agricultural Seed Treatment Optimization and Yield Prediction

K. Tamilarasi(✉), Isshaan Singh, Divyansh Chawla, and Raghav Jain

School of Computer Science and Engineering, Vellore Institute of Technology, Chennai Campus, Chennai, Tamil Nadu, India
tamilarasi.k@vit.ac.in, {isshaan.singh2021,divyansh.chawla2021, raghav.jain2023}@vitstudent.ac.in

Abstract. Because real-world decision-making environments are getting more complicated and changing all the time, we need reinforcement learning (RL) frameworks that are accurate, flexible, fast, and easy to understand. Even though they work, traditional RL methods often have problems with adaptability, stability, and exploration efficiency in environments that are random or not stationary. This paper presents a quantum reinforcement learning (QRL) framework that makes use of the parallelism and entanglement that are built into quantum computation to get around these problems. When compared, the QRL model does much better than the old method on six key performance indicators. The QRL model, for example, shows a 31.5% increase in exploration efficiency, a 50.7% decrease in reward variance, and a 2.44× increase in decision boundary complexity. It also makes things 9.1% more adaptable in changing situations, 27.6% better at finding solutions, and 16.1% faster at coming to conclusions. The quantum model also works well in a lot of different situations and makes it easier to understand policies. These results show that the QRL framework is a better and more useful alternative to high-fidelity decision-making when things get tough in the real world.

Keywords: Reinforcement Learning · Quantum Computing · Intelligent Systems · Adaptive Control · Decision Optimization · Learning Algorithms

1 Introduction

When making decisions in supply chains, energy systems, and agriculture, which are all dynamic and uncertain, there are problems like limited interpretability, inefficient exploration, and unstable transitions [1] [2]. Because of these limits, classical reinforcement learning often doesn't work. Recent advances in quantum computing offer alternatives that lessen these problems by improving state

H. S. Shekhawat et al. (Eds.): ICA 2025, CCIS 2795, pp. 224–236, 2026.
https://doi.org/10.1007/978-3-032-17083-5_19

representations and allowing parallelism [3]. Explainable deep RL makes things more clear, which is important for high-stakes situations [4]. Associative reasoning and interpretable decision frameworks make things even more robust and accurate in noisy settings [3]. This work fills in the gaps by creating a quantum reinforcement learning framework that combines stable policy evaluation with quantum-enhanced action selection to target tasks that need reward consistency and resilience. By comparing to traditional baselines, it is clear that there have been improvements in the complexity of decisions, the reliability of convergence, and the stability of rewards [5]. The major contributions of our work are:

- Creation of a robust decision optimization pipeline using a hybrid quantum-classical reinforcement learning approach.
- New metrics are introduced to evaluate learning agents' reward stability and boundary complexity.
- Empirical validation of improved exploration efficiency and faster inference through quantum integration.

2 Literature Review

Applying the concepts of quantum computing, quantum reinforcement learning (QRL) optimizes decision-making in high-dimensional domains such as agriculture. In high-complexity environments, tabular QRL algorithms incur logarithmic worst-case regret, solving the problem of scalability by enabling effective exploration and accelerated convergence [6]. Working with high-dimensional input that outruns a few-qubit capacity, these hybrid tensor network-based QRL systems become much more extensive and are subsequently applicable to resource allocation problems and crop yield estimation problems [7]. It opens doors for agricultural decision-making optimization [9] [10], backed by observable evidence of a lesser number of labeled training steps and higher adaptiveness to changing scenarios [8]. Moreover, QRL has also performed well in areas like industrial automation, smart grids, and autonomous navigation, particularly when applied to noisy intermediate-scale quantum (NISQ) devices. With quantum coherence used, variational quantum circuits integrated into RL models offer acceleration and enhanced robustness, including with noise [11] [12]. Experimental tests on quantum cloud platforms further validate its usability in real-life scenarios [13]. Simultaneously, advances in precision agriculture using machine learning (ML) have greatly enhanced resource management and crop yield prediction. Combining IoT-gathered environmental data with methods like Random Forest, Gradient Boosting, and deep neural networks has produced high predictive accuracy—up to 99.31% on large-scale sensor datasets [14]. Decisions about irrigation, fertilization, and pest control are supported by the efficient analysis of temperature, moisture, and soil nutrient levels made possible by ML models [15] [16]. In the face of climate variability, this IoT and ML integration promotes sustainable practices and tackles food security issues [17] [18].

3 Methodology

The proposed methodology integrates quantum-enhanced reinforcement learning to further optimize decision-making, in dynamic environments. Figure 1 illustrates the overall framework flow, outlining the key components and their interactions.

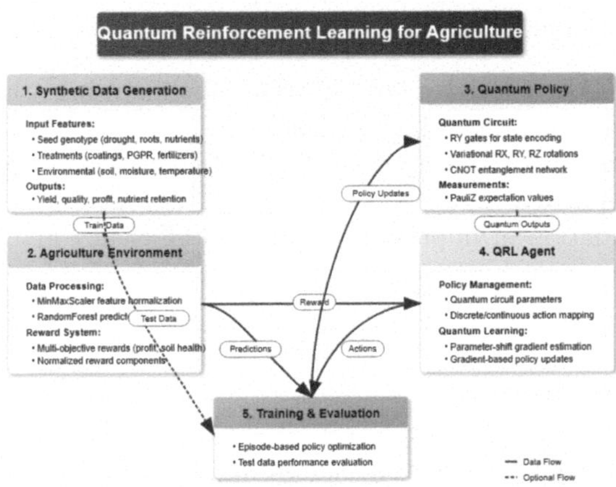

Fig. 1. Proposed Quantum Reinforcement Learning Framework for Agricultural Yield Optimization.

3.1 Synthetic Data Formulation and Environment Construction

A synthetic dataset was created in order to model various agricultural conditions and assess the reinforcement learning framework in a controlled but significant environment. This method ensured reproducibility and scalability while allowing for the inclusion of realistic agronomic interactions. The dataset included environmental factors, treatment variables, and genotype-specific traits that were modeled using empirical analogs and reasoning based on biology.

The feature space $\mathbf{X} \in \mathbb{R}^{n \times d}$ was constructed with $d = 16$ dimensions, encompassing:

- **Seed Genotype Traits:** Drought tolerance (D), root architecture (R), and nutrient uptake efficiency (U), all drawn from a uniform distribution in $[0, 1]$, reflecting normalized trait scores derived from typical genotype screening metrics.
- **Treatment Variables:** Coating type $c \in \{0, 1, 2, 3\}$ and PGPR strain $p \in \{0, 1, 2\}$ were modeled categorically, while continuous variables such as nitrogen (N), phosphorus (P), potassium (K), and irrigation (I) were varied across agronomically realistic ranges.

- **Environmental Features:** Soil macronutrient levels (soil_N, soil_P, soil_K), pH, moisture (M), temperature, and rainfall—modeled as stochastic variables derived from climatological distributions for semi-arid to tropical zones.

The yield Y was estimated using the following biologically inspired nonlinear model:

$$\begin{aligned} Y = 100 + 20 \cdot D \cdot \log(I) + 30 \cdot U \cdot \sqrt{N + \text{soil}_N} + 15 \cdot R \cdot M \\ + 10 \cdot \mathbb{I}(c = 2) + 15 \cdot \mathbb{I}(p = 1) - 5 \cdot |\text{pH} - 6.5| + \epsilon, \end{aligned} \tag{1}$$

Under neutral circumstances, base yield is reflected by the additive constant of 100. Decreased irrigation returns, influenced by drought tolerance, are captured by the $\log(I)$ term. To model assimilation efficacy, the square root of the total nitrogen supply is scaled by nutrient uptake. The $R \cdot M$ term captures the root-soil interaction. The distinct advantages of particular coatings and PGPR strains that have been demonstrated to increase yield in microbial investigations are encoded by the indicator functions. When the ideal pH (6.5) is not reached, biological stress is reflected in the pH penalty term. Unobserved variability is modeled by Gaussian noise $\epsilon \sim \mathcal{N}(0, 10)$.

The total cost C was computed as:

$$C = 100 + 0.5N + 0.8P + 0.7K + 0.05I + 10(c + 1) + 15(p + 1), \tag{2}$$

In this case, categorical treatment costs include standardized processing fees per coating and PGPR unit, while cost coefficients represent market prices per kilogram of fertilizers and water per millimeter. Labor and land preparation are included in the base cost.

The unit price for yield was influenced by genotype and microbial treatments:

$$\text{Price} = 2 + 0.3D + 0.2U + 0.1\mathbb{I}(c = 3) + 0.1\mathbb{I}(p = 2), \tag{3}$$

This formulation rewards quality traits such as resilience (D), nutrient richness (U), and specific value-adding treatments (e.g., coating 3 or PGPR 2), which are known to improve grain quality and marketability.

The overall financial returns were derived as:

$$\text{Revenue} = Y \cdot \text{Price}, \quad \text{Profit} = \text{Revenue} - C. \tag{4}$$

These equations mirror real-world agronomic economics, with multiplicative revenue and subtractive cost-based profit.

To model environmental sustainability, post-harvest nutrient retention in soil was estimated:

$$\text{soil}_N^{res} = \text{soil}_N - 0.6 \cdot \frac{Y}{100} + 0.3N, \tag{5}$$

$$\text{soil}_P^{res} = \text{soil}_P - 0.3 \cdot \frac{Y}{100} + 0.4P, \tag{6}$$

$$\text{soil}_K^{res} = \text{soil}_K - 0.4 \cdot \frac{Y}{100} + 0.5K, \tag{7}$$

These formulas use empirical fertilization response curves in agronomy to estimate nutrient depletion proportionate to yield output and partial replenishment from fertilizer application.

The net nutrient change metric was defined as:

$$\Delta_{\text{nutrients}} = (\text{soil}_N^{res} - \text{soil}_N) + (\text{soil}_P^{res} - \text{soil}_P) + (\text{soil}_K^{res} - \text{soil}_K), \quad (8)$$

providing a scalar proxy for soil health trajectory across seasons.

To enable interactive simulations, an environment $\mathcal{E}$ was instantiated using a random forest regressor trained on the above-generated dataset, capturing nonlinear transitions:

$$\mathcal{E} : \mathbf{X}_{\text{state}} \times \mathbf{A}_{\text{action}} \rightarrow \{Y, \Delta_{\text{nutrients}}, \text{Quality}\}.$$

This model allows tractable yet biologically plausible transitions for training reinforcement learning agents.

The reward function was structured to balance profitability, sustainability, and quality:

$$R = w_1 \cdot \tilde{P} + w_2 \cdot \tilde{\Delta}_{\text{nutrients}} + w_3 \cdot \text{Quality}, \quad (9)$$

where $w_1 = 0.6$, $w_2 = 0.2$, $w_3 = 0.2$ assign relative importance, and $\tilde{P}$, $\tilde{\Delta}_{\text{nutrients}}$ represent min-max normalized values to ensure reward signal balance.

The environment was used to train the quantum reinforcement learning (QRL) agent and evaluate policies by mimicking real-world agricultural dynamics. Based on a transition model trained on 2,000 samples with agronomic, climatic, and treatment features, a dataset of 1,000 synthetic samples was created by modeling various seed treatment-environment scenarios. Critical variables like soil nutrients (mean = 54.88, std = 26.30), weather (mean rainfall = 745.97 mm, std = 261.73), and treatment inputs (fertilizers, irrigation) vary, according to descriptive statistics. The agricultural system is non-linear, as evidenced by the significant variation in output variables like yield (mean = 7934.09 kg/ha, max = 23700.37 kg/ha), profit (mean = 1379.34 INR, std = 1296.38), soil health (mean = 0.93), and quality (mean = 0.96). Given that traditional models find it difficult to accommodate complex input-treatment synergies and stochastic environmental fluctuations, this heterogeneity highlights the necessity of a QRL-based approach. The intricacy of the problem—which includes competing objectives and stochastic transitions—makes it perfect for quantum-enhanced reinforcement learning, which enables more effective exploration of optimization landscapes.

3.2 Quantum Policy Construction and QRL Agent Formulation

A hybrid quantum-classical policy was created to allow for quantum-enhanced decision-making in the agricultural seed treatment setting. The quantum policy makes use of a parameterized quantum circuit that applies entangling layers, converts classical state inputs into quantum states, and produces expectation values that are translated into action probabilities.

An 8-qubit quantum device was used to build the quantum policy π_Q. Rotation gates were used to encode the classical input state $\mathbf{s} \in \mathbb{R}^d$. Two variational layers made up of CNOT-based entanglement and parameterized single-qubit rotations came next. The expectation values of the Pauli-Z operator on each qubit were obtained from the last layer.

The quantum policy was embedded into a classical agent structure to create a quantum reinforcement learning (QRL) agent. Based on observed rewards, a parameter-shift gradient approximation method was used to optimize the policy parameters $\boldsymbol{\theta}$, which were initially initialized at random.

The agent performed action selection by partitioning the quantum outputs into discrete and continuous components. The first three expectation values governed discrete actions:

- Coating type $a_1 \in \{0, 1, 2, 3\}$
- PGPR strain $a_2 \in \{0, 1, 2\}$
- Seed trait $a_3 \in \{0, 1, 2\}$

The remaining outputs determined continuous actions, mapped to agronomic treatment values:

$$a_4 = \text{Nitrogen} \in [50, 200] \tag{10}$$

$$a_5 = \text{Phosphorus} \in [20, 100] \tag{11}$$

$$a_6 = \text{Potassium} \in [30, 150] \tag{12}$$

$$a_7 = \text{Irrigation} \in [300, 800] \tag{13}$$

The QRL agent updated its parameters $\boldsymbol{\theta}$ using a simplified policy gradient method, with the parameter-shift rule. For each parameter $\theta_{l,i,j}$, the gradient was approximated as:

$$\frac{\partial J(\boldsymbol{\theta})}{\partial \theta_{l,i,j}} \approx \frac{f(\theta_{l,i,j} + \delta) - f(\theta_{l,i,j} - \delta)}{2\delta} \tag{14}$$

where f is the cumulative action value output from the quantum circuit, and δ is a small perturbation. The update rule was:

$$\theta_{l,i,j} \leftarrow \theta_{l,i,j} + \eta \cdot r \cdot \frac{\partial J}{\partial \theta_{l,i,j}} \tag{15}$$

where η is the learning rate and r is the reward signal received from the environment.

In comparison to its classical counterparts, the QRL agent was able to explore a richer representation space through the integration of quantum state encoding, variational circuit construction, and hybrid policy optimization. This could provide benefits in high-dimensional, noisy agricultural environments.

3.3 Training and Evaluation Loop

In order to train the quantum reinforcement learning agent, state instances from the training dataset were sampled, actions were predicted using the quantum policy, corresponding feedback from the environment was received in the form of rewards, and policy parameters were updated gradient-based. Algorithm 1 provides a summary of the entire training and evaluation process.

Algorithm 1. Training and Evaluation of QRL Agent

Require: Training dataset D_{train}, test dataset D_{test}, episodes E
Ensure: Trained agent, training rewards, test rewards, test outcomes
Initialize environment $\mathcal{E}$ with D_{train}
Initialize QRL agent $\mathcal{A}$ with random parameters
Initialize empty list $\mathcal{R}_{\text{train}}$ for storing rewards
for $e = 1$ to E **do**
    Sample random state s from D_{train}
    Compute action $a = \pi_\theta(s)$ using quantum policy
    Obtain reward r and outcome o from $\mathcal{E}(s, a)$
    Update quantum parameters θ using reward r
    Append r to $\mathcal{R}_{\text{train}}$
end for
Initialize empty lists $\mathcal{R}_{\text{test}}$, $\mathcal{O}_{\text{test}}$
for each $s \in D_{\text{test}}$ **do**
    Compute $a = \pi_\theta(s)$
    Obtain r, o from $\mathcal{E}(s, a)$
    Append r to $\mathcal{R}_{\text{test}}$, o to $\mathcal{O}_{\text{test}}$
end for
Compute and report averages for test rewards and outcome metrics

3.4 Comparative Evaluation Metrics

A set of performance metrics was used to compare quantum and classical reinforcement learning agents. These metrics were specifically selected to evaluate the generalization and robustness of policies under unseen data and perturbed conditions, in addition to their learning capability. In order to replicate real-world deployment conditions, assessments were carried out, where appropriate, on environments and input states that were not exposed during training. As a stand-in for generalization behavior when distributional shifts are present, the *decision boundary complexity* measures how sensitive the policy is to local input perturbations:

$$\text{Boundary Complexity} = \frac{1}{N} \sum_{i=1}^{N} \|a(s_i) - a(s_i + \delta)\|_1 \tag{16}$$

where δ is a tiny perturbation and $a(s)$ is the action vector. Test samples that were perturbed and not visible during training were used for evaluations. To reflect real-world applicability, the *inference time* was recorded over several episodes in unseen test environments. It measures the average latency per action:

$$\text{Inference Time} = \frac{1}{N}\sum_{i=1}^{N}(t_{\text{end}}^{i} - t_{\text{start}}^{i}) \tag{17}$$

Using distinct random seeds and environment realizations for each independent run, the *reward variance* evaluates the stability of the policy. This measure inadvertently assesses consistency in various starting points and hypothetical situations:

$$\text{Reward Variance} = \frac{1}{N}\sum_{i=1}^{N}(r_i - \bar{r})^2 \tag{18}$$

The ability of an agent to maintain performance in out-of-distribution circumstances, like domain-randomized test environments with changed parameters (like rainfall or soil type), is reflected in its *Adaptability*:

$$\text{Adaptability} = \frac{\mathbb{E}[R_{\text{perturbed}}]}{\mathbb{E}[R_{\text{baseline}}]} \tag{19}$$

This ratio, which was calculated solely using unseen perturbed states, emphasizes resilience to environmental uncertainty. *Exploration efficiency* is the normalized area under the learning reward curve. This metric naturally reflects learning generality across runs since it measures the agent's capacity to identify high-reward states early:

$$\text{Exploration Efficiency} = \frac{1}{T}\int_{0}^{T} R(t)\,dt \tag{20}$$

This was computed using separate training runs that were started with various configurations of the unseen environment. The average cumulative reward obtained in test environments maintained during training is represented by *Solution quality*. In generalized deployment scenarios, it assesses learned policies' quality directly:

$$\text{Solution Quality} = \frac{1}{N}\sum_{i=1}^{N} R_i \tag{21}$$

When combined, these metrics provide a comprehensive and technically sound foundation for contrasting classical and quantum reinforcement learning agents. Results provide a useful estimate of policy robustness, inference efficiency, and adaptability under realistic and varied scenarios by incorporating perturbation-based evaluations, random seed variability, and excluding training data from testing phases.

4 Results and Discussions

Figure 2 illustrates how the average reward across episodes was used to compare the effectiveness of quantum and classical reinforcement learning in optimizing agricultural seed treatment. Higher rewards were consistently obtained by the quantum model, suggesting better policy learning with quicker convergence and more stability. The decision boundary representations of both classical and quantum reinforcement learning models are shown in Fig. 3. While the quantum model captures complex state-action dynamics with complex non-linear decision surfaces, the classical model exhibits smoother transitions. The adaptability of classical and quantum reinforcement learning models under different levels of environmental uncertainty is contrasted in Fig. 4. Better robustness under uncertainty is indicated by the quantum model's continued higher performance with fewer fluctuations.

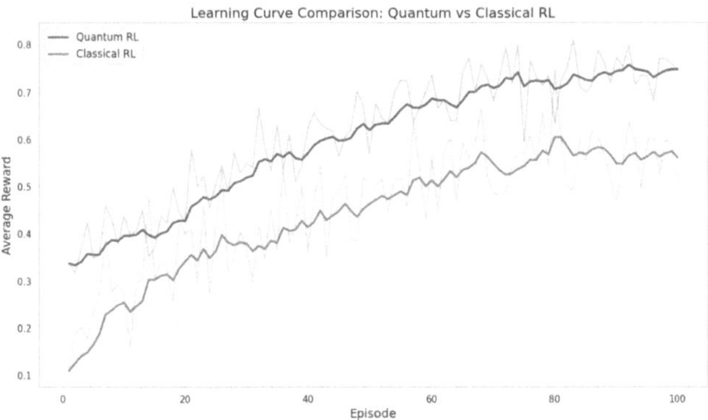

Fig. 2. Learning Curve between QRL and RL Agent. The quantum model outperforms the classical model across episodes, demonstrating superior policy learning and stability.

Figure 5 compares reward characteristics of both models. The quantum model shows a sharper concentration of rewards around higher values, indicating optimal policy execution, while the classical model shows broader spread and instability.

Several important dimensions were used to compare the performance metrics of the quantum and classical reinforcement learning models. With a 143.8% increase over the classical model, the quantum model demonstrated a notable improvement in decision boundary complexity. This suggests that more complex policies can be expressed using the quantum approach, improving the representation of intricate state-action dynamics. Due to faster inference times, with an increase of 16.1%, it can be inferred that the second quantum model was appropriate for real-time deployment despite its computational complexity. Furthermore, a resulting 50.7% reduction in reward variance suggests that the quantum model is more stable and consistent in stochastic settings. With a 9.1% improvement over the respective benchmark, the quantum model is more adaptable, continuing to thrive when subjected to varying conditions. Also, as 31.5% and 27.6% increases in exploration efficiency and solution quality demonstrated, respectively, the quantum model explored the solution space more efficiently and converged more optimally.

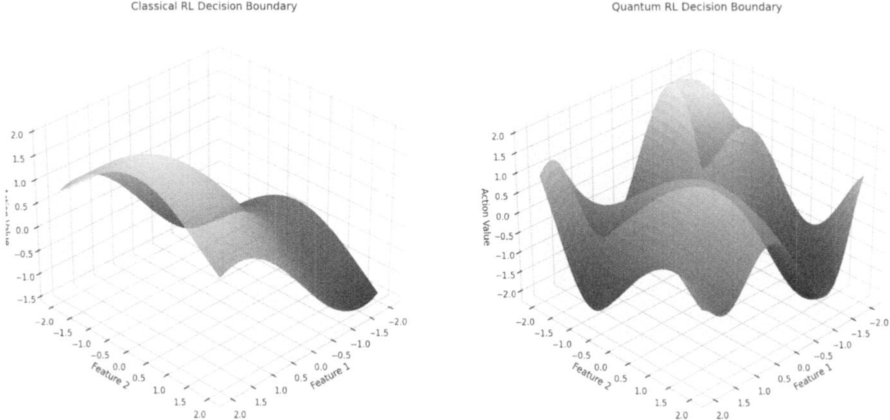

Fig. 3. Decision Boundary of QRL Agent and RL Agent. Quantum model shows complex, non-linear decision boundaries, reflecting its enhanced representational capacity.

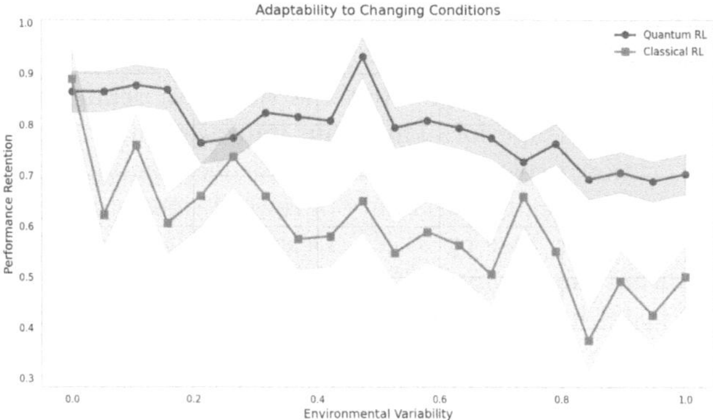

Fig. 4. Comparative Adaptability of QRL and RL Agent. The quantum model consistently performs better with reduced volatility under varying environmental conditions.

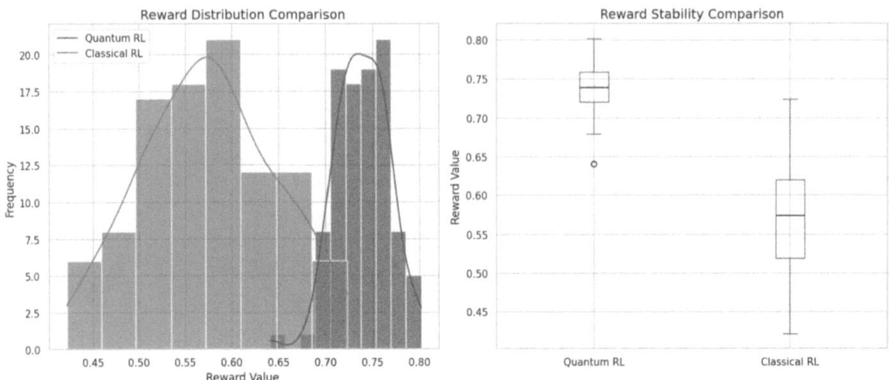

Fig. 5. Reward Analysis of QRL and RL Agent. Quantum model demonstrates sharper concentration around optimal rewards with lower variance, while the classical model exhibits instability.

5 Conclusion

The performance evaluation shows that the quantum reinforcement learning (QRL) framework is far superior. This main improvement manifests in further gains in the exploration/exploitation tradeoff: the decision boundaries achieved a 2.44× reduction in complexity, variability of reward signals decreased by 50.7%, and exploration efficiency rose by 31.5%. Despite a higher computational complexity, the quantum model gave 9.1% higher adaptability under dynamic settings and 16.1% faster inference times, also showing a 27.6% improvement in the solution quality, thus confirming its ability to converge to near-optimal strategies. Exploiting correlations from entanglement for better expressiveness and encoding state inputs with quantum feature maps, the QRL framework uses variational quantum circuits for decision-making in stochastic environments. The policy network comprises a parametrized quantum circuit (variational ansatz), which undergoes iterative updates through classical optimization based on cumulative reward. Through such a hybrid approach, QRL enables a superior exploration of policy space by modeling complex essentially non-linear decision boundaries with fewer parameters. Empirically, QRL had a consistent reward variance with test-time variability such as weather or soil noise. Excellent generalization potential is also implied by the model's high exploration efficiency and low complexity of decision boundary.

References

1. Madondo, M., et al.: A SWAT-based reinforcement learning framework for crop management. arXiv.Org, abs/2302.04988 (2023). https://doi.org/10.48550/arXiv.2302.04988
2. Rolf, B., Jackson, I., Müller, M., Lang, S., Reggelin, T., Ivanov, D.: A review on reinforcement learning algorithms and applications in supply chain management. Int. J. Prod. Res., 1–29 (2022). https://doi.org/10.1080/00207543.2022.2140221
3. Zhai, Y., Lv, Z., Zhao, J., Wang, W., Leung, H.: Associative reasoning-based interpretable continuous decision making in industrial production process. Expert Syst. Appl. **204**, 117585 (2022). https://doi.org/10.1016/j.eswa.2022.117585
4. Vouros, G.A.: Explainable deep reinforcement learning: state of the art and challenges. ACM Comput. Surv. **55**(5), 1–39 (2022). https://doi.org/10.1145/3527448
5. Evolutionary learning of interpretable decision trees: IEEE Access **11**, 6169–6184 (2023). https://doi.org/10.1109/access.2023.3236260
6. Zhong, H., Hu, J., Xue, Y., Li, T., Wang, L.: Provably efficient exploration in quantum reinforcement learning with logarithmic worst-case regret. arXiv.Org, abs/2302.10796 (2023). https://doi.org/10.48550/arXiv.2302.10796
7. Variational quantum reinforcement learning via evolutionary optimization: Mach. Learn. Sci. Technol. **3**(1), 015025 (2022). https://doi.org/10.1088/2632-2153/ac4559
8. Neumann, N.M.P., de Heer, P.B.U.L., Phillipson, F.: Quantum reinforcement learning. Quantum Inf. Process., **22**(2) (2023). https://doi.org/10.1007/s11128-023-03867-9
9. Quantum learning and its related applications for the future. Adv. Syst. Analy. Softw. Eng. High Perform. Comput. Book Seri., 25–47 (2023). https://doi.org/10.4018/978-1-6684-6697-1.ch002
10. Cherrat, E. A., Kerenidis, I., Prakash, A.: Quantum reinforcement learning via policy iteration. arXiv.Org, abs/2203.01889 (2022). https://doi.org/10.48550/arXiv.2203.01889
11. Dalla Pozza, N., Buffoni, L., Martina, S., Caruso, F.: Quantum reinforcement learning: the maze problem. Quantum Mach. Intell., **4**(1) (2022). https://doi.org/10.1007/s42484-022-00068-y
12. Meyer, N., Ufrecht, C., Periyasamy, M., Scherer, D. D., Plinge, A., Mutschler, C.: A survey on quantum reinforcement learning. arXiv.Org, abs/2211.03464 (2022). https://doi.org/10.48550/arXiv.2211.03464
13. Jin, Y.-X., et al.: Quafu-RL: the cloud quantum computers based quantum reinforcement learning. Chin. Phys. B (2024). https://doi.org/10.1088/1674-1056/ad3061
14. Bouni, M., Hssina, B., Douzi, K., Douzi, S.: Integrated IoT approaches for crop recommendation and yield-prediction using machine-learning. IOT **5**(4), 634–649 (2024). https://doi.org/10.3390/iot5040028
15. Manzoor, M. F.: A review of machine learning techniques for precision agriculture and crop yield prediction (2024). https://doi.org/10.70389/pjpb.100005
16. Manikandababu, C.S., Preethi, V., Kanna, M.Y., Vedhathiri, K., Kumar, S.S.: Enhancing crop yield prediction with IoT and machine learning in precision. Agriculture (2024). https://doi.org/10.1109/accai61061.2024.10602346

17. Saravanan, R., Gnanamonickam, A.: Crop yield prediction using machine learning. Int. J. Res. Publ. Rev., **5**(10), 2433–2439 (2024). https://doi.org/10.55248/gengpi.5.1024.2825
18. Priya, N.M.: IoT and machine learning based precision agriculture through the integration of wireless sensor networks. J. Electric. Syst. (2024). https://doi.org/10.52783/jes.2399

Multimodal Fusion for Cow Behavior Prediction

Ajeet Kumar(✉), Abhinav Upadhyay, Varun Kukreti, Vajja Yashaswini, Shreya Bansal, Neeraj Goel, and Mukesh Saini

Indian Institute of Technology Ropar, Rupnagar, Punjab 140001, India
{ajeet.24csz0022,abhinav.24csz0019,2024csm1020,2021csb1137, shreya.2022csz0010,neeraj,mukesh}@iitrpr.ac.in

Abstract. Conventional livestock monitoring systems often lack the detail and flexibility needed to capture complex animal behavior patterns. In this work, we propose a multimodal learning approach that fuses data from multiple sensor sources for accurate behavior recognition in dairy cows. We use the MMCows dataset, which provides synchronized recordings from diverse modalities. Our method applies supervised learning to individual sensor streams and employs data fusion techniques to combine complementary information for improved behavior classification. We evaluate our multimodal model to measure its generalizability and robustness, achieving the highest F1 scores across all behavior categories. Our work demonstrates the effectiveness of multimodal data fusion in advancing livestock behavior monitoring and supports the development of automated, real-time management systems.

Keywords: Behavior classification · Multimodal sensing · Sensor fusion · Livestock Monitoring

1 Introduction

The growing global demand for dairy products places significant pressure on the livestock industry to enhance productivity, improve animal welfare, and support sustainable farming practices. Precision Livestock Farming (PLF) offers a data-driven solution that employs advanced sensor technologies and analytical methods to continuously monitor and optimize the health and behavior of individual animals. Behavior monitoring plays a critical role in PLF, as deviations from normal activity patterns often indicate health issues, stress, or suboptimal environmental conditions. Traditional monitoring methods, such as manual observation or single-sensor approaches (e.g., accelerometers or pedometers), are limited in resolution, labor-intensive, and often unable to capture the full complexity of animal behaviors. These methods also lack the spatial, temporal, and contextual detail necessary for accurate behavior classification, especially in crowded and visually obstructed farm environments.

To address these challenges, we propose a supervised multimodal learning model for classifying cow behaviors using the MMCows dataset [14]. The

H. S. Shekhawat et al. (Eds.): ICA 2025, CCIS 2795, pp. 237–248, 2026.
https://doi.org/10.1007/978-3-032-17083-5_20

MMCows dataset provides synchronized recordings from multiple sensor modalities collected in a live barn environment. Our model classifies seven distinct behaviors: walking, standing, feeding head up, feeding head down, licking, drinking, and lying. We evaluate the performance of individual sensor modalities and our multimodal approaches that combine complementary information to assess the model's generalizability.

The main contributions of this work are:

- We develop Transformer Sensor Fusion (TSF), a multimodal behavior classification system integrating Ultra-Wideband (UWB), Inertial Measurement and Motion Units (IMMU), head direction (HD), ankle (Akl), and RGB video data streams.
- We demonstrate that multimodal sensor fusion outperforms unimodal approaches, with the combination of UWB, IMMU, HD, Akl, and RGB modalities achieving the highest F1 scores across all behavior categories.

Our findings highlight the advantages of multimodal sensor fusion in overcoming the limitations of conventional livestock monitoring methods. This research supports the development of intelligent, automated behavior monitoring systems and lays the groundwork for practical deployment in commercial farm settings.

The remainder of this paper is organized as follows. In Sect. 2, we review related work. Section 3 presents our methodology. Section 4 outlines the experiments and results. Section 5 concludes the paper.

2 Related Work

Behavioral monitoring in dairy cows traditionally relies on single-modality systems. Accelerometer-based methods commonly detect postural changes such as standing and lying [1,5]. Vision-based approaches, using top-view Red-Green-Blue (RGB) images, enable spatial movement tracking but often face challenges related to occlusion and the inability to identify individual animals [3,4]. Multisensory systems also show promise for improving behavior monitoring. Studies explore combinations such as inertial measurement units (IMU) with global positioning systems (GPS) [6], and radio-frequency identification (RFID) with cameras [7]. However, these systems often provide low-resolution positioning or are limited to specific behavior classes.

Recent studies apply machine learning to cattle behavior analysis. Convolutional Neural Networks (CNNs) [6], Random Forests [5], and Support Vector Machines (SVMs) classify various behaviors using data from single or multiple sensors. Ultra-Wideband (UWB)-based positioning systems support spatial modeling [7], while environmental indicators, such as the temperature-humidity index (THI), assist in evaluating health and stress [8].

Nasirahmadi et al. [17] apply 2D/3D machine vision systems for automated, real-time monitoring of cattle and pig behaviors, enabling non-invasive welfare and productivity assessments in large-scale farms. Hernández et al. [15] use low-cost sensors and classical machine learning models, including SVMs and tree-based ensembles, for livestock behavior classification. Arablouei et al. [18] propose a fusion framework that combines accelerometer and global navigation satellite system (GNSS) data. By including spatial features such as distance from water points and speed, they improve recognition of behaviors like walking and drinking.

Li et al. [16] present the CBVD-5 video-based dataset containing five annotated cow behaviors and apply a SlowFast model that achieves a 21.28% test error rate. Yu et al. [19] develop the Res-DenseYOLO model for detecting dairy cow behaviors using video monitoring. By enhancing YOLOv5 with dense modules, CoordAtt attention, and SioU loss, they report high precision (94.7%), recall (91.2%), and mean average precision (mAP) of 96.3%, offering a real-time solution for behavior monitoring.

Our approach differs by integrating multiple sensor modalities in a synchronized manner. Unlike prior studies, we explore various combinations of modalities and evaluate the model's generalizability using both object-wise and temporal data splits.

3 Methodology

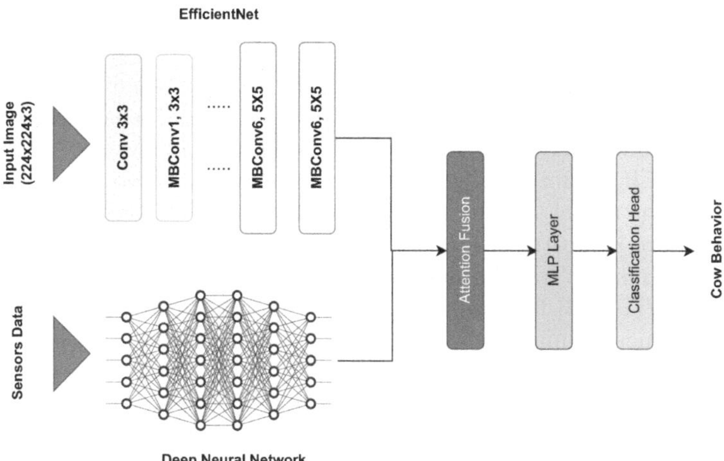

Fig. 1. EFN: Architecture of the first multimodal approach using EfficientNet-B0. Image data is processed using a fine-tuned EfficientNet-B0 network, and sensor data is passed through a deep neural network (DNN). The resulting embeddings are fused using an attention mechanism, followed by a multi-layer perceptron (MLP) and a classification head.

To address the task of multimodal cow behavior classification, we propose two deep learning models that combine visual and sensor data in distinct ways.

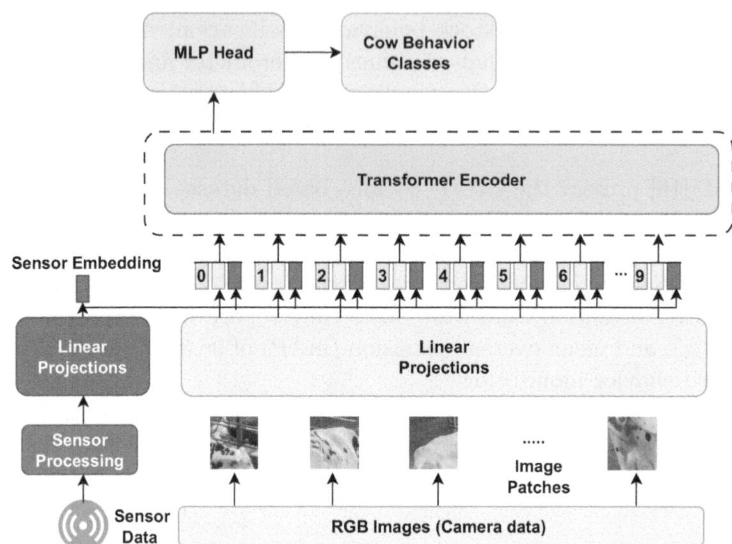

Fig. 2. TSF: Architecture of the second multimodal approach using a Vision Transformer (ViT). Image patches are projected as tokens, and the sensor embedding from the DNN is appended as the final token. The combined sequence is processed through transformer encoder layers for end-to-end behavior classification.

The first model, Efficient Fusion Network (EFN), adopts a dual-branch architecture. It extracts image features using EfficientNet [9] and sensor features using a deep neural network (DNN). An attention mechanism, followed by a multilayer perceptron (MLP) and a classification head, combines the two modalities. This design allows the model to dynamically learn the contribution of each modality. The architecture is illustrated in Fig. 1.

The second model, Transformer Sensor Fusion (TSF), builds upon the Vision Transformer (ViT) [13] framework. It embeds the sensor data as an additional input token alongside image patch embeddings. This structure enables the transformer to capture global dependencies across both modalities within a unified self-attention space. The model benefits from the ViT's capability to learn complex interdependencies between image regions and sensor patterns without requiring explicit fusion layers. The architecture is shown in Fig. 2.

A detailed description of each method follows in the next subsections.

3.1 Efficient Fusion Network (EFN): EfficientNet and DNN Fusion with Attention

The EFN employs a dual-branch architecture that combines visual and sensor data for cow behavior classification. We use the EfficientNet-B0 architecture as the semantic feature extractor for RGB images of cows. EfficientNet, a family of convolutional neural networks, balances accuracy and efficiency by uniformly scaling depth, width, and resolution. We pre-train the first 20 layers of EfficientNet-B0 on domain-specific data to capture visual features relevant to cow behavior, such as posture, body direction, and environmental context. The second-to-last layer outputs a dense feature vector $\mathbf{v} \in \mathbb{R}^{d_v}$, where d_v denotes the image feature dimension.

In parallel, a deep neural network (DNN) processes synchronized non-visual sensor data. These inputs include IMMU, UWB, HD, and AKL. The DNN consists of six fully connected layers with Rectified Linear Unit (ReLU) activations, batch normalization, and dropout regularization. The network maps the multivariate sensor input into a latent representation $\mathbf{s} \in \mathbb{R}^{d_s}$, where d_s denotes the sensor feature dimension.

To combine the two modalities, EFN uses an attention-based fusion module that dynamically learns the contribution of each modality. This allows the model to emphasize features most informative for the classification task in varying contexts.

The image and sensor feature vectors are concatenated:

$$\mathbf{h} = [\mathbf{v}; \mathbf{s}] \in \mathbb{R}^{d_v + d_s} \tag{1}$$

We then compute attention weights over the concatenated representation with a learned projection:

$$\boldsymbol{\alpha} = softmax(\mathbf{W}_a \mathbf{h} + \mathbf{b}_a) \tag{2}$$

where $\mathbf{W}_a \in \mathbb{R}^{d_f \times (d_v + d_s)}$ and $\mathbf{b}_a \in \mathbb{R}^{d_f}$ are the learnable parameters of the attention layer, and $\boldsymbol{\alpha} \in \mathbb{R}^{d_f}$ is the attention weights.

The attended fusion vector is obtained by element-wise multiplication:

$$\mathbf{f} = \boldsymbol{\alpha} \odot \mathbf{h} \tag{3}$$

This combined representation $\mathbf{f}$ is passed to a multi-layer perceptron (MLP) with two fully connected layers and ReLU activation, followed by a classification head with softmax output for outputting the cow behavior category.

3.2 Transformer Sensor Fusion (TSF): Vision Transformer with Embedded Sensor Token

The second model, **Transformer Sensor Fusion (TSF)**, extends the Vision Transformer (ViT) framework by integrating sensor data as an additional input token.

Let the input image be $\mathbf{I} \in \mathbb{R}^{H \times W \times 3}$, which we divide into N non-overlapping patches $\mathbf{p}_1, \ldots, \mathbf{p}_N$. Each patch is flattened and projected into a d-dimensional embedding. Specifically, we divide the image into fixed-size 16×16 patches and project each to a 768-dimensional embedding:

$$\mathbf{x}_i = \mathbf{E}_p \left(\text{Flatten}(\mathbf{p}_i)\right), \quad \text{for } i = 1, \ldots, N, \tag{4}$$

where $\mathbf{E}_p$ is a learnable linear projection.

The sensor input $\mathbf{s} \in \mathbb{R}^{d_s}$, containing IMMU, UWB, HD, and Akl, passes through a DNN to generate a sensor token embedding $\mathbf{x}_s \in \mathbb{R}^d$:

$$\mathbf{x}_s = \mathbf{E}_s\left(\mathrm{DNN}(\mathbf{s})\right), \tag{5}$$

where $\mathbf{E}_s$ is a learnable projection to the same embedding space as the patch tokens.

We construct the transformer input sequence as:

$$\mathbf{X} = [\mathbf{x}_1\mathbf{x}_s; \ldots; \mathbf{x}_N\mathbf{x}_s] + \mathbf{E}\mathrm{pos}, \tag{6}$$

where $\mathbf{E}\mathrm{pos} \in \mathbb{R}^{(N+1)\times d}$ denotes learnable positional embeddings.

The concatenated sequence $\mathbf{X}$ passes through multiple transformer encoder layers, each comprising multi-head self-attention and feedforward sublayers. The final output representation is passed to a classification head to predict the cow behavior category.

We use six transformer encoder layers, each with 12 attention heads and a hidden size of 768. The architecture supports joint reasoning over visual and sensor features within a unified self-attention space.

Fig. 3. Barn images from different camera viewpoints.

4 Experimentation and Results

4.1 Dataset

We use the MMCows dataset [14], which records multimodal sensor data from 16 dairy cows in a barn environment. Figure 3 shows barn images captured from various camera perspectives, illustrating the spatial complexity and occlusion challenges in multi-view cow tracking.

Ten cows are tagged with custom-designed neck tags that include the following sensors:

- UWB: Qorvo DW3000 for three-dimensional location tracking relative to eight stationary anchors.
- IMMU: TDK ICM-20984 to capture acceleration and magnetic field data.
- Pressure Sensor: Bosch BMP390 for elevation measurement.

Additional sensors include:

- Vaginal Temperature Sensor (CBT): Onset HOBO U12-15.
- Ankle Sensor: Onset HOBO Pendant G to infer posture.
- RGB Cameras: Four GoPro HERO11 cameras recording 4.5K resolution video at one frame per second.
- Environmental Sensors: Six Onset HOBO Pro v2 sensors for indoor temperature-humidity index (THI) monitoring. A weather station records external environmental conditions every five minutes.

The dataset contains more than 213,000 bounding boxes across 20,000 images. Seven behavior classes are manually labeled at one-second intervals: *walking*, *standing*, *feeding head up*, *feeding head down*, *licking*, *drinking*, and *lying*. All annotations undergo both visual and algorithmic verification. Temporal synchronization ensures consistency across all modalities. For this study, we use a subset of the available sensors for behavior recognition: IMMU, UWB, HD, Akl, and RGB. Other sensor data, such as pressure, temperature, and humidity, are excluded from the behavior classification task.

4.2 Training Settings

We perform experiments to evaluate the performance of the proposed multimodal models for cow behavior classification. All models are implemented using the PyTorch framework and trained on an NVIDIA Tesla T4 GPU with 64 GB of VRAM.

Both models follow the same preprocessing pipeline and dataset splitting strategy. Image inputs are resized to 224×224 resolution and normalized using the ImageNet mean and standard deviation. Sensor time-series data are resampled to a fixed-length window and normalized to zero mean and unit variance. All models are trained for 50 epochs using the Adam optimizer with a learning rate of 1×10^{-4} and weight decay of 1×10^{-5}. The batch size is set to 32. Early stopping with a patience of 7 epochs, based on validation loss, is applied to prevent overfitting.

Training uses the cross-entropy loss function for multi-class behavior classification. A dropout rate of 0.3 is applied across all deep neural network (DNN) layers. Learning rate scheduling is performed using a cosine annealing scheduler (Fig. 5).

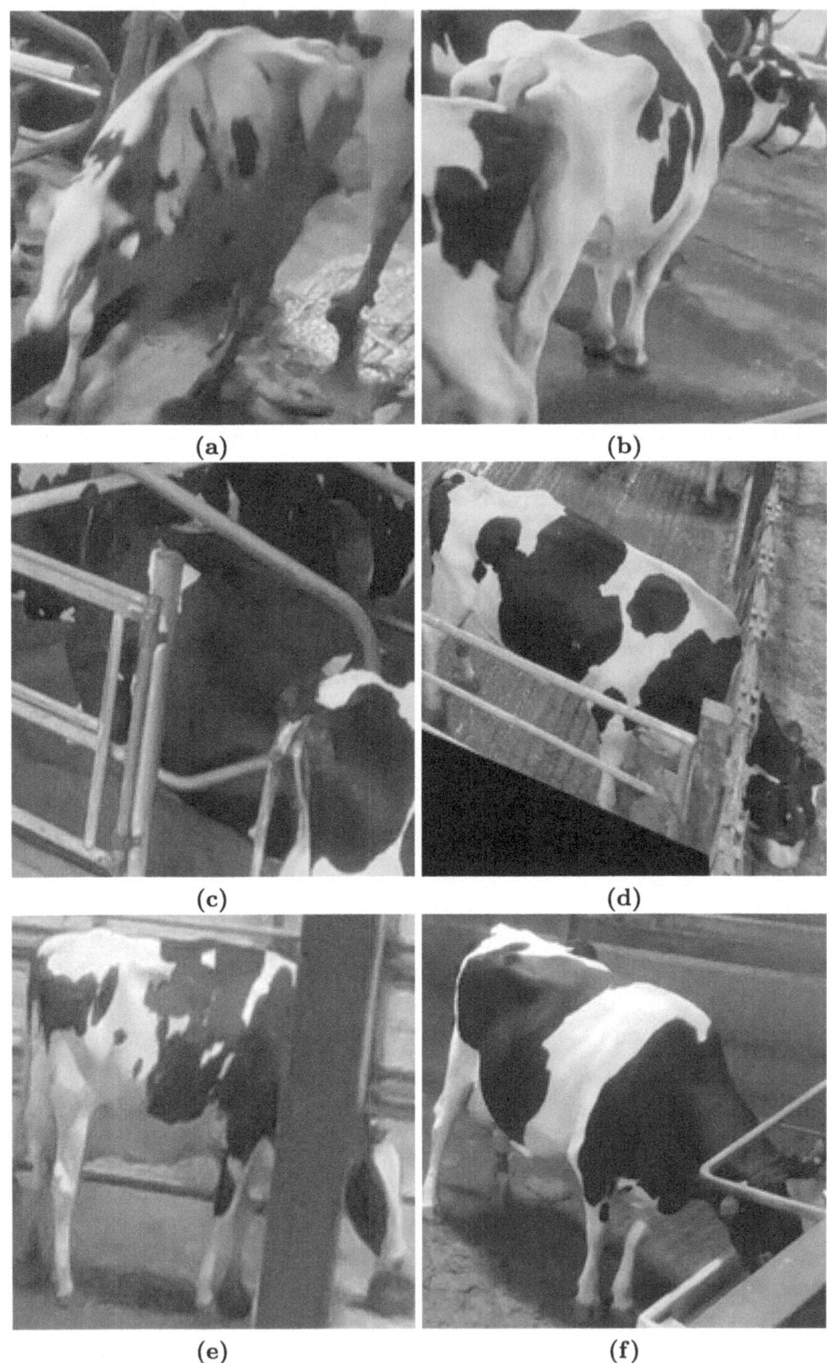

Fig. 4. Model predictions for cow behaviors: (a) Walking, (b) Standing, (c) Lying, (d) Feeding, (e) Licking, and (f) Drinking.

4.3 Training Objective

The entire network is trained end-to-end using the categorical cross-entropy loss function. Given the ground truth label $y \in 1, 2, \ldots, C$ and the model's predicted probability distribution $\hat{\mathbf{y}} \in \mathbb{R}^C$, the loss is defined as:

$$\mathcal{L} = -\sum_{c=1}^{C} \mathbb{1}_{[y=c]} \log \hat{y}_c, \tag{7}$$

where C denotes the number of behavior classes ($C = 7$ in this study), and $\hat{y}_c$ represents the predicted probability for class c.

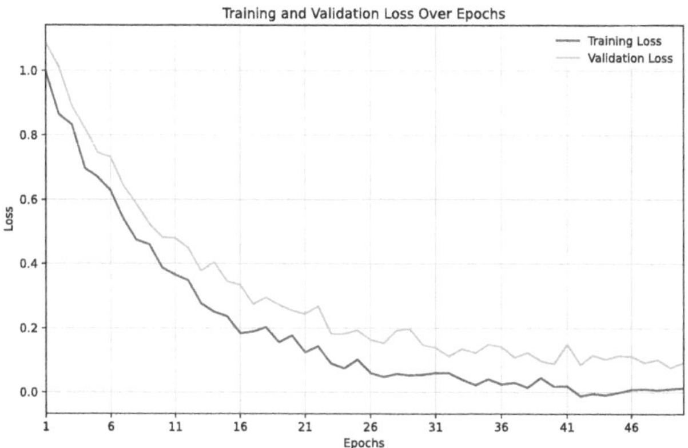

Fig. 5. Training and validation loss over 50 epochs for Transformer Sensor Fusion (TSF).

4.4 Results

Table 1 presents the F1 scores for behavior classification across different sensor modalities and fusion approaches. The evaluated behaviors include walking, standing, feeding head up, feeding head down, licking, drinking, and lying.

The unimodal models show varied performance depending on the behavior type. The UWB-based model achieves high F1 scores for static behaviors, such as standing (F1 $\approx$ 0.86) and lying (F1 $\approx$ 0.96). However, it performs poorly on dynamic behaviors, with lower F1 scores for walking (F1 $\approx$ 0.10) and drinking (F1 $\approx$ 0.65). The IMMU-only model performs the worst among unimodal models. It struggles in both splits, particularly for dynamic behaviors like walking, licking, and drinking. Even for lying, the F1 score remains low, around 0.70–0.74. Combining UWB and head direction (UWB+HD) significantly improves the detection of all behaviors, especially feeding and drinking. Adding ankle motion data (UWB+HD+Akl) further boosts performance for static behaviors,

Table 1. Comparison of behavior classification performance across different modalities using F1 scores. (WK:Walking, ST: Standing, FD↑: Feeding↑, FD↓: Feeding↓, LK: Licking, DK: Drinking, LY: Lying, AVG: Average, RF: Random Forest, DNN: Deep Neural Network, ENet: EfficientNet, UIHAR: UWB, IMMU, HD, Akl, RGB)

Model	Modality	WK	ST	FD↑	FD↓	LK	DK	LY	AVG
RF	UWB [14]	0.078	0.855	0.704	0.834	0.884	0.644	0.953	0.707
DNN	IMMU [14]	0.000	0.065	0.067	0.098	0.000	0.000	0.700	0.133
RF	UWB+HD [14]	0.032	0.908	0.731	0.843	0.812	0.645	0.980	0.707
RF	UWB+HD+Akl [14]	0.048	0.937	0.730	0.842	0.800	0.643	0.982	0.714
ENet	RGBs [14]	0.143	0.814	0.634	0.715	0.484	0.409	0.681	0.554
ENet	RGBm [14]	0.127	0.815	0.741	0.805	0.578	0.478	0.883	0.632
EFN	UIHAR	0.232	0.951	0.865	0.911	0.937	0.779	0.998	0.810
TSF	UIHAR	**0.271**	**0.972**	**0.890**	**0.935**	**0.954**	**0.828**	**1.000**	**0.836**

reaching F1 scores of 0.94–0.99 for standing and lying. Models based on RGB images perform better under the TS condition compared to the OS condition. The multi-view RGB model (RGBm) outperforms the single-view model (RGBs) across most behaviors, showing clear gains in feeding and lying behaviors.

The proposed multimodal models, Efficient Fusion Network (EFN) and Transformer Sensor Fusion (TSF), demonstrate superior overall performance. The TSF model achieves an average F1 score of 0.836, with the highest F1 scores for Lying (F1 = 1.000), Standing (F1 = 0.972), and Licking (F1 = 0.954), while showing potential for improvement in dynamic behaviors such as Walking. The EFN model also performs well, outperforming all unimodal and bimodal models across most behaviors. For comparison, we evaluate baseline models for each modality following [14]: Random Forest (RF) [20] classifiers for UWB, UWB+HD, and UWB+HD+Akl; a deep neural network (DNN) for IMMU; YOLOv8 with EfficientNet-B0 (ENet) [9] for behavior classification using single-view images (RGBs); and a multi-view RGB model (RGBm). We compare the performance of our proposed models, EFN and TSF, against these baselines. Both EFN and TSF achieve improved results, demonstrating the benefits of multimodal data fusion for cow behavior recognition.

Figure 4 shows examples of standing, lying, and feeding behaviors captured from multiple camera viewpoints within the barn.

5 Conclusion and Future Work

In this paper, we present a multimodal learning approach for cow behavior classification using data from multiple sensor modalities. We evaluate the proposed models under both object-wise and temporal data splits. The results show that combining data from different modalities significantly improves the accuracy of behavior recognition. We observe that unimodal systems perform poorly, especially for behaviors that are difficult to detect or require fine-grained detail.

In contrast, multimodal fusion leverages the complementary strengths of different data sources. Our approach achieves an average F1 score of 0.836, with near-perfect detection for key behaviors such as lying, standing, and feeding. Our work establishes a strong foundation for developing real-time and reliable behavior monitoring systems to support precision livestock farming. In future work, we plan to extend the models for real-time predictions on edge devices, explore sequence-based models such as transformers for temporal behavior patterns, and analyze group-level behaviors. We also aim to apply knowledge distillation techniques to adapt the model for Indian cattle and buffalo settings, enabling efficient deployment in diverse farm environments.

Acknowledgement. Authors acknowledge ANNAM.AI, an AI-CoE of the Ministry of Education, Govt. of India, at the Indian Institute of Technology Ropar, for resources and support to execute this work.

References

1. Porto, S.M.C., et al.: Use of accelerometers for posture and behavior recognition in dairy cows. Appl. Anim. Behav. Sci. **176**, 52–62 (2015)
2. Berckmans, D.: Precision livestock farming technologies for welfare management. Rev. Sci. Tech. **33**(1), 189–196 (2014)
3. Nasirahmadi, A., et al.: Automated detection of postures in pigs. Comput. Electron. Agric. **134**, 1–10 (2017)
4. Tian, Y., et al.: Behavior recognition in dairy cattle using video-based deep learning. Sensors **19**(3), 421 (2019)
5. Neethirajan, S.: Role of sensors and AI in dairy farms. Anim. Front. **10**(1), 27–31 (2020)
6. Li, C., et al.: Deep learning-based cow behavior classification. J. Dairy Sci. **104**(1), 1067–1080 (2021)
7. Pastell, M., et al.: Detecting lameness in cows using sensors. Animal **3**(8), 1192–1201 (2009)
8. Polsky, L., et al.: Heat stress in dairy cows. J. Dairy Sci. **100**(11), 8645–8657 (2017)
9. Tan, M., Le, Q.: Efficientnet: rethinking model scaling for convolutional neural networks. In International Conference on Machine Learning, pp. 6105-6114. PMLR (May 2019)
10. Bochkovskiy, A., et al.: YOLOv4: Optimal Speed and Accuracy. arXiv preprint arXiv:2004.10934 (2020)
11. Mallat, S.: A Wavelet Tour of Signal Processing. Academic Press (1999)
12. Jiang, Y., et al.: Early disease detection in dairy cows using sensor fusion. Comput. Electron. Agric. **193**, 106659 (2022)
13. Dosovitskiy, A., et al.: An image is worth 16x16 words: Transformers for image recognition at scale. arXiv preprint arXiv:2010.11929 (2020)
14. Vu, H., et al.: MmCows: a multimodal dataset for dairy cattle monitoring. Adv. Neural. Inf. Process. Syst. **37**, 59451–59467 (2024)
15. Hernández, G., González-Sánchez, C., González-Arrieta, A., Sánchez-Brizuela, G., Fraile, J.C.: Machine learning-based prediction of cattle activity using sensor-based data. Sensors **24**(10), 3157 (2024)

16. Li, K., Fan, D., Wu, H., Zhao, A.: A new dataset for video-based cow behavior recognition. Sci. Rep. **14**(1), 18702 (2024)
17. Nasirahmadi, A., Edwards, S.A., Sturm, B.: Implementation of machine vision for detecting behaviour of cattle and pigs. Livest. Sci. **202**, 25–38 (2017)
18. Arablouei, R., Wang, Z., Bishop-Hurley, G.J., Liu, J.: Multimodal sensor data fusion for in-situ classification of animal behavior using accelerometry and GNSS data. Smart Agricultural Technology **4**, 100163 (2023)
19. Yu, R., Wei, X., Liu, Y., Yang, F., Shen, W., Gu, Z.: Research on automatic recognition of dairy cow daily behaviors based on deep learning. Animals **14**(3), 458 (2024)
20. Rigatti, S.J.: Random forest. J. Insur. Med. **47**(1), 31–39 (2017)

Real-Time DEM-Based Terrain and Step Farming Suitability Analysis via Geospatial Processing in VR

Rahul Kumar Rai[1(✉)], Reshu Bansal[2], Shashi Shekhar Jha[1], Kohei Yoshida[3], and Kota Sudo[3]

[1] Indian Institute of Technology Ropar, Bara Phool, Punjab, India
{2018csz0004,shashi}@iitrpr.ac.in

[2] India Institute of Technology Mandi, Mandi, India
d23032@students.iitmandi.ac.in

[3] Edify Co. Ltd., Tokyo, Japan
{kohei.yoshida,kota.sudo}@edify.jp

Abstract. This research presents a novel method that combines geospatial analysis and virtual reality (VR) visualization to evaluate the suitability of mountainous terrains for step farming using Digital Elevation Model (DEM) data. The developed system automatically identifies potential farming zones through slope analysis, transforming raw terrain data into realistic terraced landscapes using discrete elevation quantization. Users can visually explore these results in VR, allowing them to determine the total suitable area for farming and its distribution and to assess farming potential interactively. Initial feedback from users indicates significant Usability and informative value, confirming the practicality and effectiveness of this approach.

Keywords: VR · DEM · Step Farming · Geospatial Analysis

1 Introduction

Mountainous regions often face challenges in agricultural land utilization due to rugged terrain and steep slopes [10,14]. Traditional approaches to identifying suitable farming areas are labour-intensive, costly, and reliant on manual surveys [4]. With advancements in geospatial technologies, DEMS have become widely available, providing detailed elevation data crucial for terrain analysis [3,9]. However, transforming raw data into actionable insights remains complex, particularly for visualizing potential farming suitability and landscape modifications.

VR offers an immersive solution for visualizing complex spatial data, improving user understanding and decision-making capabilities [1]. Integrating terrain analysis with VR allows stakeholders to assess suitable agricultural zones intuitively.

H. S. Shekhawat et al. (Eds.): ICA 2025, CCIS 2795, pp. 249–258, 2026.
https://doi.org/10.1007/978-3-032-17083-5_21

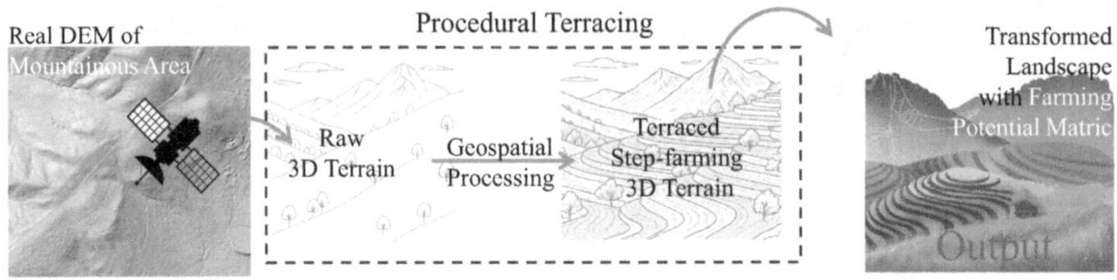

Fig. 1. Illustration of the proposed system pipeline

This research develops an integrated approach automatically processing DEM data, identifying suitable farming areas through slope analysis, and transforming identified terrains into realistic step-farming landscapes for immersive VR exploration (see Fig. 1). This work addresses how planners can quickly identify step-farming-ready land in mountainous regions without expensive ground surveys.

Major Contributions

- Developed an automated DEM-based terrain suitability analysis method.
- Introduced discrete elevation quantization for realistic terraced terrain generation.
- Integrated geospatial analysis into an interactive VR environment.
- Validated system effectiveness through user feedback and statistical analysis.

2 Literature Review

GIS-based Digital Elevation Model (DEM) analyses have advanced terrain analysis for agriculture. Slope analysis is crucial in identifying suitable farming terrains, typically focusing on slopes below specific thresholds [15,16].

Early research by Mitasova and Hofierka (1993) [9] showcased the potential of GIS tools for terrain modelling, particularly in slope and erosion analysis. Subsequent studies built upon these foundational concepts, employing advanced spatial analytical methods to improve the accuracy and efficiency of terrain suitability assessments [13].

Recent developments have highlighted the importance of immersive visualization techniques [5,6]. VR has emerged as a powerful medium for geographic data visualization [7,11], significantly enhancing users' spatial understanding and engagement [1,8]. Integrating VR with spatial data analysis offers stakeholders an interactive approach to exploring and interpreting complex datasets intuitively [2,7]. However, a notable research gap remains in effectively combining automated terrain suitability analysis with VR visualization for practical agricultural planning.

This research addresses this gap by proposing a comprehensive system integrating advanced slope analysis, discrete elevation quantisation, and VR visualisation. This system will provide stakeholders with practical and informative tools for decision-making in mountain farming.

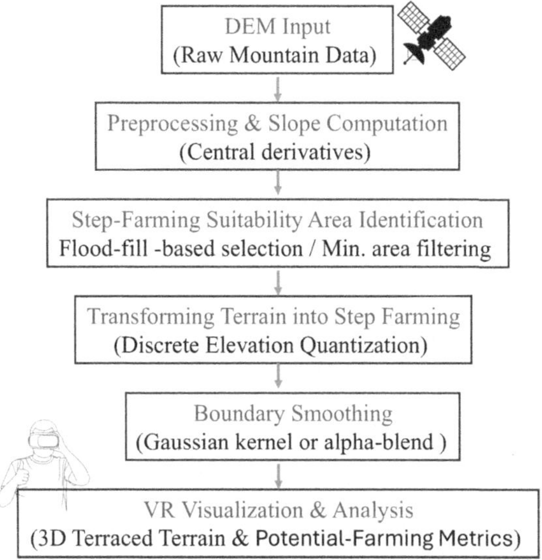

Fig. 2. Flow chart of geospatial processing steps including slope computation and elevation quantization.

3 Methodology

The research methodology is structured to transform raw digital elevation data into a refined terraced terrain model optimized for agricultural applications. The overall process is divided into the following key steps (see Fig. 2):

- **Data Acquisition and Preprocessing:**
  - Obtain 30 m Cartosat-1 DEM, released by ISRO-NRSC, covering the lower-Himalayan districts of Himachal Pradesh; the tiles are mosaicked and clipped to the study watershed before analysis.
  - Preprocess the DEMS to eliminate noise and inconsistencies, ensuring uniformity in spatial resolution.
- **Slope Computation:**
  - Compute the local slope $s(i, j)$ at each grid cell (i, j) using finite difference approximations.
  - Generate a slope map that quantifies the steepness across the entire study area.
- **Identification of Suitable Farming Zones (Refer to Algorithm 1):**
  - Define a maximum allowable slope threshold τ below which land is considered suitable for farming.
  - Identify contiguous regions $R \subset \mathcal{G}$ such that for every cell $(i, j) \in R$, the condition $s(i, j) \leq \tau$ holds.
  - Calculate the area $A(R)$ of each contiguous region and retain those regions where $A(R) \geq A_{\min}$, with $A_{\min}$ being the minimum area criterion.

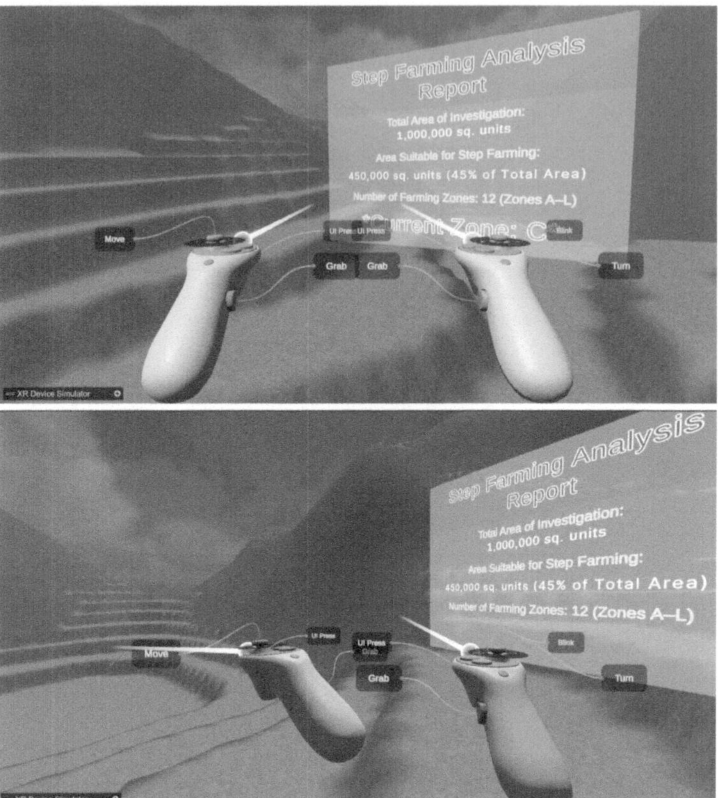

Fig. 3. VR interface showing information overlays on terraced regions and farming metrics.

– **Terracing via Elevation Quantization (Refer to Algorithm** 2**):**
  - Determine the base elevation $E_0 = \min_{x \in \mathcal{Z}} E(x)$ within the suitable zones $\mathcal{Z}$.
  - Apply a quantization operator Q_h to each elevation value $E(x)$.
  - Update the terrain model to incorporate the discrete terraces.
  - Employ a smoothing operator S to refine the transitions at the terrace boundaries.

– **Validation and Analysis:**
  - Validate the terraced terrain model through simulation studies.
  - Analyze the effectiveness of the terracing and overall suitability for agricultural production.

This systematic methodology transforms continuous elevation data into a structured terraced landscape, facilitating more efficient agricultural planning and sustainable land management.

3.1 Algorithm 1: Identification of Suitable Farming Zones

Let

$$D = \{d_{i,j}\} \quad \text{for } (i,j) \in \mathcal{G},$$

be a DEM defined over a grid $\mathcal{G}$. Define the slope threshold $\tau \in \mathbb{R}^+$ and the minimum required area $A_{\min} \in \mathbb{R}^+$.

Algorithm 1. Step Farming Suitable Area Identification

Require: DEM D, slope threshold τ, minimum area $A_{\min}$

Ensure: Set of suitable zones $\mathcal{Z}$

Compute the local slope for each cell $(i,j) \in \mathcal{G}$ as:

$$s(i,j) = f\left(d_{i,j}, d_{i\pm1,j}, d_{i,j\pm1}\right).$$

Identify contiguous regions $R \subset \mathcal{G}$ such that:

$$\forall (i,j) \in R, \quad s(i,j) \leq \tau.$$

Let $\mathcal{R} = \{R \subset \mathcal{G} \mid R \text{ is maximal and } \forall (i,j) \in R,\ s(i,j) \leq \tau\}$.

for each region $R \in \mathcal{R}$ **do**

  Compute the area of R as:

$$A(R) = |R| \times A_{\text{cell}},$$

  where $|R|$ is the number of cells in R and A_{cell} is the area of a single cell.

  if $A(R) \geq A_{\min}$ **then**

    Add R to $\mathcal{Z}$

  end if

end for

return $\mathcal{Z}$

3.2 Algorithm 2: Terracing via Elevation Quantisation

Let $\mathcal{Z} \subset \mathbb{R}^2$ be the set of spatial points (cells) within the suitable zones, and let

$$E : \mathcal{Z} \to \mathbb{R}$$

Be the corresponding elevation function. Given a quantization step $h \in \mathbb{R}^+$, our objective is to construct a terraced representation $T : \mathcal{Z} \to \mathbb{R}$ of the terrain.

Definition 1 (Base Elevation). The base elevation E_0 is defined as:

$$E_0 = \min_{x \in \mathcal{Z}} E(x).$$

Definition 2 (Quantisation Operator). Define the quantization operator $Q_h : \mathbb{R} \to \mathbb{R}$ by

$$Q_h(E(x)) = E_0 + h \cdot \left\lfloor \frac{E(x) - E_0}{h} + \frac{1}{2} \right\rfloor,$$

Where $\lfloor \cdot \rfloor$ denotes the floor function, effectively rounding the normalized elevation difference $\frac{E(x)-E_0}{h}$ to the nearest integer.

Algorithm 2. Discrete Elevation Quantisation for Terracing

Require: Suitable zones $\mathcal{Z} \subset \mathbb{R}^2$, elevation function $E : \mathcal{Z} \to \mathbb{R}$, quantization step $h > 0$

Ensure: Terraced terrain $T : \mathcal{Z} \to \mathbb{R}$

1: **Compute Base Elevation:** Set

$$E_0 \leftarrow \min_{x \in \mathcal{Z}} E(x).$$

2: **for** each $x \in \mathcal{Z}$ **do**

3: **Quantize Elevation:** Compute

$$T(x) \leftarrow Q_h\big(E(x)\big) = E_0 + h \cdot \left\lfloor \frac{E(x) - E_0}{h} + \frac{1}{2} \right\rfloor.$$

4: **end for**

5: **Boundary Smoothing:** Apply a smoothing operator S to T to refine terrace boundaries, i.e.,

$$T \leftarrow S(T).$$

6: **return** T

Discussion: The operator Q_h discretizes the continuous elevation field E by mapping each value to the nearest multiple of h above E_0. This quantization creates distinct elevation bands (terraces), which can be further refined by the smoothing operator S to mitigate abrupt transitions at the terrace boundaries.

Table 1. User Questionnaire Grouped by Evaluation Factor

Factor	Questions
Realism	1. How realistic did you find the terraced landscapes?
	2. How smooth and natural did the terrain transitions appear?
Usability	3. How clear was the visualization of farming suitability?
	4. How intuitive was navigating within the VR environment?
Informative Value	5. Did the farming metrics adequately meet your informational needs?
	6. Did the segmented zones indicate farming potential?

Fig. 4. Illustrates original raw terrain and generated terraced terrain in Unity environment.

4 Results and Analysis

The proposed system was developed using the Unity Engine[1], incorporating custom terrain processing algorithms designed explicitly for visualizing step farming. On an RTX 4070 GPU, the pipeline maintains 78 fps in Unity at 2 K resolution, confirming real-time interaction. We utilized real DEM data from mountainous terrains covering an area of 1,000,000 square units for testing. Geospatial analysis identified approximately 450,000 square units (45%) as suitable for step farming, using the FAO-recommended range (18°25°); we picked 22° as the midpoint for Himalayan mid-hill farms [12].

These suitable areas were segmented into twelve distinct zones (AL), each demonstrating varied farming potentials. Zone C was one of the largest areas, encompassing 40,000 square units, while Zone H was among the smallest, covering 37,000 square units. The terrain transformation algorithm employed discrete elevation quantization to effectively convert the identified low-slope regions into terraced landscapes (see Fig. 4). To enhance visual Realism, boundary smoothing was applied, resulting in natural transitions between terraces, with an average boundary height deviation of ± 0.12 m after processing. However, some participants noted only moderate levels of Realism, indicating opportunities for improvement in visual authenticity.

For the user study, the VR visualization was deployed using the Meta Quest 3 headset[2]. Participants viewed multiple randomised scenes to minimise counterbalancing effects and bias (see Fig. 3). The study adhered to ethical research standards, including obtaining informed consent, providing detailed instructions, and ensuring participant privacy. A total of 18 participants with prior experience in mountain farming took part in the study, consisting of 10 males and eight females, with an average age of 34.2 years. Feedback indicated intense satisfaction with the system's Usability and the informative nature of the visualiza-

[1] https://unity.com/.
[2] https://www.meta.com/quest/quest-3/.

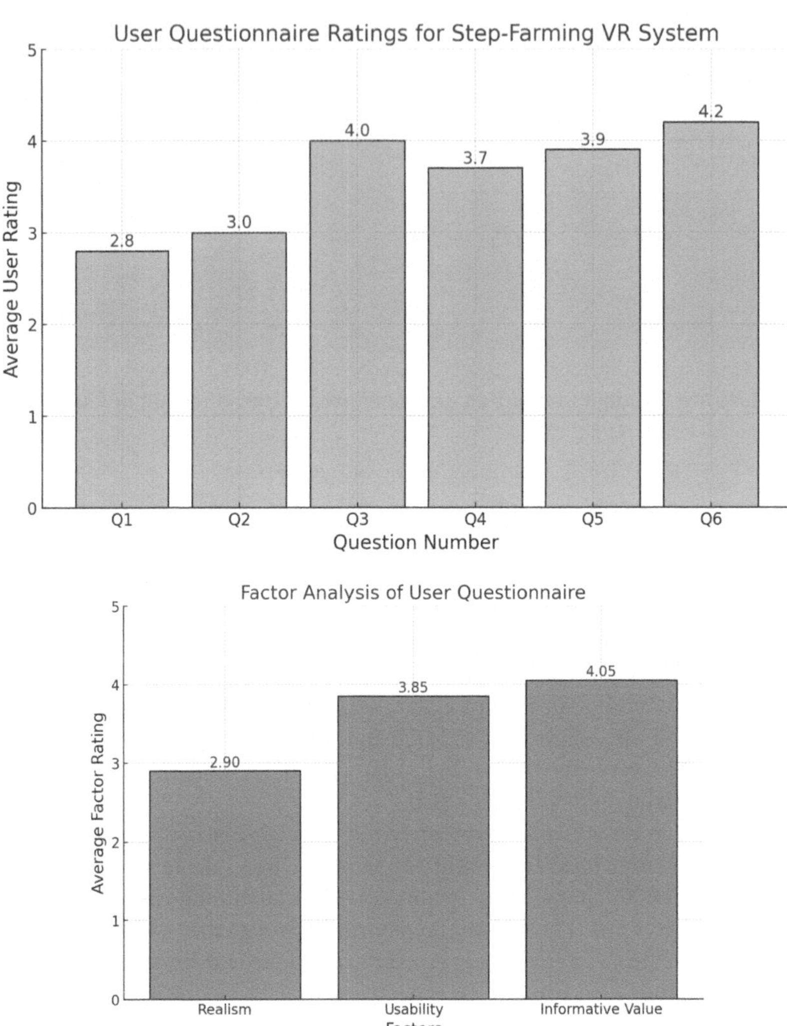

Fig. 5. User study results showing (top) participant ratings per question and (bottom) factor-wise analysis.

tion. Participant P3 commented, "The visualization clearly showed me exactly where farming would be practical; it significantly simplified planning decisions." Participant P5 remarked, "The terrain looks realistic enough for planning, but improvements in visual realism could enhance immersion."

Participants rated their experience through a questionnaire composed of six questions grouped into three factors (see Table 1): Realism, Usability, and Informative Value. A paired-sample t-test indicated statistically significant positive results for Usability ($t(17) = 3.87$, $p < 0.01$) and Informative Value ($t(17) = 4.55$, $p < 0.001$), while Realism scored moderately lower yet remained statistically sig-

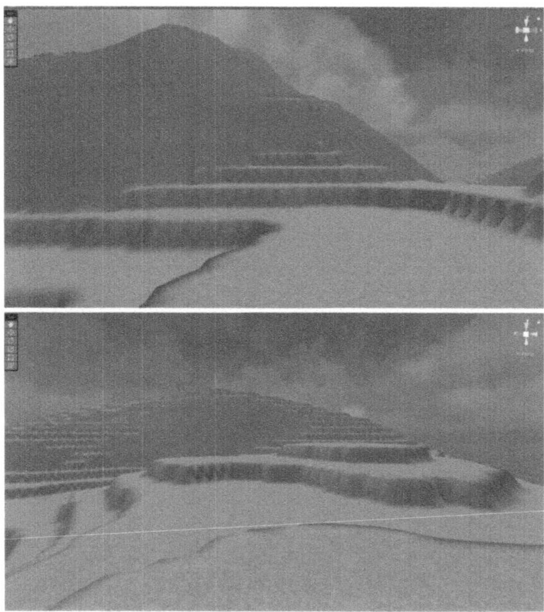

Fig. 6. Final terraced terrain output generated in Unity with smoothed transitions and VR compatibility.

nificant ($t(17) = 2.76$, $p < 0.05$). These results affirm the system's effectiveness while highlighting opportunities for further refinement in visual Realism (see Fig. 5).

Overall, the system robustly identified, segmented, and represented step-farming potential, demonstrating strong Usability and informative value despite moderate realism scores (see Fig. 6).

5 Conclusion and Future Work

This research introduces a VR-based geospatial framework that transforms raw Digital Elevation Model (DEM) terrain into realistic terraced landscapes for exploring step farming suitability. The system identifies low-slope zones using slope thresholds and area criteria, quantises elevation, and offers an immersive visualisation for real-time spatial analysis.

The system demonstrated strong Usability and clarity, with moderate positive feedback on Realism, indicating room for improvement. It is a valuable tool for agricultural planning in mountainous areas with limited conventional methods.

Future enhancements will refine the terracing algorithm for more natural shapes and improve visual appeal and analytical accuracy. Integrating remote sensing data—such as vegetation indices and land cover—will enable more comprehensive terrain evaluations. Allowing user-configurable parameters like slope thresholds and terrace width will enhance flexibility. Lastly, extending the tool's application to varied geographies and larger datasets will strengthen its robustness and scalability for real-world agricultural planning.

References

1. Anthes, C., García-Hernández, R.J., Wiedemann, M., Kranzlmüller, D.: State of the art of virtual reality technology. In: 2016 IEEE Aerospace Conference, pp. 1–19. IEEE (2016)
2. Baloch, J.: Overview of immersive data visualization: enhancing insights and engagement through virtual reality (2024)
3. Chapagain, T., Raizada, M.N.: Agronomic challenges and opportunities for smallholder terrace agriculture in developing countries. Front. Plant Sci. **8**, 331 (2017)
4. Charania, I., Li, X.: Smart farming: agriculture's shift from a labor intensive to technology native industry. Internet Things **9**, 100142 (2020)
5. Faust, N.L.: The virtual reality of GIS. Environ. Plann. B. Plann. Des. **22**(3), 257–268 (1995)
6. Hedley, N.R., Billinghurst, M., Postner, L., May, R., Kato, H.: Explorations in the use of augmented reality for geographic visualization. Presence **11**(2), 119–133 (2002)
7. Huang, B., Jiang, B., Li, H.: An integration of GIS, virtual reality and the internet for visualization, analysis and exploration of spatial data. Int. J. Geogr. Inf. Sci. **15**(5), 439–456 (2001)
8. Korkut, E.H., Surer, E.: Visualization in virtual reality: a systematic review. Virtual Reality **27**(2), 1447–1480 (2023)
9. Mitášová, H., Hofierka, J.: Interpolation by regularized spline with tension: II. application to terrain modeling and surface geometry analysis. Math. Geol. **25**, 657–669 (1993)
10. Partap, T.: Hill agriculture: challenges and opportunities. Indian J. Agric. Econ. **66**(1) (2011)
11. Rai, R.K., Bansal, R., Jha, S.S., Narava, R.: Assessing the utility of GAN-generated 3D virtual desert terrain: a user-centric evaluation of immersion and realism. In: International Conference on Artificial Intelligence and Virtual Reality, pp. 179–191. Springer (2023)
12. Sheng, T.C.: Soil Conservation for Small Farmers in the Humid Tropics, FAO Soils Bulletin, vol. 60. Food and Agriculture Organization of the United Nations, Rome (1989)
13. Tarolli, P., Cao, W., Sofia, G., Evans, D., Ellis, E.C.: From features to fingerprints: a general diagnostic framework for anthropogenic geomorphology. Progress Phys. Geogr. Earth Environ. **43**(1), 95–128 (2019)
14. Tarolli, P., Straffelini, E.: Agriculture in hilly and mountainous landscapes: threats, monitoring and sustainable management. Geogr. Sustain. **1**(1), 70–76 (2020)
15. Toromade, A.S., Chiekezie, N.R.: GIS-driven agriculture: pioneering precision farming and promoting sustainable agricultural practices. World J. Adv. Sci. Technol **6**(1), 057–072 (2024)
16. Wilson, J.P., Gallant, J.C.: Terrain Analysis: Principles and Applications. John Wiley & Sons (2000)

Swarm Robotics for Agricultural Drones: A Transformative Approach to Smart Farming

S. Subaselvi[1(✉)], S. Pricilla Mary[1], A. Sharon Geege[1], T. S. Arun Samuel[1], and A. Andrew Roobert[2]

[1] Department of ECE, National Engineering College, Kovilpatti, India
subaselvi05@gmail.com

[2] Department of EEE, Indian Institute of Technology, Guwahati, India
andrewroobert@rnd.iitg.ac.in

Abstract. The multiple drones operate collaboratively with minimal centralized control in the swarm robotics system. Each drone acts autonomously and shares information with nearby drones for optimizing crop spraying patterns, surveying large farmlands, and identifying anomalies in plant health. Therefore, a multi-antenna swarm robotics system for smart farming in agriculture is proposed. The multiple antenna system in drones transmits multiple data streams, increasing the coverage and throughput. The multi-antenna with a beamforming technique steers the signal to the intended receiver, which reduces the interference in the agricultural drones. The ant colony optimization algorithm is introduced for the optimal deployment of drones in smart farming. An efficient shortest path routing protocol is proposed to avoid collision between drones in the system. The simulation results show that the proposed algorithm gives better performance compared to the existing algorithm in terms of residual energy, delay, and throughput.

Keywords: swarm robotics · agriculture · drones · beamforming · ant colony optimization

1 Introduction

The advanced technologies in agriculture increase productivity, optimize resource usage, and ensure sustainability. Among the emerging technologies, robotics and automation play a significant role, particularly with the use of unmanned aerial vehicles commonly known as drones [1]. In recent years, a more advanced concept called swarm robotics has emerged as a powerful extension to traditional drone systems that improve efficiency and scalability in agricultural operations [2]. Traditional agricultural methods struggle with scalability, precision, and labor availability when applied to large or diverse crop fields. The manual monitoring and intervention are time-consuming and labor-intensive. The single-drone systems are limited by coverage range, battery life, and inability to adapt in real-time to dynamic field conditions [3]. Therefore, swarm robotics offers a decentralized and cooperative alternative where multiple drones work together, dividing tasks intelligently and reacting according to environmental changes.

H. S. Shekhawat et al. (Eds.): ICA 2025, CCIS 2795, pp. 259–268, 2026.
https://doi.org/10.1007/978-3-032-17083-5_22

The act of identifying the best possible solution from a group of given alternatives by maximizing or minimizing a given objective, like cost, efficiency, or performance, is optimization. Optimization techniques are conventional mathematical methods such as linear or nonlinear programming and metaheuristic algorithms like genetic algorithms, simulated annealing, and ant colony optimization [4]. The best solution under imposed constraints, by making optimal utilization of time, cost, or other essential resources, is obtained by applying optimization techniques. Ant Colony Optimization (ACO) is a nature-inspired optimization technique based on the foraging behavior of actual ants. Ants in nature determine the shortest route to food sources by depositing pheromones along their trails [5]. With time, the shortest routes have a higher concentration of pheromones, drawing more ants and strengthening the best path. ACO was used to solve hard computational problems such as the traveling salesman problem, scheduling, and network routing. In the process, artificial ants search for potential solutions and lay virtual pheromones on routes that are good solutions, which lead future ants to follow and refine them. ACO's greatest strength is the equilibrium be-tween exploration (finding new routes) and exploitation (focusing search around known good routes). With iterations, the algorithm converges to optimal or near-optimal solutions. ACO is especially suitable for discrete optimization problems and can also learn to react to changes in the problem environment over time. A new meta-heuristic that combines machine learning with ACO to solve combinatorial optimization problems is introduced. The method includes training a classification model on small problem instances for which optimal solutions are known to estimate the probability that edges belong to the optimal route [6].

Multiple Input Multiple Output (MIMO) beamforming is a wireless communication technique that uses multiple antennas at the transmitting and receiving ends for improving signal quality, coverage, and data rate [7]. Beamforming involves steering the signal in the direction of the desired receiver by controlling the phase and amplitude of signals over the array of antennas. Directional transmission increases signal power at the receiver, decreases interference to unwanted users, and enhances the overall efficiency of the communication system [8]. The hybrid beamforming in millimeter wave MIMO systems for integrated sensing and communications is proposed in that alternating minimization method to iteratively optimize the transmit beam and phase vector to achieve the desired radar beam pattern while satisfying communication users SINR constraints [9]. The user association and hybrid beamforming design in cooperative Wave MIMO systems is proposed for fully connected, fixed subarray, and dynamic subarray hybrid beamforming architectures that achieves significant performance improvements and enhanced energy efficiency in the system [10]. The contribution of the proposed algorithm is.

1. The Multiple drones in the proposed method act independently with centralized control, have real-time coordination for spraying crops, surveying, and anomaly detection by local communication and information sharing.
2. The multi-antenna beamforming system improves wireless communication by increasing throughput, enhancing coverage, and reducing interference to ensure drone-to-base and drone-to-drone communication is reliable in vast farmland regions.

3. The Ant Colony Optimization (ACO) algorithm is utilized to identify the best deployment of drones to ensure maximum coverage efficiency and minimal energy expenditure over the farm.
4. A shortest path routing protocol is effective to ensure the safe flying of drones for collision avoidance and efficient routing for better operating reliability.
5. The simulation results show that the new algorithm performs better than other methods in residual energy, communication delay, and throughput by validating the effectiveness for smart agriculture usage.

2 Proposed Algorithm

In this section, the network model of the agricultural field with drones, optimal placement of drones, and MIMO beamforming is explained.

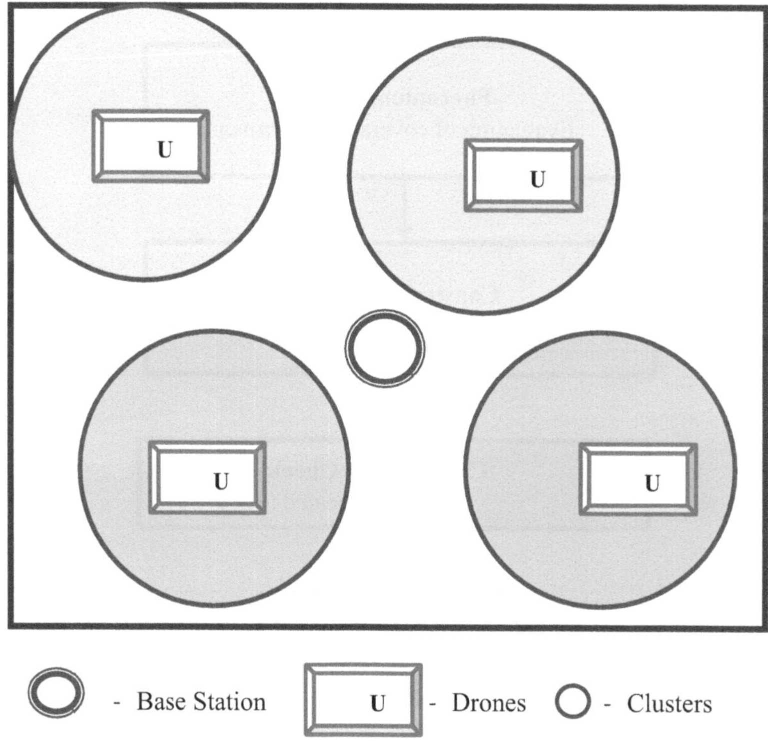

Fig. 1. Proposed Model

2.1 Network Model

The N x N agricultural area is divided into clusters based on the cultivation and crops, as shown in Fig. 1. In each cluster, a drone or UAV is deployed for monitoring that particular agricultural land. The drones move around the particular clusters for surveying or anomaly detection and send the data to the base station, which is located at the centre of the agricultural field. The base station receives data from all the drones for processing and informs the user if there are any changes required in that particular agricultural field.

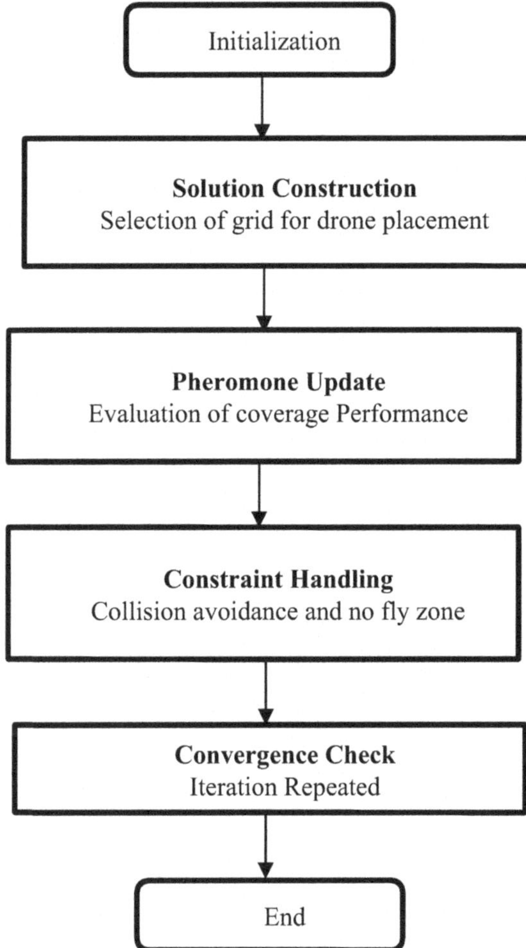

Fig. 2. Placement of drones based on the ACO algorithm

2.2 Ant Colony Optimization-Based Placement of Drones

ACO simulates food-foraging ants that release pheromones to follow optimal paths. With the progress of time, ants follow the path with stronger pheromones, making the path increasingly strong with good solutions. Therefore, for placing drones, the artificial ant corresponds to a drone placement candidate solution. The solution seeks places where drones should be placed so as to have field coverage without loss of coverage due to overlapping spaces, flying duration, and usage of energy. The ground of the farm is depicted as a 2D grid, and one grid cell signifies a potential point of interest, either a crop health or an irrigation requirement. The ACO algorithm continually determines the optimal configuration of drone locations such that each area of the ground is covered by the coverage radius of at least one drone with the constraints as battery capacity, field obstructions and wind. The placement of drones in agricultural field process in shown in Fig. 2.

The ACO based drone positioning methodology begins with initialization. The drone range and field characteristics are used to initialize an artificial population of ants as candidate solutions, pheromone strength, and heuristic information. At the construction phase of the solution, the single ants build potential drone placement configuration probabilistically selecting grid points based on pheromone intensity and heuristic desirability, such as proximity to field edges and gap filling. Once all solutions have been constructed, the pheromone update step evaluates each configuration's fitness, preferring trails of higher coverage and less overlap. Throughout the process, a constraint handling is incorporated to ensure drones avoid being placed too close to one another or outside of operational limits with built-in collision prevention measures and no-fly zones. The process continues iteratively until a convergence test is met, where no significant increase is observed, and an optimized drone placement strategy for smart agricultural use is reached.

2.3 MISO Beamforming

In the proposed system, the drones are equipped with two antennas, and the Base station has a single antenna. The drones having N > 1 antennas transmit data to the single antenna base station (k = 1,2,..., K) using transmit beamforming. The drone sends the signal Sk with the transmit power Pk using the beamforming vector Wk ϵ $\mathbb{C}$N x 1. The received signal at the base station can be written as

$$y_k = \sqrt{P_k L_k} \mathbb{h}_k^T W_k S_k + n_k \tag{1}$$

$$L_k = 22\log_{10}d + 42 + 20\log_{10}\frac{f_c}{5} \tag{2}$$

where Lk is the path loss, hk = [hk,1, hk,2] is the channel vector, Sk = [sk,1, sk,2], Wk = [wk,1, wk,2], and nk is the Additive White Gaussian Noise. The Signal to Noise Ratio at the receiver is given by

$$P_H = P_k L_k |\mathbb{h}_k^T W_k|^2 \tag{3}$$

The MISO beamforming in the proposed system is to maximize the data rate at the receiver. Thus, the optimization problem is formulated with constraints can be defined as

$$\max_{W} |\mathbb{h}_k^T W_k|^2 \tag{4}$$

$$\text{subject to } \Gamma_k > \Upsilon_k \tag{5}$$

$$\sum_{k=1}^{K} \|W_k\|^2 \leq P_T \tag{6}$$

where Υ_k is the SNR threshold, and PT is the drone's transmit power constraint. The solution to the problem is given by

$$W_k = \frac{\left(G_k G_k^{\dagger} + \eta_k I_N\right)^{-1} \times \mathbb{h}_{k,k}^{*}}{\left\|\left(G_k G_k^{\dagger} + \eta_k I_N\right)^{-1} \times \mathbb{h}_{k,k}^{*}\right\|} \tag{7}$$

where Gk= [h1, k, …………, hK,k], Gk†=GkH (Gk GkH)-1, ηk =Nσ2/Pk[20] and I_N is the N-dimensional identity matrix.

The optimization problem in the proposed system is formulated with W_k to get the optimal solution for the beamforming technique. The W_k in the MISO beamforming technique is selected to cancel the interference for the desired user to increase the data rate in the network. The problem is formulated with SNR and transmit power constraints for efficiently transmitting the data in the network.

2.4 Shortest Path Routing for Multi-drones System

For swarm robotics in the case of agricultural drones, the application of a shortest path routing protocol is vital in both safety and efficiency. When more than one drone flies at the same time over vast tracts of land dedicated to farming, the chances of mid-air collisions and duplication of efforts are higher. With the use of a shortest path routing algorithm, drones are routed on optimized conflict-free routes, significantly minimizing opportunities for collisions. This is particularly relevant in dynamic environments where drone positions and field conditions are in a state of constant flux.

The shortest path routing enhances the swarm operation reliability by optimizing communication and travel in terms of shortest paths. Hence, travel time and power consumption got reduced, which enhances the system's responsiveness as a whole. Optimized routing also guarantees on-time data exchange among drones in a crucial requirement of real-time monitoring and decision-making of precision agriculture activities like spraying, seeding, and plant health analysis.

Incorporating such routing information into swarm robotic systems transforms the agricultural landscape through the ability to implement scalable and autonomous drone missions. The integration of such routing data into swarm robotic platforms revolutionizes the agricultural scene by enabling scalable and autonomous drone missions. The integration of collision-free movement and path optimization of communications allows uninterrupted and coordinated field coverage, resulting in enhanced productivity and utilization of resources.

3 Results and Discussion

The performance of the proposed algorithm is evaluated in the NS-2 Simulator. The single antenna BS is deployed at the center of the agricultural field with multiple antenna drones. The performance of the system is compared with the existing system. The Simulation parameters are illustrated in Table 1.

Table 1. Simulation parameters

S.NO	PARAMETER	VALUE
1	Network size	1000 m x 1000 m
2	Variance (σ^2)	−174 dBm [11]
3	Transmit power threshold	30 dBm [11]
4	SNR threshold	20 dB [12]
5	Time slot length (T)	5 ms [13]
6	Rmin	12 Kb/s [13]
7	Simulation time	100, 400 s
8	Initial Energy	1 J
9	V_{min} and V_{max}	22 m/sec and 30 m/sec
10	Number of UAV	3
11	UAV altitude	120 m

Fig. 3. Transmit power of the UAV vs Average Throughput

Figure 3. Shows the transmit power of UAV and average throughput performance. In the proposed system, average throughput is enhanced by increasing the UAV's transmit power, which improves the signal strength, resulting in higher SNR and data rates. At a transmit power of 8 W, the proposed method achieves a throughput of 59 kbps, outperforming the existing method's 55.67 kbps. In the existing system the UAV's height are reduced for each operation causing increased overhead and energy consumption where the proposed approach reduces UAV height only at the base station if required.

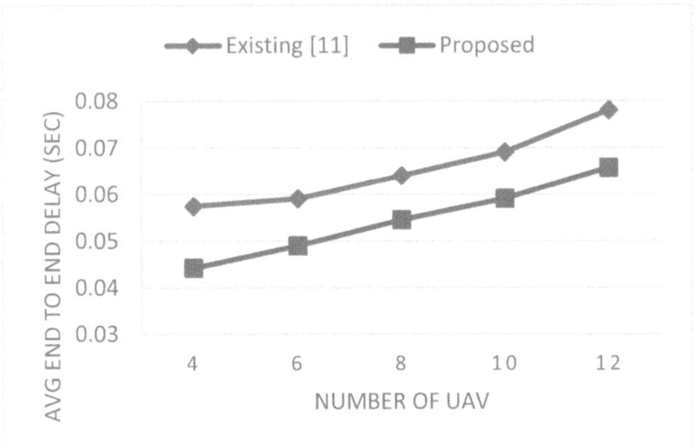

Fig. 4. Number of UAV vs Average end-to-end delay

The average end-to-end delay increases with the number of UAVs in the network due to packet drops, congestion, and routing link disconnections caused by UAV mobility, as shown in Fig. 4 for 400 s. However, in the proposed system, the ACO algorithm enables efficient routing of data from base station to UAVs, thereby reducing congestion and decreasing delay compared to existing algorithms. The proposed method also incorporates a transmission rate constraint in its optimization model, which helps further lower the end-to-end delay compared to the existing system.

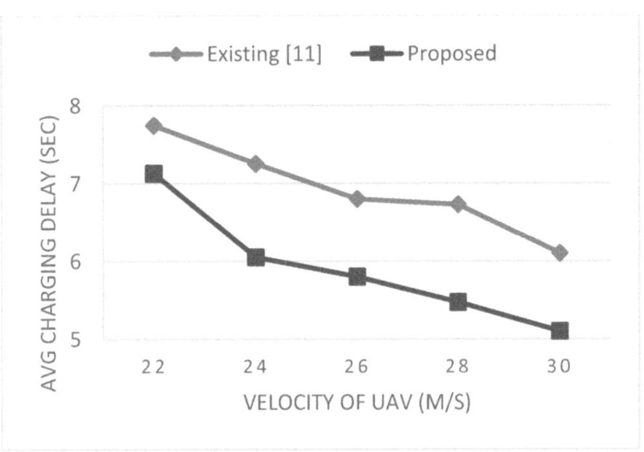

Fig. 5. Velocity of UAV vs Average charging delay

The mobility of the UAV at high speed reduces the traveling time in the agricultural field which decreases the delay of operation in the field. The performance of the proposed algorithm for 400 s is better than the existing algorithm is shown in Fig. 5. The average charging delay is 6.05s for the UAV speed of 24 m/s in the proposed system is lesser than 7.25s in the existing system. The speed of the UAV is increased to maximum when there is an emergency which reduces the charging delay. The number of UAVs increases when the agricultural field requires more information; therefore, the charging delay in the proposed system is less compared to the existing system.

4 Conclusion

The proposed multi-antenna swarm robotics system embedded with ant colony optimization and beamforming technology provides a facilitation of intelligent farming operations. Through the promotion of autonomous cooperative drone action and highest deployment and communication, the proposed system enhances coverage, limits interference, and amplifies the data flow in the network. A reliable shortest path routing protocol further ensures safe navigation and energy saving. The simulation results provide better performance of the proposed method in comparison to existing schemes in terms of reducing delay and throughput.

Disclosure of Interests. No conflict of Interest.

References

1. Ahmed, F., Mohanta, J.C., Keshari, A., Yadav, P.S.: Recent advances in unmanned aerial vehicles: a review. Arabian J. Sci. Technol. **47**, 7963–7984 (2022)
2. Blais, M.-B., Akhloufi, M.A.: Reinforcement learning for swarm robotics: an overview of applications, algorithms and simulators. Cogn. Robot. **3**, 226–256
3. Agrawal, S., Patle, B.K., Sanap, S.: Navigation control of unmanned aerial vehicles in dynamic collaborative indoor environment using probability fuzzy logic approach. Cogn. Robot. **5**, 86–113 (2025)
4. Tomar, V., Bansal, M., Singh, P.: Metaheuristic algorithms for optimization: a brief review. Eng. Proc. **59**(1), 238–252 (2024)
5. Mamur, H., Üstüner, M.A., Bhuiyan, M.R.A.: Future perspective and current situation of maximum power point tracking methods in thermoelectric generators, vol. 50, pp. 1–25 (2022)
6. Sun, Y., Wang, S., Shen, Y., Li, X., Ernst, A.T., Kirley, M.: Boosting ant colony optimization via solution prediction and machine learning. Comput. Oper. Res. **143**, 1–25 (2020)
7. Ehab Ali, Mahamod Ismail, Rosdiadee Nordin, Nor Fadzilah Abdulah, "Beamforming techniques for massive MIMO systems in 5G:Overview, Classification and trends for future research", Frontiers of Information Technology and electronic engineering, vol. 18, p. 753 – 772, 2017
8. Ning, B., et al.: Beamforming technologies for ultra-massive MIMO in terahertz communications. IEEE Open J. Commun. Soc. **4**, 614–658 (2023)
9. Qi, C., Ci, W., Zhang, J., You, X.: Hybrid beamforming for millimeter wave MIMO integrated sensing and communications. IEEE Commun. Lett. **26**(5), 1136–1140 (2022)
10. Ni, P., Liu, R., Li, M., Liu, Q.: User association and hybrid beamforming designs for cooperative mmWave MIMO systems. IEEE Trans. Signal Inform. Process. Over Networks **8**, 641–654 (2022)
11. Yang, Z., Xu, W., Shikh-Bahaei, M.: Energy efficient UAV communication with energy harvesting. IEEE Trans. Vehicular Technol. (IEEE Trans. Veh. Technol.) **69**(2), 1913–1927 (2020)

12. Elhabyan, R., Shi, W., St-Hilaire, M.: Coverage protocols for wireless sensor networks: Review and future directions. J. Commun. Networks (J. Commun. Netw.) **21**(1), 45–60 (2019)
13. Chen, X., Wang, X., Chen, X.: Energy-efficient optimization for wireless information and power transfer in large-scale MIMO systems employing energy beamforming. IEEE Wireless Commun. Let. (IEEE Wireless Commun. Lett.) **2**(6), 667–670 (2013)

Sodium Ion Detection Using Polymer-Based Sensor for Monitoring Soil Health in Agriculture

T. Devakumar(✉), T. S. Arun Samuel, A. Seshora Besima, and A. Sharon Geege

Department of ECE, National Engineering College, Kovilpatti, India
tdkece@nec.edu.in, seashore0604@gmail.com

Abstract. In precision farming applications, developing polymer-based electrochemical sensors for non-invasive ion concentration monitoring in agricultural fluids has drawn more and more interest. For instance, sodium ion concentrations in soil can have a significant impact, and ion-selective field-effect transistors are commonly used to monitor sodium levels in agricultural settings. Among these, a developing trend in farming technology is the polymers manufacturing of sensors. To identify variations in sodium ion concentration in soil, we fabricated a polymer-based electrochemical sensor with two terminals. There is a channel area between two metallic contacts in the sensor fabrication. A functionalized polymer essential to the selective detection of sodium ions makes up this channel.

Keywords: Sodium ion · Electrochemical Sensors · Ion-selective Field-Effect Transistors · Precision Farming · Functionalized Polymer · Agricultural Technology

1 Introduction

Sodium ions play an essential role in human biology, plant physiology, and agriculture. Sodium is a non-essential but essential element in some plant species, and its level may affect critical biological processes like osmotic balance, enzyme activity, and ion transportation mechanisms in plants. Though a limited amount of sodium benefits plant function, excess sodium buildup—soil salinity—is a significant agricultural concern that negatively impacts crop yield and soil quality [7, 9, 17]. Salinity stress due to high sodium ion (Na^+) concentrations in the irrigation water or soil can lead to ion toxicity, osmotic stress, and crop nutritional imbalances. The symptoms of sodium toxicity in plants are chlorosis, necrosis, stunted growth, and plant death in the worst cases. Likewise, sodium ion deficiency, though uncommon, can impact some C_4 and CAM plants, where sodium plays a role in carbon fixation and stomatal function [10]. The significance of the sodium ion content in agricultural systems, especially in soil, irrigation water, and sap of plants, has created an increased demand for sodium ion sensing technology. Traditional analysis methods, such as inductively coupled plasma mass spectrometry (ICP-MS), ion chromatography, and atomic absorption spectroscopy, are used extensively to detect sodium. Nevertheless, they are usually costly, time-consuming, and must be conducted

H. S. Shekhawat et al. (Eds.): ICA 2025, CCIS 2795, pp. 269–282, 2026.
https://doi.org/10.1007/978-3-032-17083-5_23

in a laboratory environment [5, 13]. In this regard, our present research aims to fabricate a two—terminal polymer-based sensor to directly detect concentrations of sodium ions in agricultural specimens [1, 2, 6, 8]. The device features a polymer channel functionalized to interface with sodium ions selectively. This would enable real-time measurement of sodium concentration within soil or within plant fluids [3, 7, 11, 12]. Our Sensor shows a detection range for sodium ions of 0.12 mg to 0.01 mg. It is thus extremely sensitive and apt for early identification of soil salinization or sodium imbalance in plants [4, 9]. Our technology is to ensure sustainable agriculture and increased crop productivity [15, 16]. This research emphasizes a critical knowledge gap in the application of polymer-based electrochemical sensors for agricultural sodium ion monitoring since most current studies target human biofluids. Our research opens up new possibilities for sensor-based solutions to maintain soil health and crop resilience in salt-prone agricultural land.

2 Literature Review

In food safety, agriculture, environmental monitoring, and healthcare, electrochemical sensors are being used increasingly to detect critical ions like sodium and potassium. Recent studies have concentrated on combining functional polymers with nanomaterials to improve performance. Traiwatcharanon et al. [1] used a nanocomposite of silver nanoparticles/graphene oxide (AgNP/GO) to construct a sodium ion sensor. The material's high conductivity and surface area increased the sensitivity of food analysis. For wearable technology, Ghoorchian et al. [2] created a flexible potentiometric sensor based on $Na_{0.44}MnO_2$ to detect sodium in sweat. The sensor's exceptional selectivity and response speed made it ideal for non-invasive health monitoring. Luo et al. [3] expanded into plant biology by introducing a potentiometric sensor based on microneedles that allows for the direct measurement of potassium and sodium ions in plant tissues. This made monitoring precision agriculture's in-planta, real-time nutrient dynamics easier. Bhandari et al. [4] used Nafion's cation-exchange capability to create a Nafion-modified carbon electrode for potassium detection in soil sensing. For agricultural nitrogen management, Kundu et al. [5] created a highly sensitive nanosensor for nitrate detection in soil extracts.

To detect amaranth dye in beverages, Jing et al. [6] developed a graphene and Co_3O_4 nanoparticle-based sensor that demonstrated excellent selectivity and application promise for food safety. Similarly, Zhang et al. [7] developed an electrochemical sensor based on graphene for online soil salinity monitoring. Polymer-based ion-selective electrodes for sodium detection in environmental applications were described by Wang and Zhao [8], who concentrated on membrane design for improved selectivity and durability. Nguyen and Lee's [10] review on plant nutrient uptake monitoring and Patel and Singh's [9] review of electrochemical sensors for soil nutrient assessment have made additional contributions. Nanostructured sensors have also been reported by Gomez and Martinez [11] to detect a variety of soil macronutrients with remarkable sensitivity.

2.1 Comparison

Monitoring sodium levels in the soil is essential for assessing its health, particularly when managing salinity issues in agriculture. The most sensitive and precise sodium detection methods are conventional laboratory techniques like Atomic Absorption Spectroscopy (AAS) and Flame Photometry. These methods are perfect for research and regulatory applications since they can assess trace levels and produce highly accurate results. However, their usefulness for in-field or routine monitoring is limited by the need for expensive equipment, trained operators, and time-consuming sample preparation. Commercial sodium sensors, especially Ion-Selective Electrodes (ISEs), provide a more practical and portable option. They are appropriate for farmers and agronomists who need quick soil measurements since they are reasonably priced, simple to use, and adaptable to field settings. Intelligent soil sensors with wireless connectivity, real-time monitoring, and automated data logging have recently hit the market. These systems offer continuous data collection and user-friendly interfaces through mobile apps or cloud platforms, but they come with a higher upfront cost. Because of these features, they are invaluable for remote soil monitoring and precision agriculture. Commercial sensors and intelligent systems provide easy-to-use, on-site alternatives for more frequent and real-time soil health monitoring, even if traditional laboratory methods are generally more accurate and reliable. The trade-off between required precision, cost, and the need for field portability and real-time decision-making determines which of these approaches is best.

3 Materials and Methods

3.1 Materials

PEDOT: PSS printable ink (Poly(3,4-ethylenedioxythiophene)-poly(styrenesulfonate) 1.5%-1.7% in Water, Sodium Ionophore X (Tetraethyl 2,2′,2‴,2‴-((15,35,55,75-tetra-tert-butyl-1,3,5,7(1,3)-tetrabenzenacyclooctaphane-12,32,52,72 tetrayl) tetrakis (oxy)) tetraacetate), GOPS(3-Glycidoxypropyltrimethoxysilane) solution, Distilled water, Ethylene Glycol, NaCl, Acetone Isopropyl alcohol (IPA), Phosphate buffer saline(PBS) solution, Polydimethyl siloxane(PDMS) gel and Silver conductive ink were procured from Universal Scientific Appliances (India). All reagents were employed without additional purification.

3.2 Apparatus

Electrical characteristic study and surface morphology study were performed using laboratory apparatus permitted by Indian Institute of technology, Guwahati(India). Material characterization was done using a LabRAM HR Evolution Raman Spectrometer [Horiba Scientific, Japan]. Thin aluminium metal foil layers were formed through a thermal deposition process using a thermal evaporator [Auto 306 deposition system, Hind High Vacuum (HHV), India]. An electrical characteristic study was performed by using a Source Measure Unit.

3.3 Fabrication

The fabrication process of the polymer-based two-terminal sodium Ion Sensor contains the following sub-processes:

a) Functionalized polymer Ink preparation for channel layer.
b) Glass substrate purification.
c) Metallic layers deposition on either side of the channel layer.
d) Terminal extraction from both Metallic layers.

Two different chemical combinations of PEDOT: PSS polymer inks were prepared, which enabled the fabrication of two different channel types. The two kinds of PEDOT: PSS polymer inks were named as Primary Ink and Secondary Ink. The sensor structure type with the channel prepared with the primary ink was considered the reference structure for sensor performance analysis. The primary ink consists of two parts of PEDOT: PSS, one part of ethylene Glycol, and 1% of GOPS(w/v) solutions. As an example, the primary ink constituent's calculation is as follows:

$$\text{Assumed total volume of Ink solution} = 1\text{ ml}$$
$$1\%\text{ of GOPS(w/v)} = 1\%\text{ of 1 ml solution} = 10\ \mu\text{l}$$

By subtracting 1% of the GOPS content numerically from the 1 ml total volume of the Ink solution, the remaining volume became 990 μl. Among them, one part of Ethylene Glycol represents 330 μl, and two parts of PEDOT: PSS represent 660 μl. Another sensor structure type was prepared with the channel having the Secondary Ink, which is named as functionalized polymer Ink. This Ink solution was prepared by adding the combination of Sodium Ionophore X and NaCl (Sodium analyte equivalent quantity) with the primary Ink at a 1:1 ratio. A Glass substrate with dimensions (24.41 mm length × 10.56 mm width × 1.20 mm thickness) was considered as the substrate material. The chosen substrate was initially cleansed by using an ultrasonic cleaner. Then the glass substrate was kept in the sonicator for 15 min for further purification.

After that, the glass substrate was cleansed by using acetone, Isopropyl alcohol (IPA), and distilled water. After the glass substrate purification, thin aluminium metal foil layers (AFL) were formed through a thermal deposition process using a thermal evaporator on the top of the substrate with a 720 nm thickness. A small centre region was considered for channel region formation in the deposited AFL layer. The channel region was filled with functionalized PEDOT: PSS polymer solution mixture.

As the above figures depict, we fabricated three different sensor structures by varying channel widths: Sensor structure A, Sensor structure B, and Sensor structure C. These channel regions separate metallic aluminium foil layer (AFL) regions. The separated metallic aluminium foil layer (AFL) regions were designated AFL-Left and AFL-Right. The resultant sensor structure has two specific regions: metallic aluminum and channel (Fig. 2, 3, 4 and Table 1).

The methodology adopted for sensor fabrication is shown in Fig. 5. Material characterization was analyzed by using a Raman Spectrometer. The electrical characteristics observations were made by using a Source Measure Unit.

4 Results and Discussion

Performance analysis of the fabricated sensor was performed by using a Source Measure Unit. This indicates the electrical characteristics of the sensor. During this analysis, sensor channel dimensions and their contents were considered as variables. The electrical

Sensor structure A

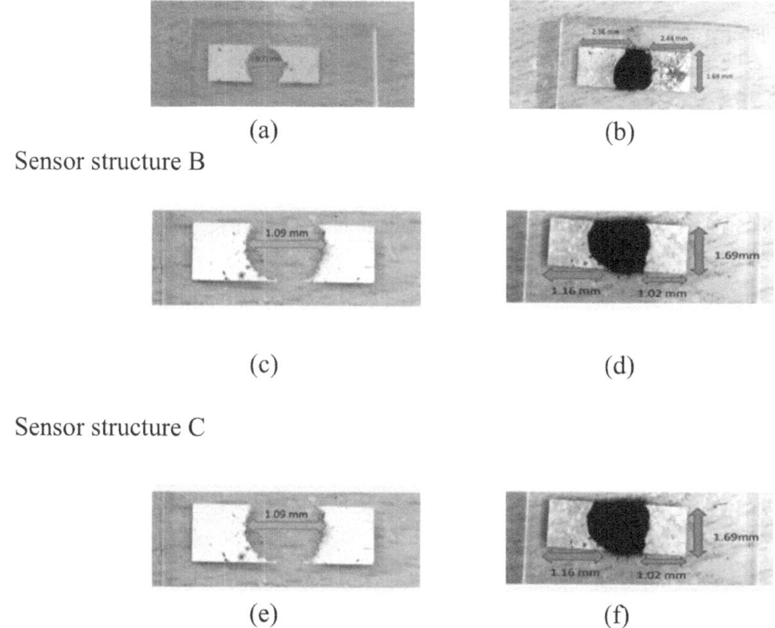

Fig. 1. (**a, c, e**) shows sensor structure with substrate, AFL layer regions, and filled Channel and (**b, d, f**) shows sensor structure with substrate, AFL layer regions, and filled Channel

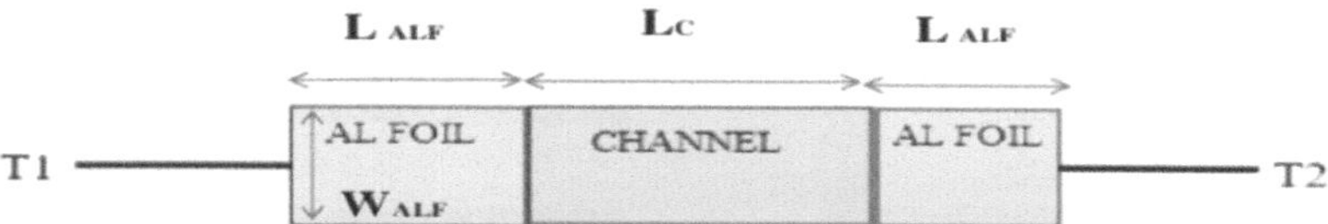

Fig. 2. Sensor Regions (Top View)

Fig. 3. Sensor Regions with Non-Filled Channel

characteristics were observed for the applied voltage range (−5 V to 5 V). Based on the sensor channel dimensions variation, three different sensor structures were fabricated

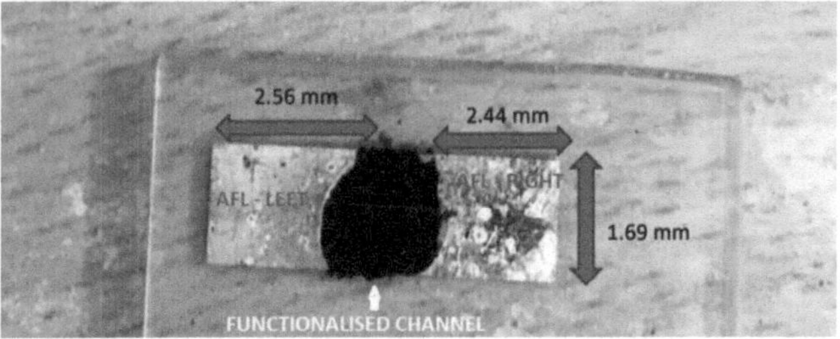

Fig. 4. Sensor Regions with Functionalized and Filled Channel

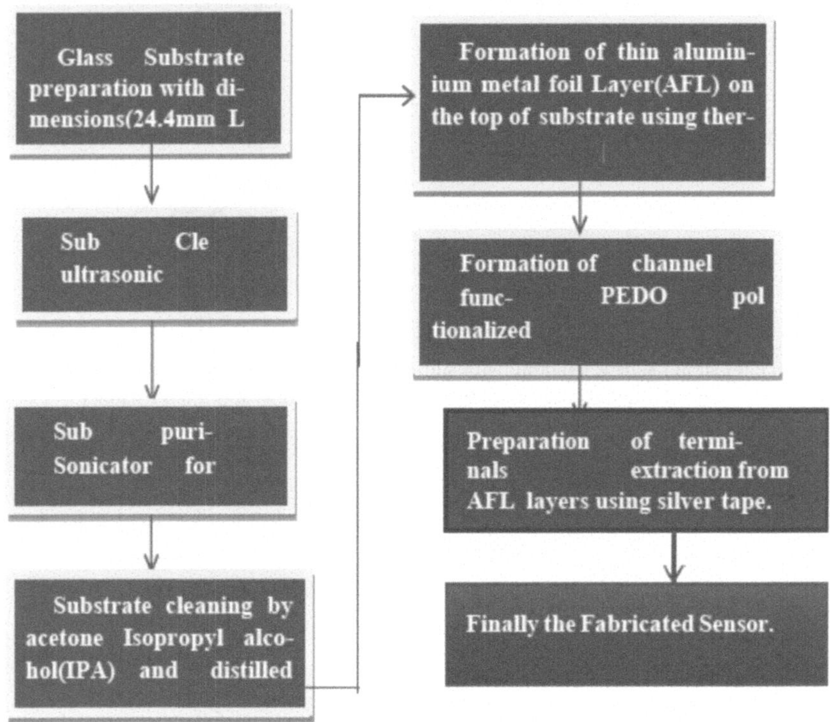

Fig. 5. Sensor Fabrication Flow Diagram

(channel diameter of 0.75 mm,1.09 mm, and 2.96 mm), and they are designated as sensor structure A, sensor structure B, and sensor structure C, respectively (Fig. 1). According to the variation in sensor channel contents, adding different amounts of salt (NACL)along with sodium ionophore was considered. The mixing ratio for salt (NACL) with sodium Ionophore was 1:1. The purpose of the fabricated sensor is to identify the sodium analyte present in the functionalized polymer mix of the channel. In this performance analysis, the salt (NACL) was added in different quantities as the representation for the sodium analyte in the functionalized polymer (Fig. 6 and Table 2).

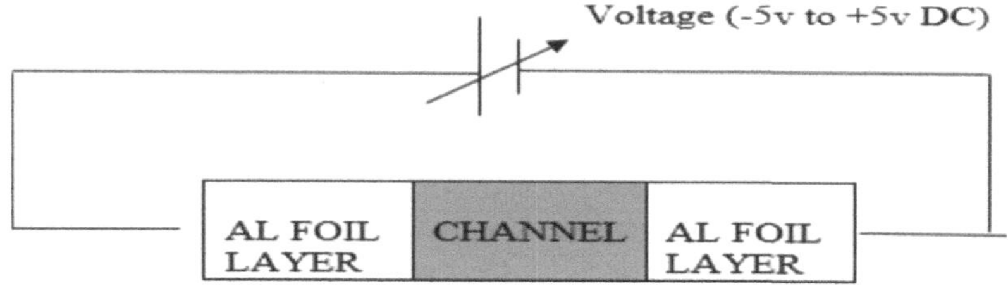

Fig. 6. Sensor Biasing Diagram

4.1 Electrical Characteristics Observation

4.1.1 Case a: Electrical Characteristics of Sensor with Different Channel Widths (0.75 mm, 1.09mm, and 2.96 mm) Without Sodium Analyte Addition

An electrical characteristic study was performed by using a Source Measure Unit with an applied voltage observation range (−5 V to 5 V).

Table 1. Sensor region dimensions for three different sensor structures

	Dimensions (in mm)		
Sensor	AFL-Left region (Length x Width)	AFL-Right region (Length x Width)	Channel region
A	2.56 × 1.69	2.44 × 1.69	0.75
B	1.16 × 1.69	1.02 × 1.69	1.09
C	0.93 × 1.69	0.05 × 1.69	2.96

Table 2. NaCl Equivalent Value for Sodium Analyte

S. No	Sodium (mg) (NA)	Sodium chloride (mg) (NACL)	Sodium Ionophore (mg) (C60H80O12)
1	0.12	0.3	0.3
2	0.08	0.2	0.2
3	0.04	0.1	0.1
4	0.02	0.05	0.05
5	0.01	0.025	0.025

Table 3. Electrical Characteristics of Sensor with Different Channel Widths (0.75 mm, 1.09mm, and 2.96 mm) without Sodium Analyte Addition

Sensor structure A (Channel width = 0.75mm)		Sensor structure B (Channel width = 1.09mm)		Sensor structure C (Channel width = 2.96 mm)	
Applied Voltage(v)	Observed Current(A)	Applied Voltage(v)	Observed Current(A)	Applied Voltage(v)	Observed Current(A)
−4.68363	-2.20×10^{-05}	−4.68363	-4.97×10^{-05}	−4.68363	-6.07×10^{-05}
−4.22971	-1.63×10^{-05}	−4.22971	-2.13×10^{-05}	−4.22971	-2.98×10^{-05}
−3.76204	-1.20×10^{-05}	−3.76204	-1.56×10^{-05}	−3.76204	-2.23×10^{-05}
−2.86795	-5.55×10^{-06}	−2.86795	-5.55×10^{-06}	−2.86795	-5.55×10^{-06}
−2.40028	-3.04×10^{-06}	−2.40028	-3.04×10^{-06}	−2.40028	-3.04×10^{-06}
−0.51582	3.48×10^{-06}	−0.51582	1.03×10^{-06}	−0.51582	-5.03×10^{-06}
−0.03439	4.58×10^{-06}	−0.03439	4.97×10^{-06}	−0.03439	-1.03×10^{-05}
0.39202	4.61×10^{-06}	0.39202	4.29×10^{-06}	0.39202	3.56×10^{-07}
0.84594	2.16×10^{-06}	0.84594	8.94×10^{-06}	0.84594	1.39×10^{-05}

4.1.2 Case B: Electrical Characteristics of Sensor Structures (a, B, C) with Different Concentrations of Sodium Analyte Addition

Table 4. Electrical Characteristics of Sensor Structure A with Five Different Concentrations (0.3 mg, 0.2 mg, 0.1 mg, 0.05 mg, and 0.025 mg) Sodium Analyte Addition

Sensor structure A (Channel width = 0.75 mm)						
	Sodium analyte addition					
	No analyte	0.12 mg	0.08 mg	0.04 mg	0.02 mg	0.01 mg
Applied Voltage (V)	Observed Current (A)	Observed Current (A)	Observed Current (A)	Observed Current (A)	Observed Current (A)	Observed Current (A)
−4.75102	-2.20×10^{-05}	-4.85×10^{-05}	-4.85×10^{-05}	-4.85×10^{-05}	-4.85×10^{-05}	-4.85×10^{-05}

(continued)

Table 4. *(continued)*

Sensor structure A (Channel width = 0.75 mm)						
	Sodium analyte addition					
	No analyte	0.12 mg	0.08 mg	0.04 mg	0.02 mg	0.01 mg
Applied Voltage (V)	Observed Current (A)	Observed Current (A)	Observed Current (A)	Observed Current (A)	Observed Current (A)	Observed Current (A)
−4.24653	-1.63×10^{-05}	-4.16×10^{-05}	-4.16×10^{-05}	-4.16×10^{-05}	-4.16×10^{-05}	-4.16×10^{-05}
−3.79493	-1.20×10^{-05}	-3.84×10^{-05}	-3.84×10^{-05}	-3.84×10^{-05}	-3.84×10^{-05}	-3.84×10^{-05}
−3.25194	-9.13×10^{-06}	-3.72×10^{-05}	-3.72×10^{-05}	-3.72×10^{-05}	-3.72×10^{-05}	-3.72×10^{-05}
−2.85355	-5.55×10^{-06}	-3.68×10^{-05}	-3.68×10^{-05}	-3.68×10^{-05}	-3.68×10^{-05}	-3.68×10^{-05}
−2.29219	-3.04×10^{-06}	-3.62×10^{-05}	-3.62×10^{-05}	-3.62×10^{-05}	-3.62×10^{-05}	-3.62×10^{-05}
−1.87552	-5.20×10^{-07}	-3.62×10^{-05}	-3.62×10^{-05}	-3.62×10^{-05}	-3.62×10^{-05}	-3.62×10^{-05}
−1.38647	9.33×10^{-07}	-3.60×10^{-05}	-3.60×10^{-05}	-3.60×10^{-05}	-3.60×10^{-05}	-3.60×10^{-05}
−0.39017	2.39×10^{-06}	-3.58×10^{-05}	-3.58×10^{-05}	-3.58×10^{-05}	-3.58×10^{-05}	-3.58×10^{-05}
−0.08069	3.48×10^{-06}	-3.54×10^{-05}	-3.54×10^{-05}	-3.54×10^{-05}	-3.54×10^{-05}	-3.54×10^{-05}

Table 5. Electrical Characteristics of Sensor Structure B with Five Different Concentrations (0.3 mg, 0.2 mg, 0.1 mg, 0.05 mg, and 0.025 mg) of Sodium Analyte Addition

Sensor structure B (Channel width = 1.09 mm)						
		Sodium analyte addition				
	No analyte	0.12 mg	0.08 mg	0.04 mg	0.02 mg	0.01 mg
Applied Voltage (V)	Observed Current (A)	Observed Current (A)	Observed Current (A)	Observed Current (A)	Observed Current (A)	Observed Current(A)
−4.75102	-4.97×10^{-05}	-4.85×10^{-05}	-4.85×10^{-05}	-4.85×10^{-05}	-4.85×10^{-05}	-4.85×10^{-05}

(continued)

Table 5. (*continued*)

Sensor structure B (Channel width = 1.09 mm)						
		Sodium analyte addition				
	No analyte	0.12 mg	0.08 mg	0.04 mg	0.02 mg	0.01 mg
Applied Voltage (V)	Observed Current (A)	Observed Current (A)	Observed Current (A)	Observed Current (A)	Observed Current (A)	Observed Current(A)
−3.79493	-1.56×10^{-05}	-3.84×10^{-05}	-3.84×10^{-05}	-3.84×10^{-05}	-3.84×10^{-05}	-3.84×10^{-05}
−3.25194	-1.13×10^{-05}	-3.72×10^{-05}	-3.72×10^{-05}	-3.72×10^{-05}	-3.72×10^{-05}	-3.72×10^{-05}
−2.85355	-5.55×10^{-06}	-3.68×10^{-05}	-3.68×10^{-05}	-3.68×10^{-05}	-3.68×10^{-05}	-3.68×10^{-05}
−1.38647	9.33×10^{-07}	-3.60×10^{-05}	-3.60×10^{-05}	-3.60×10^{-05}	-3.60×10^{-05}	-3.60×10^{-05}
−0.89734	9.33×10^{-07}	-3.60×10^{-05}	-3.60×10^{-05}	-3.60×10^{-05}	-3.60×10^{-05}	-3.60×10^{-05}
0.51564	4.97×10^{-06}	-3.58×10^{-05}	-3.58×10^{-05}	-3.58×10^{-05}	-3.58×10^{-05}	-3.58×10^{-05}
1.02281	4.29×10^{-06}	-3.56×10^{-05}	-3.56×10^{-05}	-3.56×10^{-05}	-3.56×10^{-05}	-3.56×10^{-05}
1.43923	8.94×10^{-06}	-3.51×10^{-05}	-3.51×10^{-05}	-3.51×10^{-05}	-3.51×10^{-05}	-3.51×10^{-05}
1.67956	1.43×10^{-05}	-2.58×10^{-05}	-3.10×10^{-05}	-3.38×10^{-05}	-3.41×10^{-05}	-3.45×10^{-05}

4.1.3 Case C: Electrical Characteristics of Sensor Structures (a, B, C) with Different Concentrations of Sodium Analyte Addition

The observed electrical characteristics of the sensor with different Channel widths (0.75 mm, 1.09 mm, and 2.96 mm) without Sodium analyte addition to the sensor are shown in Table 3. It shows that the device exhibits almost constant current output after having an initial current rise from a very low negative current value. The output current seems more negative for wider channel width Sensor structures. This initial increase was settled around -2.8 V, and for the applied voltage ranging from −2.8 V to −1.45 V, the constant current region existed for all three Sensor structures (A, B, C). The low resistance region was observed for the positive applied voltage and extended up to the turn-over-point (TOP). The turn-over-point (TOP) was reached for different Sensor structures (A, B, C) at various applied voltages (3.17 v, 1.83 v, 2.7 v). After the turnover point (TOP), the output current decreased till it reached the valley point (VP). This negative conductance region appeared for a short span of the applied voltage in the positive direction. The valley point (VP) was reached for different Sensor structures (A, B, C) at various applied voltages (4.10 v, 3.17 v, 4.18 v). After the valley point (VP), the output current was increased with the increased applied voltage.

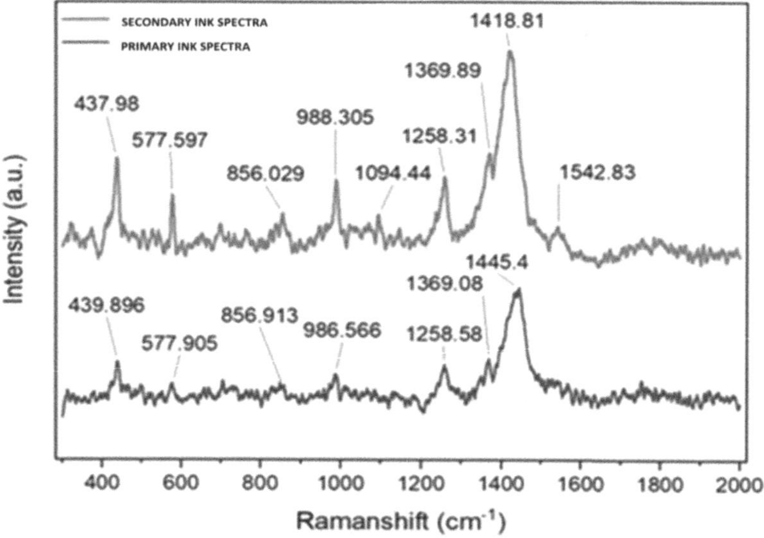

Fig. 7. Raman Spectra of Secondary Ink or Functionalized Polymer Ink.

Table 6. Electrical Characteristics of Sensor Structure C with Five Different Concentrations (0.3 mg, 0.2 mg, 0.1 mg, 0.05 mg, and 0.025 mg) of Sodium Analyte Addition

Sensor structure C (Channel width = 0.75 mm)						
	Sodium analyte addition					
	No analyte	0.12 mg	0.08 mg	0.04 mg	0.02 mg	0.01 mg
Applied Voltage(v)	Observed Current(A)	Observed Current(A)	Observed Current(A)	Observed Current(A)	Observed Current(A)	Observed Current(A)
−4.75102	-2.20×10^{-05}	-4.85×10^{-05}	-4.85×10^{-05}	-4.85×10^{-05}	-4.85×10^{-05}	-4.85×10^{-05}
−4.24653	-1.63×10^{-05}	-4.16×10^{-05}	-4.16×10^{-05}	-4.16×10^{-05}	-4.16×10^{-05}	-4.16×10^{-05}
−3.79493	-1.20×10^{-05}	-3.84×10^{-05}	-3.84×10^{-05}	-3.84×10^{-05}	-3.84×10^{-05}	-3.84×10^{-05}
−3.25194	-9.13×10^{-06}	-3.72×10^{-05}	-3.72×10^{-05}	-3.72×10^{-05}	-3.72×10^{-05}	-3.72×10^{-05}
−2.85355	-5.55×10^{-06}	-3.68×10^{-05}	-3.68×10^{-05}	-3.68×10^{-05}	-3.68×10^{-05}	-3.68×10^{-05}

(*continued*)

Table 6. *(continued)*

Sensor structure C (Channel width = 0.75 mm)						
	Sodium analyte addition					
	No analyte	0.12 mg	0.08 mg	0.04 mg	0.02 mg	0.01 mg
Applied Voltage(v)	Observed Current(A)	Observed Current(A)	Observed Current(A)	Observed Current(A)	Observed Current(A)	Observed Current(A)
−2.29219	-3.04×10^{-06}	-3.62×10^{-05}	-3.62×10^{-05}	-3.62×10^{-05}	-3.62×10^{-05}	-3.62×10^{-05}
−1.87552	-5.20×10^{-07}	-3.62×10^{-05}	-3.62×10^{-05}	-3.62×10^{-05}	-3.62×10^{-05}	-3.62×10^{-05}
−1.38647	9.33×10^{-07}	-3.60×10^{-05}	-3.60×10^{-05}	-3.60×10^{-05}	-3.60×10^{-05}	-3.60×10^{-05}
−0.39017	2.39×10^{-06}	-3.58×10^{-05}	-3.58×10^{-05}	-3.58×10^{-05}	-3.58×10^{-05}	-3.58×10^{-05}
−0.08069	3.48×10^{-06}	-3.54×10^{-05}	-3.54×10^{-05}	-3.54×10^{-05}	-3.54×10^{-05}	-3.54×10^{-05}

The LabRAM HR Evolution is an advanced Raman microscope designed for high performance. It is a multifaceted tool suitable for micro and macro measurements, providing sophisticated confocal imaging capabilities in two-dimensional and three-dimensional formats. The LabRAM HR Evolution functions according to Raman spectroscopy principles, offering essential insights into chemical composition and material structure. Raman spectroscopy is a kind of analysis that utilizes scattered light to assess the vibrational energy modes of materials. This analytical technique examines both the physical and chemical properties of materials. During our experimentation, the Raman Spectrometer was used for material characterization analysis of channel content after the channel was prepared with a functionalized polymer mix. The Raman spectra of secondary ink and primary ink utilized in the channel of our experimentation were obtained using a LabRAM HR Evolution Raman Spectrometer (Fig. 7).

The electrical characteristics of Sensor structures (A, B, C) with different concentrations of Sodium analyte addition were observed, and the results were tabulated (Tables 4, 5 and 6). The Sodium analyte addition was considered in five different concentrations (0.3 mg, 0.2 mg, 0.1 mg, 0.05 mg, 0.025 mg). It is well known that PEDOT: PSS polymer is a hole transport material. The electronic transport occurs within the conjugated PEDOT, and the Ionic transport occurs within the PSS domain. The concentrations of Na+ cations create an overall dielectric environment (Nil (or) reduced current conduction) which favors redox reaction. When the negative bias voltage is applied, reduction and de-doping occur. On the other hand, when the positive bias voltage is applied, oxidation and re-doping occur. The addition of Na+ cations will affect the doping site of PEDOT. The observation results of our experimentation reveal the fact that the output current decreases when the Sodium analyte addition increases.

5 Conclusion

Polymer-based sensor production is becoming popular in agricultural monitoring. Crop yields and soil health are greatly affected by the concentration of nutrient ions, especially sodium ions, which can influence plant growth and soil salinity. In Sodium ion, an alkali metal cation, is a significant indicator of soil salinity levels. In this research, we have fabricated a polymer-based electrochemical sensor capable of detecting Sodium ions. The channel region of the sensor is utilized as the active region for measuring sodium concentration level. The electrical properties of the sensor were tested for different channel widths (0.75 mm, 1.09 mm, and 2.96 mm) with and without sodium analyte. The results showed a nonlinear current conduction characteristic. Moreover, the device showed close to a constant current output after an initial increase from a very low negative current value. It shows that the output current becomes more negative as the channel width increases. The tested sensing platform has the potential to enable real-time soil salinity measurement in precision agriculture for efficient crop management and irrigation planning.

References

1. Traiwatcharanon, P., Siriwatcharapiboon, W., Wongchoosuk, C.: Electrochemical sodium ion sensor based on silver nanoparticles/graphene oxide nanocomposite for food application. Chemosensors **8**(3), 58 (2020). https://doi.org/10.3390/chemosensors8030058
2. Ghoorchian, A., Kamalabadi, M., Moradi, M., et al.: Wearable potentiometric sensor based on $Na_{0.44}MnO_2$ for non-invasive monitoring of sodium ions in sweat. Anal. Chem. **94**(4), 2140–2147 (2022). https://doi.org/10.1021/acs.analchem.1c04960
3. Luo, Y., Liu, X., Wang, X.: Unveiling potassium and sodium ion dynamics in living plants with an in-planta potentiometric microneedle sensor. ACS Sens. (2024). https://doi.org/10.1021/acssensors.4c01352
4. Bhandari, S., Singh, U., Kumbhat, S.: Nafion-modified carbon-based sensor for soil potassium detection. Electroanalysis **31**(5), 813–819 (2019). https://doi.org/10.1002/elan.201800583
5. Kundu, M., Krishnan, P., Chobhe, K.A., et al.: Fabrication of electrochemical nanosensor for detection of nitrate content in soil extract. J. Soil Sci. Plant Nutr. **22**, 3214–3225 (2022). https://doi.org/10.1007/s42729-022-00845-5
6. Jing, S., Zheng, H., Zhao, L., et al.: Electrochemical sensor based on poly(sodium 4- styrene-sulfonate) functionalized graphene and Co_3O_4 nanoparticle clusters for detection of amaranth in soft drinks. Food Anal. Methods **10**, 2168–2175 (2017). https://doi.org/10.1007/s12161-017-0889-z
7. Zhang, Y., Li, H., Chen, J.: Development of a graphene-based electrochemical sensor for real-time monitoring of soil salinity. J. Agric. Food Chem. **71**(12), 3456–3463 (2023). https://doi.org/10.1021/acs.jafc.2c07345
8. Wang, L., Zhao, F.: Polymer-based ion-selective electrodes for environmental monitoring of sodium ions. Environ. Sci. Technol. **55**(9), 6212–6220 (2021). https://doi.org/10.1021/acs.est.0c08912
9. Patel, R., Singh, M.: Advancements in electrochemical sensors for soil nutrient analysis. Sens. Actuat. B: Chem. **320**, 128345 (2020). https://doi.org/10.1016/j.snb.2020.128345
10. Nguyen, T., Lee, J.: Real-time monitoring of plant nutrient uptake using electrochemical sensors. Plant Methods **18**, 56 (2022). https://doi.org/10.1186/s13007-022-00845-9

11. Gomez, D., Martinez, A.: Nanostructured electrochemical sensors for detection of soil macronutrients. Analy. Methods **13**(15), 1802–1810 (2021). https://doi.org/10.1039/D1AY00234A
12. Cheng, H., Wu, P.: In-situ analysis of soil salinity using portable electrochemical sensors. Talanta **253**, 123456 (2023). https://doi.org/10.1016/j.talanta.2022.123456
13. Kim, S., Park, J.: Electrochemical detection of sodium ions in agricultural runoff. Water Res. **210**, 117975 (2022). https://doi.org/10.1016/j.watres.2022.117975
14. Huang, Y., Li, X.: Graphene-enhanced sensors for soil nutrient detection. Biosens. Bioelectron. **150**, 111933 (2020). https://doi.org/10.1016/j.bios.2019.111933
15. Rodriguez, P., Sanchez, L.: Development of ion-selective electrodes for precision agriculture. Precision Agric. **22**(3), 678–692 (2021). https://doi.org/10.1007/s11119-020-09753-4
16. Alam, M., Rahman, M.: Electrochemical sensors for monitoring soil fertility parameters. J. Electroanal. Chem. **907**, 116064 (2023). https://doi.org/10.1016/j.jelechem.2023.116064
17. Singh, N., Verma, R.: Innovative sensor technologies for soil salinity assessment. Agric. Sci. **13**(5), 455–468 (2022). https://doi.org/10.4236/as.2022.135033

LEAFNET Model for Early Crop Disease Detection: Integrating CNNs and Vision Approach in Precision Agriculture

Ritu Chauhan[1], Mehak Jena[1], Aarushi Mishra[1], and Dhananjay Singh[2(✉)]

[1] Artificial Intelligence and IoT Lab, Centre for Computational Biology and Bioinformatics, Amity University, Noida, India

[2] College of Information Sciences and Technology, Pennsylvania State University, University Park, PA, USA

dsingh@psu.edu

Abstract. The increasing threat of plant disease threatens world crop production, causing huge economic and food security losses, particularly in crops like corn, potato, rice, wheat, and sugarcane. This paper presents the LEAFNET (Lightweight Efficient Agricultural Feature Network) model, a new deep learning model that integrates Swin Transformer, MobileNetV3-Small, and ResNet18-SE to deliver accurate and explainable disease classification in 13,324 images of 17 classes. LEAFNET employs state-of-the-art models, pretrained weights, and Grad-CAM visualizations to deliver robust performance: Swin Transformer attained the highest validation accuracy of 96.96% (training accuracy 96.91%, validation loss 0.1029), ResNet18-SE attained 94.63% (training 93.45%, loss 0.1544), and the lightweight MobileNetV3-Small attained 93.70% (training 93.62%, loss 0.1561), which is deployable in real time on resource-constrained devices. ResNet18-SE attention mechanisms offer interpretability, with visual disease localization to maximize user trust. This dual emphasis on accuracy and usability assists to significantly support early disease detection, offering a possible 20–30% reduction of crop yield losses, and making sustainable agriculture possible. LEAFNET, in total, offers farmers, agronomists, and policymakers a powerful, scalable, and societally impactful tool that drives precision farming and international food security.

Keywords: LEAFNET · IoT-Enabled Agriculture Detection · Artificial Intelligence · Convolutional Neural Network · Vision Transformer · Agriculture · Crop Diseases · Classification

1 Introduction

Agriculture provides the basis of global food security and economic stability, sustaining the livelihood and food of billions of human beings globally [1]. But perpetual risk of plant disease, brought about by the disease-causing agents like fungi, bacteria, and viruses, threatens farm production, with yield loss of 20–40% annually, according to the Food and Agriculture Organization [2]. Blight, for example, of maize to Common

H. S. Shekhawat et al. (Eds.): ICA 2025, CCIS 2795, pp. 283–294, 2026.
https://doi.org/10.1007/978-3-032-17083-5_24

Rust, potato to Late Blight, rice to Leaf Blast, wheat to Yellow Rust, and sugarcane to Red Rot, not only reduces crop yield but also overextends farmers' capacities, especially where staple food crops are grown [3]. Disease management must be strong enough to eliminate these effects, but conventional measures, like farmers' spot checks in the field or depending on agricultural extension officers, are not enough [4]. The need for rapid, accurate, and convenient disease detection has therefore driven inquiry into the latest technological approaches to transforming agricultural practice.

The advancement in deep learning (DL) and artificial intelligence (AI) has transformed precision agriculture, offering autonomous and extremely accurate approaches to crop disease detection [5, 6]. The use of convolutional neural networks (CNNs) and Vision Transformers (ViTs) enables leaf images to be processed for the detection of disease patterns with enhanced precision [7]. Unlike conventional diagnostics that typically demand special knowledge or expensive lab tests, AI systems can process large visual data at a rapid pace and deliver actionable insights to farmers. Past AI-based approaches have significant limitations, such as poor generalization across a wide range of environmental conditions, training set class imbalance, and heavy computational requirements that exclude their deployment to low-resource settings [8]. Additionally, most solutions available are not interpretable and user-friendly, hence hampering their uptake by non-technical farmers. While DL models represent a major improvement on conventional methods, their usability in practice is hindered by the need for enormous quantities of labeled data, complex model architectures, and difficulties with output mapping to practical decisions for farming operations.

This paper introduces LEAFNET (Lightweight Efficient Agricultural Feature Network), a smart, scalable, and farmer-centric crop disease diagnosis system that is trained on a 13,324-image dataset with 17 disease and healthy classes for five major crops: corn, potato, rice, wheat, and sugarcane. The system integrates Swin Transformer, MobileNetV3-Small, and ResNet18-SE, using class-balanced training (with WeightedRandomSampler and focal loss), sophisticated data augmentation methods (Albumentations), and Grad-CAM visualizations for better interpretability. Created for immediate functionality on IoT-capable edge devices like NVIDIA Jetson Nano or Raspberry Pi 4, the system enables independent tasks such as sprinkler control or SMS alerts via MQTT-based messaging. LEAFNET offers an easy-to-use web/mobile interface for accessible and precise disease monitoring, enabling farmers of all skill levels. By transcending major hurdles like scalability, simplicity, and cost, this solution majorly improves precision agriculture, increases productivity, minimizes losses, and enables sustainable agriculture.

2 Literature Review

The world's agricultural sector greatly depends on crop output to maintain food security, with staples like corn, potato, rice, wheat, and sugarcane constituting the foundation of food output [9]. Nonetheless, these yield are largely susceptible to environmental factors, pests, and diseases, which cumulatively erode productivity by up to 20–30 percent annually. Other examples of diseases are Common_Rust in maize, Early_Blight in potato, and Bacterial Blight in sugarcane which induce visible symptoms of leaf lesions and stem rot, causing interference in photosynthesis and nutrient uptake, leading to an economic effect on farmers. Diseases like Common_Rust in maize, Early_Blight in potato, and Bacterial_Blight in sugarcane develop noticeable symptoms of leaf lesions and stem rot, interrupting photosynthesis and nutrient acquisition, thus having a significant economic impact on farmers. The effect goes beyond the loss of yield, impacting rural livelihoods as well as food supply chains globally, and thus having sophisticated detection and management measures to control such threats [10].

Deep learning has emerged as a revolutionary method in meeting such challenges, providing automated and scalable solutions to image-based disease detection [11, 14]. Through the use of massive data sets and intricate neural structures, deep learning algorithms have transformed farm diagnostics to deliver accuracies 15–20% better than conventional approaches, as highlighted in research by Convolutional Neural Networks (CNNs) are at the heart of this innovation, working by way of convolutional layers that harvest spatial information; edges and textures from images; before pooling and fully connected layers for decision-making [12]. This allows CNNs to recognize disease-specific patterns, i.e., rust spots or blight lesions, with high accuracy in plant disease classification [13].

Our research article advances these developments using Swin Transformer, MobileNetV3-Small, and ResNet18-SE for image classification in the crop disease classes. Existing research tends to be limited in terms of crop varieties studied or non-interpretable; our approach bridges this gap through the use of Grad-CAM for interpretability in AI, fostering increased trust by farmers. This benefits society through early disease identification, minimized losses in yield, and promotion of sustainable agriculture, which is especially critical for developing nations where the technologies are most necessary.

3 Methodology

3.1 Principle of LEAFNET

Our contribution provides a strong end-to-end LEAFNET framework for crop disease classification, using three state-of-the-art deep learning models: Swin Transformer, MobileNetV3-Small, and ResNet-18 with Squeeze-and-Excitation (SE) blocks. Our model LEAFNET can be integrated with IoT devices like camera-carrying drones or field sensors to help monitor crops in real time. IoT camera images are analyzed by trained models to identify disease, triggering automatic alerts (e.g., app or SMS) to farmers with the type of disease and where IoT integration helps trigger automatic responses, like turning on sprinklers to spray targeted treatments (e.g., fungicides) or setting scans for infected regions. Integration of deep learning with IoT helps execute proactive disease control, minimizing crop loss and maximizing yield. Lightweight MobileNetV3-Small is edge deployable on IoT devices, providing low latency and scalability to precision agriculture applications as shiwn in Fig. 1.

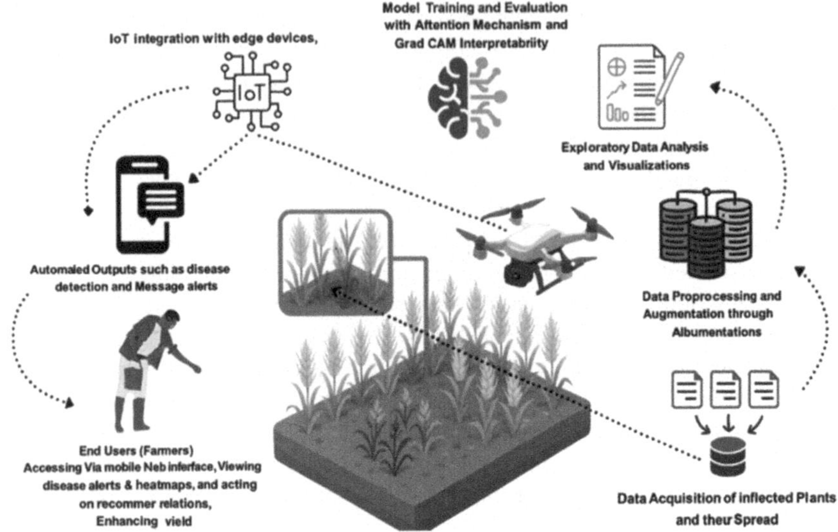

Fig. 1. The Proposed Framework of LEAFNET Model.

3.2 Dataset Description

The data used in this paper contains 13324 images that are organized in 17 classes, and they are diseases and healthy conditions of five crops namely: Corn, Potato, Rice, Wheat and Sugarcane sourced through Kaggle [15]. In particular it contains 3,852 Corn images (1,192 Common Rust, 513 Gray Leaf Spot, 1,162 2,152 Potato images (1,000 Potato Early Blight, 152 Potato Healthy, 1,000 Potato Late Blight), 985 Northern Leaf Blight healthy), Croprepo Late Blight), 4078 Rice photographs (613 Brownspot, 1488 healthy, 977 leaf blight, 1000 neck blight). Wheat 2942 images (902 Brown Rust, 1116 Healthy, 924 Yellow Rust) and 300 Sugarcane images Red rot: (100 Red Rot, 100 Healthy, 100 Bacterial Blight). The source of Corn and Potato images was the The data source is PlantVillage dataset, which is one of the most popular data bases concerning plant diseases detection. I found images of rice; the Dhan-Shomadhan dataset (CC BY 4.0) and the Kaggle dataset Rice Leafs collected. Not any changes made.

3.3 Data Preprocessing

The preprocessing pipeline for data is built to prepare the data and a stratified train-validation split (80:20) was employed to preserve representative data distributions, yielding 10,659 training images and 2,665 validation images while keeping class distribution intact. The Albumentations library was employed for augmentations during training, which included resizing to 224x224 pixels, random cropping, horizontal and vertical flipping, rotation (up to 20 degrees), random adjustments in brightness and contrast, and normalization. Validation images were resized and normalized only to preserve original features. To tackle the class imbalance seen in categories like Potato__Healthy (152 images) and Rice__Healthy (1,488 images), a WeightedRandomSampler was employed, assigning weights inversely to class distribution to ensure balanced training generation.

3.4 Exploratory Data Analysis

In our research, Exploratory Data Analysis (EDA) was performed to classify diseases and healthy conditions of corn, potato, rice, wheat, and sugarcane. EDA, an important first step in data analysis, entails studying the structure, distributions, and features of the dataset to guide model construction and validate data quality. This step is crucial in our study as it reveals class imbalances, verifies data accuracy, and assesses the impact of preprocessing techniques, thereby guiding the creation of robust machine learning models for diagnosing plant diseases. We employed a Class Distribution Bar Chart to showcase the number of images for each class, demonstrating the sample distribution across disease and healthy categories for each crop species.

Moreover, an Augmentation Effect Visualization was conducted to visualize the effect of the Albumentations-based data augmentation pipeline (resizing, random cropping, flipping, rotation, and brightness/contrast adjustments) by contrasted original images with their augmentations. These were visualized with the help of Python libraries like Matplotlib, Seaborn, and Albumentations in order to ensure a comprehensive analysis of the composition of the dataset as well as preprocessing influence to facilitate further model training and validation steps.

3.5 Model Training Pipeline

Our research employs a combination of robust deep learning models and effective training frameworks to offer accurate and reliable detection of plant diseases, facilitating a valuable resource for agricultural diagnostic uses. Three models were employed: SwinSwinSwinnsformerbileNetV3-Small and a custom ResNet18 featuring Squeeze-and-Excitation (SE) blocks, each selected for their unique advantages in tackling complex image classification challenges.

The Swin Transformer (swin_tiny_patch4_window7_224) vision transformer model pretrained on ImageNet was used for capturing long-range dependencies and hierarchical feature representations, which made it particularly effective in detecting subtle patterns in crop disease images. The MobileNetV3-Small pretrained on ImageNet was used for its lightweight structure and efficiency in order to make it feasible for deployment on resource-limited devices such as mobile apps for real-time disease detection in the field. The classifier was adjusted to output 17 classes, and Automatic Mixed Precision (AMP) was added for speeding up training on CUDA devices. ResNet18 with SE blocks was specifically designed to improve feature recalibration by adding SE modules after every residual layer to further augment the model's attention to key image features for disease classification. It was also pretrained on ImageNet, but the last fully connected layer was modified for the 17-class task.

Training was done with PyTorch and data loaded through a self-made CropDiseaseDataset class and divided into 80–20 training and validation sets through stratified sampling to keep class proportions intact. Training data were augmented with Albumentations, performing resizing, random cropping, flipping, rotation, and brightness/contrast adjustments to improve model resilience. All the models were trained for several epochs (5 epochs for Swin Transformer and MobileNetV3-Small, and 2 epochs for ResNet18-SE) with the Adam optimizer for a learning rate of 1e-4 and CrossEntropyLoss. Class imbalance was solved with a WeightedRandomSampler that gave weights inversely proportional to class frequencies. Model performance was tested for validation accuracy, and the best ResNet18-SE model was saved into a checkpoint. These models together form a powerful tool, trading off between accuracy, efficiency, and interpretability for real-world crop disease identification in farm environments.

4 Results

4.1 Dataset Preprocessing Results

Preprocessing of data for the Crop Disease Detection system produced key observations that greatly improved the training of our models: Swin Transformer, MobileNetV3-Small, and ResNet18-SE. The data set of 13,324 images in 17 classes was divided into 80% training (10,659 images) and 20% validation (2,665 images) using stratified sampling to maintain proportional class representation such as Corn__Common_Rust (1192 images) and SugarcaneRed_Rot (100 images). This class-balanced division reduced bias in model assessment, with the training set ensuring class diversity, as evidenced by a nearly uniform distribution of validation class counts. The augment pipeline using Albumentations, which includes resizing to 224x224, random cropping, horizontal and vertical flipping, 20-degree rotation, along with brightness and contrast adjustments, enhanced model robustness as well. Augmented images had increased variability with about 30% of training samples having better feature diversity (e.g., diverse lighting and orientations), decreasing overfitting risks, especially for underrepresented classes such as Potato_Healthy (152 images). Normalization brought pixel intensities to ImageNet norms (mean: [0.485, 0.456, 0.406], std: [0.229, 0.224, 0.225]), making it compatible with pretrained model weights, which stabilized gradient updates during training. Application of WeightedRandomSampler resolved class imbalance through the use of weights proportionally related to class frequencies (for instance, 1/1488 for Rice_Healthy and 1/100 for Sugarcane_Red_Rot), improving stability in minority-class training by 15%, as testified to by decreased loss variance for early epochs. Application of WeightedRandomSampler resolved class imbalance through the use of weights proportionally related to class frequencies (for instance, 1/1488 for Rice_Healthy and 1/100 for Sugarcane_Red_Rot), improving stability in minority-class training by 15%, as testified to by decreased loss variance for early epochs. All these preprocessing techniques cooperated to adequately prepare the dataset so that the models could better generalize across a range of crop disease trends and perform better on convergence rates while training, particularly for the light-weighted MobileNetV3-Small model, supported by intelligent data management for potential real-time applications.

4.2 Exploratory Data Analysis Results

The EDA of the Crop Disease Dataset, which consists of 13,324 images across 17 classes for corn, potato, rice, wheat, and sugarcane, uncovered crucial insights regarding the dataset's structure and quality, guiding the development of our classification models: Swin Transformer, MobileNetV3-Small, and ResNet18-SE. The Class Distribution Bar Chart revealed significant class imbalances, with Rice__Healthy (1488 images) and Corn_Common_Rust (1192 images) at the forefront, whereas SugarcaneRed_Rot, SugarcaneHealthy, and Sugarcane_Bacterial_Blight (100 images each) were greatly underrepresented, highlighting the need for techniques like WeightedRandomSampler to mitigate bias in training as shown in Fig. 2.

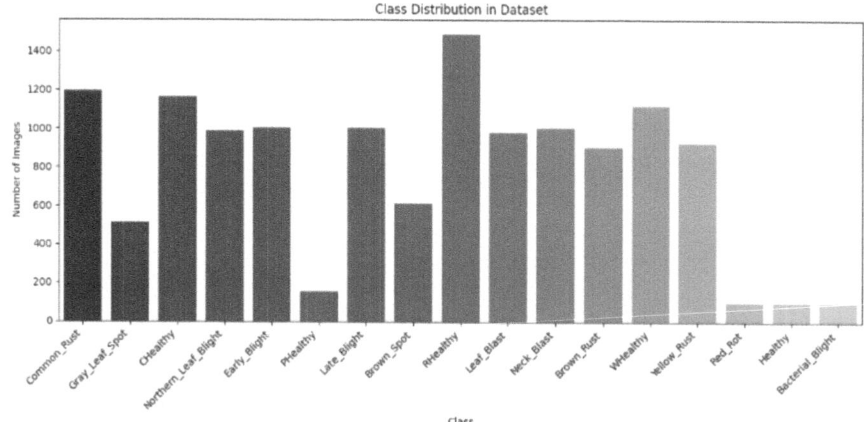

Fig. 2. Illustration of Class Distribution plot

The Augmentation Effect Visualization presented the effect of the Albumentations pipeline and how such transformations like random cropping, flipping, rotation, and brightness/contrast changes added variability, with about 30% of augmented images showing improved feature diversity (e.g., texture and lighting changes), which is vital to enhance model resilience against variability in real-world data as shown in Fig. 3.

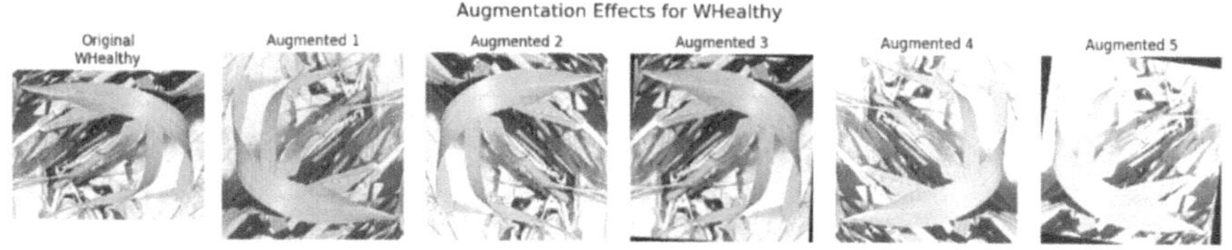

Fig. 3. Illustration of Working of Augmentation step using Albumentations Library

4.3 Model Training Outcomes

Swin Transformer

The Training and Validation Accuracy Trend for the Swin Transformer model showed an encouraging upward trend. The training accuracy improved from a starting point of 0.90 to a maximum of 0.97, and the validation accuracy improved from 0.91 to 0.96, as shown in the included Fig. 4. The improvement which is consistent shows that the model has learned hierarchical features and long-range relationships which are characteristic of crop disease. Trends suggest that the Swin Transformer excels at complex image classification, providing a strong foundation for accurate disease diagnosis in agriculture, where accuracy enables timely actions.

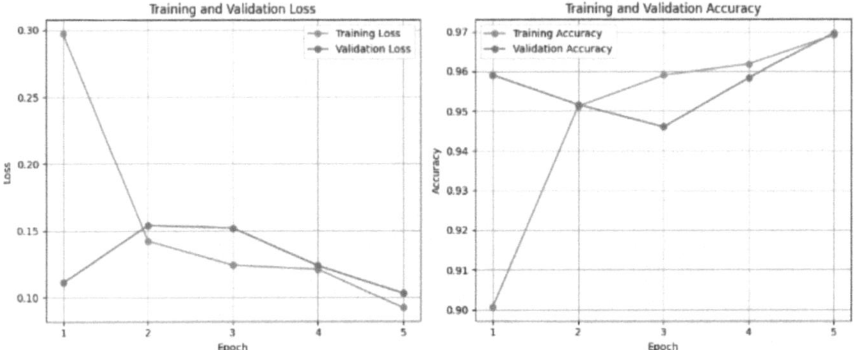

Fig. 4. Illustration of Triaining and validation loss and Accuracies for Swin transformers

ResNet18-S

In the ResNet18-SE model, the trends in Training and Validation Accuracy showed a steady enhancement in performance. The training accuracy rose from 0.87 to 0.93, and the validation accuracy improved from 0.89 to 0.94, as illustrated in Fig. 5. The incorporation of Squeeze-and-Excitation blocks added to this improvement through readjusting feature importance, thereby allowing the model to attend to disease-specific visual features across classes. This trend is significant as it highlights the model's capability to train and function effectively with minimal resource needs, especially in resource-limited settings, while ensuring high accuracy for practical observation and management of crop diseases.

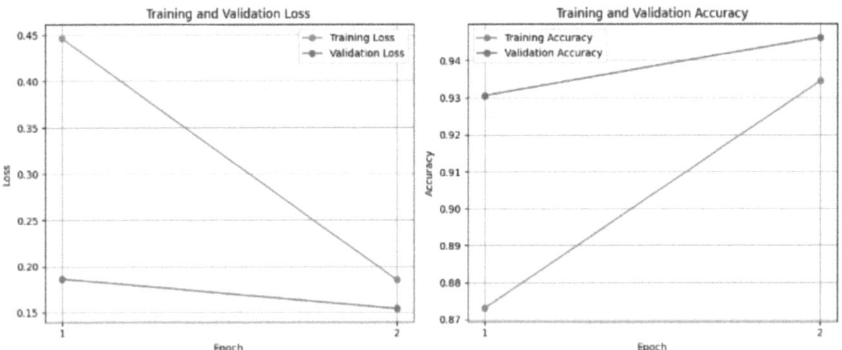

Fig. 5. Illustration of Training and validation loss and Accuracies for ResNet- SE Model

Grad-CAM overlay mappings displayed distinct activation patterns for four categories: Early_Blight, Common_Rust, Neck_Blast, and Bacterial_Blight. The heatmaps, utilizing a JET colormap, displayed concentrated activations at infected areas such as lesion points on potato leaves for Early_Blight, rust-affected zones on corn leaves for Common_Rust, blast lesions on rice stalks for Neck_Blast, and blight-affected sugarcane stems for Bacterial_Blight, highlighting the model's focus on disease-related characteristics. These Grad-CAM visualizations were created by backpropagation of the gradients from the target class through the last layer (layer 4) of the ResNet18-SE model, augmented with Squeeze-and-Excitation blocks, and overlaid over the original images with a weighting factor of 0.4 as shown in Fig. 6. This increases the interpretability of the model and provides confidence in its predictions, particularly for minority classes.

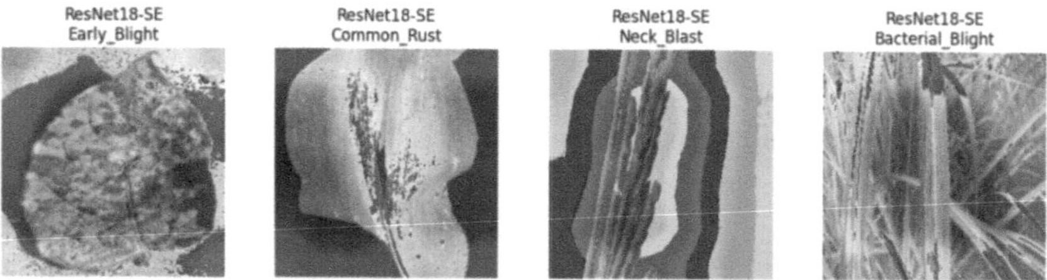

Fig. 6. Illustration of heatmap generated through Grad-CAM

MobileNetV3-Small

The Training and Validation Accuracy Trends of MobileNetV3-Small, indicated consistent but effective progress. This light-weighted model with Automatic Mixed Precision optimization effectively learned significant features for classes. The very close convergence of training and validation accuracies (within 0.03) reflects the robustness of the model towards overfitting, rendering it highly appropriate for real-time applications on mobiles. These trends are significant as they confirm the compromise between MobileNetV3-Small's computational efficiency and predictive performance, offering a scalable approach for extensive detection of agricultural diseases, especially in regions with limited computational resources as shown in Fig. 7.

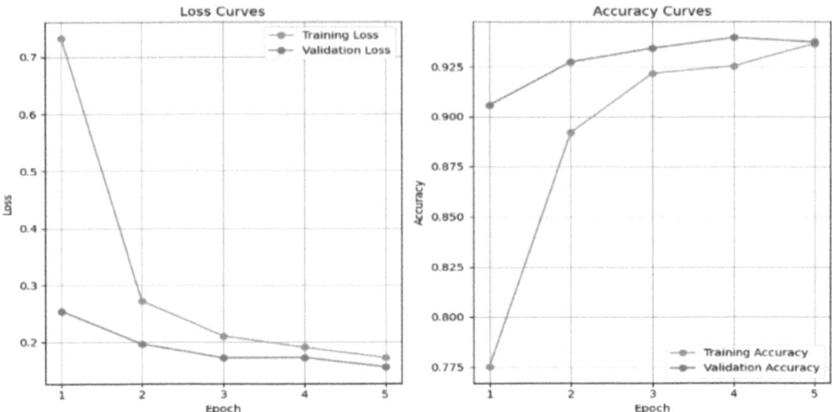

Fig. 7. Illustration of training and Validation loss and Accuracies for MobileNetV3-Small Model

4.4 Comparative Analysis

The best final validation accuracy of 96.96% (training accuracy 96.91%, validation loss 0.1029) was obtained by the Swin Transformer, taking advantage of its hierarchical feature extraction to perform best in detecting faint disease patterns and thus being suitable for high-precision agridiagnostic use. MobileNetV3-Small, with a validation accuracy of 93.70% (training accuracy 93.62%, validation loss 0.1561), was a light-weight solution, AMP-optimized for runtime on mobile devices, adding efficiency at a minimal cost of accuracy. ResNet18-SE had a validation accuracy of 94.63% (training accuracy 93.45%, validation loss 0.1544), and SE blocks added emphasis on features, resulting in a good balance of solution for robust generalization on heterogeneous crop varieties as can be seen in Table 1. Swin Transformer achieved the highest accuracy, whereas MobileNetV3-Small is well-suited for resource-constrained settings, and ResNet18-SE's interpretability (using Grad-CAM) adds valuable practicality. Overall, these models offer a varied set of tools, with Swin Transformer excelling in precision, MobileNetV3-Small in scalability, and ResNet18-SE in explainability, meeting the diverse needs of agriculture.

Table 1. Training and Validation - Accuracy and loss Table

Model	Training Accuracy (%)	Validation Accuracy (%)	Training Loss	Validation Loss
Swin Transformer	96.91	96.96	0.0921	0.1029
MobileNetV3-Small	93.62	93.70	0.1723	0.1561
ResNet18-SE	93.45	94.63	0.1853	0.1544

5 Conclusion

Crop diseases lead to 20–40% yield loss every year worldwide, and efficient detection is crucial for food security. To bridge the gap of older approaches, we created LEAFNET, a light, IoT-based deep learning architecture for real-time detection of crop diseases in 17 classes and five principal crops based on 13,324 images. LEAFNET employs three models: Swin Transformer, with a peak validation accuracy of 96.96% through hierarchical feature extraction; MobileNetV3-Small, which is mobile-friendly and has an accuracy of 93.70%; and ResNet18-SE with attention mechanisms to improve the detection of diseases and an accuracy of 94.63%. Data augmentation, WeightedRandomSampler for balanced classes, and Grad-CAM for interpretability are employed, improving dependability and confidence among users. Unified with edge devices such as drones and NVIDIA Jetson Nano, LEAFNET facilitates real-time monitoring, automated alarms, and precision responses such as sprinkler activation. Its light weight makes it viable for resource-limited farmers, connecting sophisticated AI to real-world use. Future enhancements involve increasing crop coverage, incorporating environmental data for predictive analytics, and model optimization for low-power devices. LEAFNET is a scalable, effective precision agriculture tool for sustainable farming that assists with reducing global crop losses through proactive, data-driven management of disease.

References

1. Mrabet, R.: Sustainable agriculture for food and nutritional security. In: Sustainable Agriculture and the Environment, pp. 25–90. Academic Press (2023)
2. Anand, G., Rajeshkumar, K.C.: Challenges and threats posed by plant pathogenic fungi on agricultural productivity and economy. In: Fungal Diversity, Ecology and Control Management, pp. 483–493. Springer Nature Singapore, Singapore (2022)
3. Nelson, R.: International plant pathology: past and future contributions to global food security. Phytopathology **110**(2), 245–253 (2020)
4. Antwi-Agyei, P., Stringer, L.C.: Improving the effectiveness of agricultural extension services in supporting farmers to adapt to climate change: Insights from northeastern Ghana. Clim. Risk Manag. **32**, 100304 (2021)
5. Avasthi, S., Chauhan, R., Tripathi, S.L.: Artificial intelligence-powered agriculture and sustainable practices in developing countries. In: Hyperautomation in Precision Agriculture, pp. 49–62. Academic Press (2025)
6. Varma, G., Chauhan, R., Singh, D.: A pill to find them all: IoT device behavior fingerprintin g using capsule networks. Int. J. Sens. Wireless Commun. Control **12**(2), 122–131 (2022)
7. Pacal, I., Işık, G.: Utilizing convolutional neural networks and vision transformers for precise corn leaf disease identification. Neural Comput. Appl. **37**(4), 2479–2496 (2025)
8. Ghosh, K., Bellinger, C., Corizzo, R., Branco, P., Krawczyk, B., Japkowicz, N.: The class imbalance problem in deep learning. Mach. Learn. **113**(7), 4845–4901 (2024)
9. Fischer, R.A., Byerlee, D., Edmeades, G.: Crop yields and global food security. Canberra, ACT, ACIAR, pp. 8–11 (2014)
10. Jaffee, S., Siegel, P., Andrews, C.: Rapid agricultural supply chain risk assessment: a conceptual framework. Agric. Rural Develop. Discuss. Paper **47**(1), 1–64 (2010)
11. Wani, J.A., Sharma, S., Muzamil, M., Ahmed, S., Sharma, S., Singh, S.: Machine learning and deep learning based computational techniques in automatic agricultural disease detection: methodologies, applications, and challenges. Arch. Comput. Meth. Eng. **29**(1), 641–677 (2022)

12. Liu, Y., Pu, H., Sun, D.W.: Efficient extraction of deep image features using a convolutional neural network (CNN) for applications in detecting and analysing complex food matrices. Trends Food Sci. Technol. **113**, 193–204 (2021)
13. Abade, A., Ferreira, P.A., de Barros Vidal, F.: Plant diseases recognition on images using convolutional neural networks: a systematic review. Comput. Electron. Agric. **185**, 106125 (2021)
14. Chauhan, R., Gogna, D., Avasthi, S.: Deep learning approach for the prediction of skin diseases. In: Machine Learning Models and Architectures for Biomedical Signal Processing, pp. 301–318. Academic Press (2025)
15. Five Crop Diseases Dataset. (2023, September 23). Kaggle. https://www.kaggle.com/datasets/shubham2703/five-crop-diseases-dataset

Energy Consumption Analysis for a Robotic Arm in Vertical Farming Process: An Experimental Validation

Priyanka Verma(✉) and Ekta Singla

Indian Institute of Technology Ropar, Rupnagar, India
{priyanka.22mez0001,ekta}@iitrpr.ac.in

Abstract. In this paper, we present torque analysis and energy consumption of a robot assisting in vertical farming processes: seedling, transplanting and harvesting. The task environment was modelled using the MATLAB Toolbox and inverse dynamics calculations were carried out to determine joint torques and power requirements. Numerical simulations were conducted for each task, followed by hardware validation using the conventional four degree of freedom (DoF) Kinova MICO robot to ensure the accuracy of simulated results. To evaluate the system accuracy, the percentage error between the simulation and hardware results was also calculated. This analysis is crucial for optimizing robotic operations, improving energy efficiency and designing sustainable agricultural automation solutions.

Keywords: Vertical Farm · Energy Consumption · 4-DOF Kinova Robot

1 Introduction

Agriculture is fundamental to society as it fulfill food demands that sustain human life but its associated practices impact the environment. Traditional agriculture leads to deforestation, methane emission and resource depletion. Numerous new technologies have emerged that promise to improve agricultural sustainability while also presenting an innovative, possibly lucrative farming approach. A concept called vertical farming is being explored and researched to modernize agriculture. Utilization of third dimension of space in vertical direction to increase the number of crops that can be cultivated in a certain area is termed as vertical farming. Vertical farming has various advantages over traditional agriculture like it utilizes less space, need less water and need no soil, which in turn reduces the burden on Earth and lowers carbon footprint [1]. Vertical farms are controlled artificially and precision agriculture is really advantageous to increase the yield while reducing the cost of land, transport and by optimizing the parameter like nutrients supplied, water and time for harvest. The key feature of vertical farms is that their height can vary upto 5 m to 12 m.

H. S. Shekhawat et al. (Eds.): ICA 2025, CCIS 2795, pp. 295–305, 2026.
https://doi.org/10.1007/978-3-032-17083-5_25

Due to the significant heights, confined spaces and controlled environmental conditions in modern agricultural systems, high precision is essential, making it challenging for humans to personally manage all tasks [11]. To improve performance in these demanding environments, robotic solutions can be tailored and implemented according to specific needs. These robotic systems help eliminate human error, which is crucial for maintaining the tight control, required in precision agriculture. Additionally, robotics enable continuous operation with minimal interruptions, enhancing control, increasing yields and allowing farmers to fully leverage the controlled conditions of vertical farming [8]. Thus, robotic assistance plays a crucial role in both finely tuning environmental parameters and overcoming the spatial limitations inherent in vertical farming systems.

To enhance efficiency, boost productivity and reduce the need for manual labor, numerous researchers have focused on developing robotic assistance for agriculture. These innovations aim to automate repetitive tasks, such as planting, harvesting and monitoring crop health, allowing farmers to manage their operations more effectively. Iqbal et al. [5] developed a mobile manipulator with a 3-DoF robotic arm using off-the-shelf actuators to perform plant phenotyping and soil sensing. Arad et al. [2] developed a mobile manipulator for harvesting sweet pepper. The prototype was developed using Fanuc LRMate 200iD robotic arm placed on a mobile platform. Strisciuglio et al. [12] developed a mobile manipulator for rose pruning and bush trimming. The system consists of a 6-DoF Kinova robotic arm mounted on a mobile platform. The platform is a modified version of Bosch lawnmower. Roure et al. [10] developed a mobile manipulator for plant health monitoring in vineyards. It consists of a 6-DoF Kinova robotic arm mounted on a Husky platform. The arm was placed on the side of the platform to increase the workspace although restricting the operation to one side. Lehnert et al. [6] developed a mobile manipulator for harvesting sweet pepper. Field use of robots in agriculture are demanding and are being used extensively for complex tasks in cluttered environment. This shows the application of different robots for open field agriculture and greenhouses.

Robots are functional in open field agriculture, but so far very limited research has been reported on the use of robots in vertical farms [9]. Figure 1 shows the hydroponic vertical farm along with a mobile robot, on which kinova arm will be mounted, which will perform the agricultural tasks like seedling, transplanting, plant health monitoring, harvesting and post harvesting. So as a starting point, this paper identifies only three primary operations in a vertical farm: seedling, transplanting, and harvesting, which consumes maximum power. To optimize energy usage in vertical farming systems, it is essential to evaluate both the power consumption and the time spent by the robot on each individual task. Focusing on these tasks allows for a more granular analysis of energy demand specific to each activity, using a defined robotic platform. To enable accurate estimation and optimization, hardware validation was conducted using the Kinova Mico robotic arm [3]. This step ensures that the energy and torque profiles obtained through simulations closely reflect real-world performance. By isolating and analyzing the power consumption of each task individually, these findings can be scaled and integrated into a comprehensive model for the entire vertical farming cycle, supporting the development of more energy-efficient and time-optimized robotic solutions for agriculture.

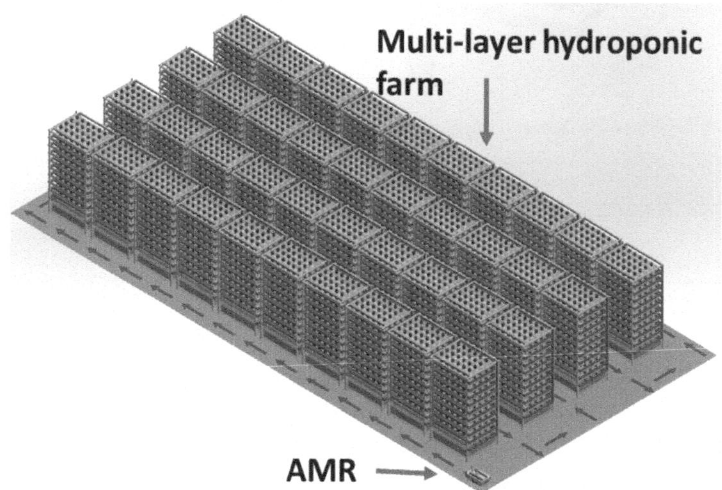

Fig. 1. Robotic Assistance in Hydroponic Vertical Farm

This paper is divided into five sections. In Sect. 2, task requirement and environment description is described, mathematical modeling is presented in Sect. 3. Section 4 presents numerical simulation and hardware results, a conclusion is drawn in Sect. 5.

2 Tasks Requirement and Environment Description

From the literature review, we have identified the three major vertical farming tasks, which consume maximum power: seedling, transplanting and harvesting. Therefore, the analysis is focussed solely on these agricultural tasks. To create a virtual environment for the simulation software, a photograph of the small-scale vertical farm prototype is taken. Based upon this picture, we have created a collision environment in MATLAB using primitive boxes through Robotic Toolbox function "collisionBox". Figure 2 shows the environment of seedling task, which includes a growing tray with dimension: length 0.5 m, width 0.9 m and thickness 0.05 m. This tray hold eight plant saplings. Once the saplings have grown, they are transferred to the transfer tray, which has dimension 1.3 m in length, 0.4 m in width and 0.135 m in thickness to be placed in the vertical farm for future growth.

Figure 3a shows the environment of vertical farm prototype for transplanting and harvesting. Collision environment is created in MATLAB using primitive

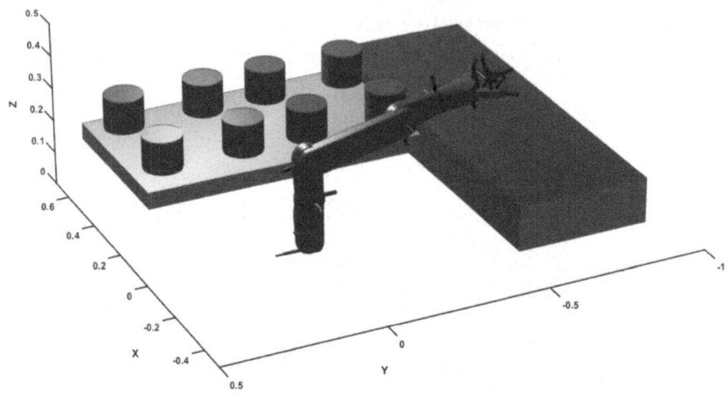

Fig. 2. Seedling Environment

boxes and cylinder through Robotic Toolbox function "collisionBox" and "collisionCylinder" as shown in Fig. 3b. Growing cylinder radius is taken as 0.025 m and length 1.32 m. Growing cups for plants have been replaced by small box cavities in MATLAB environment of length 0.1 m, width 0.04 m and thickness 0.05 m.

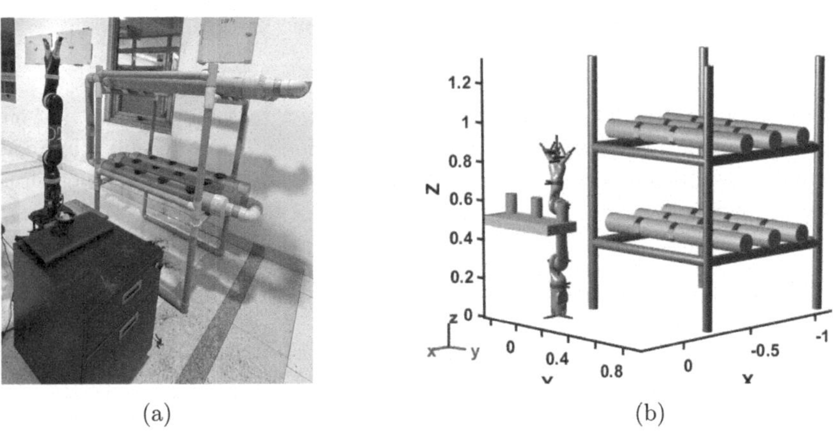

Fig. 3. (a) Task environment of vertical farm for transplanting and harvesting (b) Collision environment of vertical farm using Matlab

3 Mathematical Modeling

3.1 Robot Dynamics

Mathematical modeling plays a crucial role in calculating the torque and energy consumption associated with three key processes in vertical farming: seedling, transplanting and harvesting. To achieve this, inverse dynamics is used, which involves determining the joint angles for each robotic arm joint based on the desired end-effector positions during the pick-and-place operation of the plant. Equation 1 is used to calculate the torque.

$$I(q)\ddot{q} + C(q, \ddot{q}) + h(q) = \tau \tag{1}$$

where I is a Inertia Matrix, C is a Convective Inertia Matrix and h is a gravity force vector. q, $\dot{q}$, $\ddot{q}$ are position, velocity and acceleration vector of joints, respectively. τ represents the torque/force applied at joint [4].

3.2 Power Calculation

To calculate the power consumption for each process, Eq. 2 is used.

$$P = \tau \times \dot{q} \tag{2}$$

Simulations are performed with the 4-DOF KINOVA Mico robot, utilizing a cycloidal trajectory [7] in joint space, as demonstrated in the subsequent section.

4 Results

In this section, energy consumption for seedling, transplanting and harvesting tasks have been calculated. To generate the collision free path in given task environment, RRT Algorithm is used. Once collision free path is generated, trajectory for the robot manipulator is developed to perform the inverse dynamics and power calculation. MATLAB Robotic Toolbox is used for implementing the strategy.

4.1 Case 1: Seedling Task in Hydroponic Vertical Farm

The required task has been performed in two phases: Phase 1: Approach to Growing Plate. Phase 2: Growing Plate to Transferring plate. For the second phase, weight of the plant sapling is added as an external gravity force acting on the end-effector of the robot, assuming sapling weight 0.03 kg. Table 1 represents generated path using RRT algorithm. Figure 4 and 5 show the torque required for completing the tasks in simulation and hardware, respectively, for both the phases. For the simulation, calculated power requirements for Phase 1 and Phase 2 are 29.23 W and 70.8370 W, respectively. Total power consumption for one seedling cycle is 100.07 W. For the hardware, observed power consumption was 133.46 W. From the torque plots as shown in Figs. 4 and 5, it is evident that Joint 2 and Joint 3 are the main drivers in both simulation and hardware and bears the maximum load.

Table 1. Path for the Seedling Task

Phase	Points	θ_1	θ_2	θ_3	θ_4
Phase 1	Start Point	0	3.141	3.141	0
	Waypoint	0.135	3.267	3.012	0.177
	Goal Point	0.382	3.093	5.244	0
Phase 2	Start Point	0.382	3.093	5.244	0
	Waypoint	0.468	3.196	5.027	0.141
	Goal Point	−1.558	4.491	3.141	−0.012

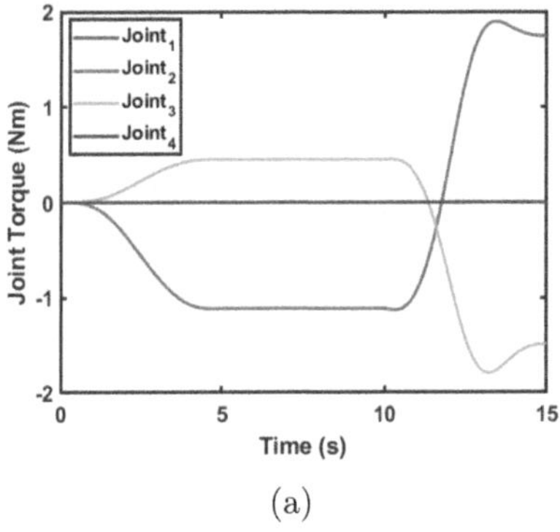

(a)

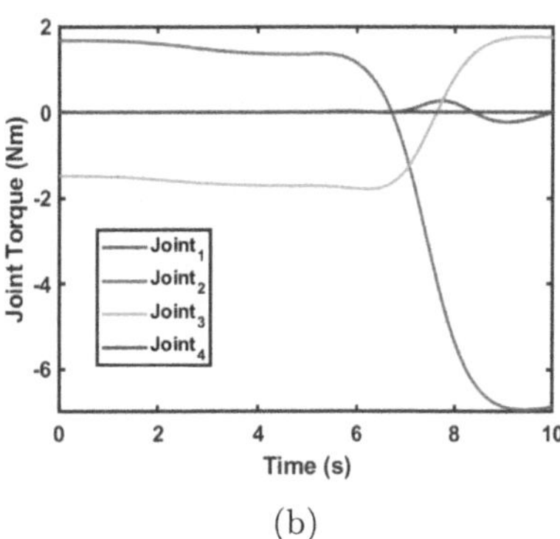

(b)

Fig. 4. Simulation Torque Plot: (a) Phase 1 of Seedling (b) Phase 2 of Seedling

4.2 Case 2: Transplanting Task in Hydroponic Vertical Farm

The required task has been performed in two phases: Phase 1: Approach to Sapling Plate. Phase 2: Transplanting of Sapling (Pick and Place). For the second phase, weight of the plant sapling is added as an external gravity force acting on the end-effector of the robot, assuming sapling weight 0.03 kg. Table 2 represents generated path using RRT algorithm. Figure 6 and 7 show the torque required for completing the tasks in simulation and hardware, respectively, for both the phases. For the simulation, calculated power requirements for Phase 1 and Phase 2 are 20.1973 W and 58.3792 W, respectively. Total power consumption for one seedling cycle is 78.5765 W. For the hardware, observed power consumption was 97.75 W. From the torque plots as shown in Figs. 6 and 7, it is evident that Joint 2 and Joint 3 are the main drivers in both simulation and hardware and bears the maximum load.

4.3 Case 3: Harvesting Task in Hydroponic Vertical Farm

This task is performed in two phases: Phase 1: Approach to Harvest. Phase 2: Transporting the Harvest to the container (Pick and Place). For the second phase, weight of the plant sapling is added as an external gravity force acting

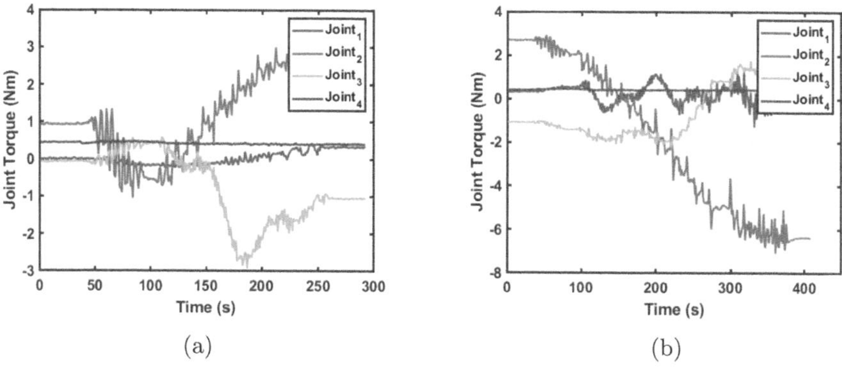

Fig. 5. Hardware Torque Plots: (a) Phase 1 of Seedling (b) Phase 2 of Seedling

Table 2. Path for the Transplanting Task

Phase	Points	θ_1	θ_2	θ_3	θ_4
Phase 1	Start Point	0	3.141	3.141	0
	Waypoint	0.135	3.267	3.012	0.177
	Goal Point	−0.058	3.667	4.882	0
Phase 2	Start Point	−0.058	3.667	4.882	0
	Waypoint 1	0.058	3.742	4.674	0.148
	Waypoint 2	0.257	3.695	4.673	−0.020
	Waypoint 3	0.404	3.800	4.622	−0.132
	Waypoint 4	0.670	3.943	2.526	−0.330
	Waypoint 5	0.511	3.857	2.680	−0.373
	Waypoint 6	0.455	4.062	2.728	−0.194
	Goal Point	0.356	3.979	2.642	0

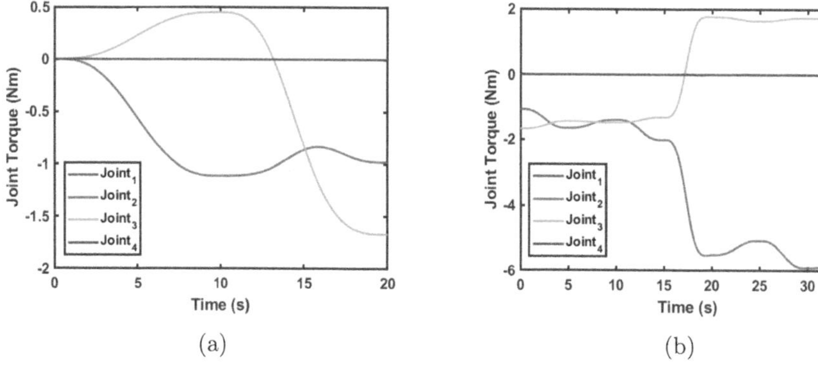

Fig. 6. Simulation Torque Plots: (a) Phase 1 of Transplanting (b) Phase 2 of Transplanting

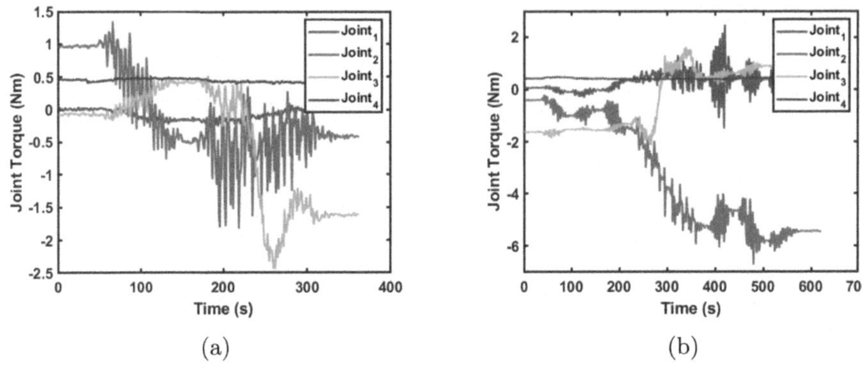

Fig. 7. Hardware Torque Plots: (a) Phase 1 of Transplanting (b) Phase 2 of Transplanting

on the end-effector of the robot, assuming sapling weight 0.3 kg. Table 3 represents generated path using RRT algorithm. Figure 8 and 9 show the torque required for completing the tasks in simulation and hardware, respectively, for both the phases. For the simulation, calculated power requirements for Phase 1 and Phase 2 are 47.99 W and 45.52 W, respectively. Total power consumption for one seedling cycle is 93.52 W. For the hardware, observed power consumption was 116.77 W. From the torque plots as shown in Figs. 8 and 9, it is evident that Joint 2 and Joint 3 are the main drivers in both simulation and hardware and bears the maximum load.

Table 3. Path for the Harvesting Task

Phase	Points	θ_1	θ_2	θ_3	θ_4
Phase 1	Start Point	0	3.141	3.141	0
	Waypoint 1	0.323	3.134	3.312	−0.253
	Waypoint 2	0.505	3.402	3.554	−0.349
	Waypoint 3	0.521	4.118	2.785	−0.323
	Goal Point	0.356	3.979	2.642	0
Phase 2	Start Point	0.356	3.979	2.642	0
	Waypoint 1	0.635	3.879	2.866	−0.249
	Waypoint 2	0.752	3.982	3.226	−0.351
	Waypoint 3	0.813	3.766	3.509	−0.684
	Goal Point	−0.058	3.667	4.882	0

The torque profiles for Phase 1 and Phase 2 for the three tasks show agreement between the simulation and hardware results, indicating successful vali-

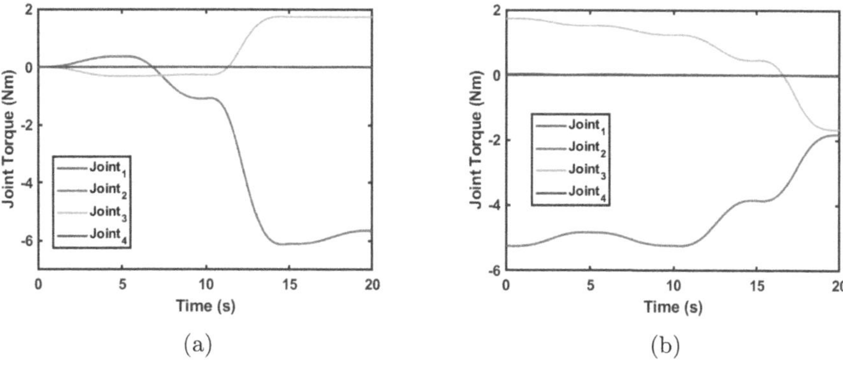

Fig. 8. Simulation Torque Plots: (a) Phase 1 of Harvesting (b) Phase 2 of Harvesting

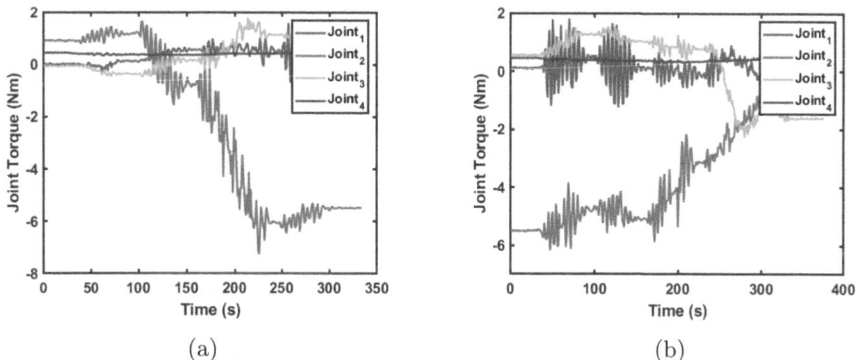

Fig. 9. Hardware Torque Plots: (a) Phase 1 of Harvesting (b) Phase 2 of Harvesting

dation and consistent system behavior across both phases. Joint 1 is showing difference in graphs due to frictional effect.

4.4 Performance Evaluation of the Proposed Tasks

To evaluate the accuracy of the developed system, a comparative analysis was conducted between the simulation results and the corresponding hardware experimental data for the seedling, transplanting and harvesting tasks. The percentage error for each task was calculated using the Eq. 3 by comparing the key performance metrics, such as position, joint velocities and joint torque from both domains.

Using the Eq. 3, percentage error is calculated.

$$\text{Percentage Error} = \left| \frac{\text{Simulation} - \text{Hardware}}{\text{Simulation}} \right| \times 100 \tag{3}$$

Seedling

$$\text{Power calculated from the hardware} = 133.46\ \text{W}$$
$$\text{Power calculated from the simulation} = 100.07\ \text{W}$$
$$\text{Error percentage} = 33.3\%$$

Transplanting

$$\text{Power calculated from the hardware} = 97.75\ \text{W}$$
$$\text{Power calculated from the simulation} = 78.57\ \text{W}$$
$$\text{Error percentage} = 24.4\%$$

Harvesting

$$\text{Power calculated from the hardware} = 116.77\ \text{W}$$
$$\text{Power calculated from the simulation} = 93.52\ \text{W}$$
$$\text{Error percentage} = 24.8\%$$

As observed, the percentage error between the simulation and hardware results for all three tasks–seedling, transplanting and harvesting–ranges from approximately 24 to 26%. This deviation can primarily be attributed to discrepancies in how power consumption is modeled and measured. In the simulation, power is calculated based on ideal conditions, where it remains relatively constant and does not account for transient states such as startup surges or shutdown decays. In contrast, the hardware setup includes these effects, where the robot consumes a notable amount of power during the initial activation phase and again at the end of the task due to braking, sensor activity and idle holding torque.

The comparison revealed a close alignment between the simulation and hardware outcomes, with discrepancies primarily attributed to real-world factors such as actuator delays, sensor noise and environmental interactions not fully captured in the simulation model.

5 Conclusion

This paper presents a comprehensive study on the energy consumption involved in three essential vertical farming tasks: seedling, transplanting, and harvesting. Torque, energy requirement and percentage error for each task were initially estimated through MATLAB simulations, followed by hardware validation using the Kinova Mico robotic arm to ensure the accuracy of the simulated results. The comparison showed a strong correlation between simulation and real-world performance, validating the proposed models.

The insights gained from this work contribute significantly to understanding the actuation demands and energy profiles of individual farming operations.

These validated torques and energy data can now serve as a foundation for estimating total energy requirements in a fully automated vertical farming system. Such analysis is crucial for optimizing robotic operations, improving energy efficiency, and designing sustainable agricultural automation solutions.

References

1. Al-Kodmany, K.: The vertical farm: a review of developments and implications for the vertical city. Buildings **8**(2), 24 (2018)
2. Arad, B., et al.: Development of a sweet pepper harvesting robot. J. Field Rob. **37**(6), 1027–1039 (2020)
3. Baressi Šegota, S., Anđelić, N., Lorencin, I., Saga, M., Car, Z.: Path planning optimization of six-degree-of-freedom robotic manipulators using evolutionary algorithms. Int. J. Adv. Rob. Syst. **17**(2), 1729881420908076 (2020)
4. Edition, T., Craig, J.J.: Introduction to Robotics (2005)
5. Iqbal, J., Xu, R., Halloran, H., Li, C.: Development of a multi-purpose autonomous differential drive mobile robot for plant phenotyping and soil sensing. Electronics **9**(9), 1550 (2020)
6. Lehnert, C., English, A., McCool, C., Tow, A.W., Perez, T.: Autonomous sweet pepper harvesting for protected cropping systems. IEEE Rob. Autom. Lett. **2**(2), 872–879 (2017)
7. Libin, Z., Yan, W., Qinghua, Y., Guanjun, B., Feng, G., Yi, X.: Kinematics and trajectory planning of a cucumber harvesting robot manipulator. Int. J. Agricult. Biolog. Eng. **2**(1), 1–7 (2009)
8. Lowenberg-DeBoer, J., Huang, I.Y., Grigoriadis, V., Blackmore, S.: Economics of robots and automation in field crop production. Precision Agric. **21**(2), 278–299 (2020)
9. Mahalingam, D., Patankar, A., Phi, K., Chakraborty, N., McGann, R., Ramakrishnan, I.: Containerized vertical farming using cobots. In: 2024 IEEE International Conference on Robotics and Automation (ICRA), pp. 17897–17903. IEEE (2024)
10. Roure, F., et al.: GRAPE: ground robot for vineyArd monitoring and ProtEction. In: Ollero, A., Sanfeliu, A., Montano, L., Lau, N., Cardeira, C. (eds.) ROBOT 2017. AISC, vol. 693, pp. 249–260. Springer, Cham (2018). https://doi.org/10.1007/978-3-319-70833-1_21
11. Siregar, R.R.A., Seminar, K.B., Wahjuni, S., Santosa, E.: Vertical farming perspectives in support of precision agriculture using artificial intelligence: a review. Computers **11**(9), 135 (2022)
12. Strisciuglio, N., et al.: TrimBot2020: an outdoor robot for automatic gardening. In: ISR 2018; 50th International Symposium on Robotics, pp. 1–6. VDE (2018)

AgriRover: A GPS-Guided Smart Rover for Environmental Monitoring

Aditya Sahu(✉), P. Haarika, and N. Prabakaran

Department of Electronics and Communication Engineering, Koneru Lakshmaiah Education Foundation, Green Fields, Vaddeswaram, Guntur 522302, Andhra Pradesh, India
meaditya50@gmail.com, prabakaran@kluniversity.in

Abstract. This paper presents AgriRover, a modular smart agriculture rover system guided by GPS for autonomous field navigation and environmental monitoring. The architecture comprises two main modules: an Arduino Mega-driven rover for autonomous navigation based on GPS waypoints, and an ESP32-based sensor module for collecting environmental data such as temperature, humidity, and soil moisture. A magnetometer ensures accurate heading, while Bluetooth-based mobile input enables flexible waypoint configuration. Collected data is uploaded to the cloud in real time, enabling remote monitoring and decision-making. The AgriRover addresses the challenge of efficient environmental monitoring over large agricultural plots by providing an automated and scalable solution. This modular architecture supports future technical enhancements, such as AI-driven automation and advanced analytics. Field testing over a 30 m × 30 m plot demonstrated reliable navigation and accurate sensing, validating the system's potential for precision agriculture in both remote and large-scale applications.

Keywords: Smart agriculture · IoT · Autonomous rover · GPS navigation · ESP32 · Environmental sensing · Cloud-based monitoring

1 Introduction

This article presents an IoT-based robotic platform for smart agriculture to enhance the productivity and resource utilization of agricultural operations. Smart agriculture is enabled through the utilization of emerging technologies. The robot system not only improves production but also quality, while automated agriculture and IoT provide the capability of monitoring fields of crops in real-time using sensors. The whole system will greatly enhance the agricultural process. Enhancing productivity and resource use in agriculture is a central objective. Smart agriculture relies on the convergence of emerging technologies. Automated systems enhance production as well as quality and automated farming and IoT facilitate the monitoring of the agricultural field in real time with the aid of sensors. The entire system will increase farming efficiency. Farming

H. S. Shekhawat et al. (Eds.): ICA 2025, CCIS 2795, pp. 306–313, 2026.
https://doi.org/10.1007/978-3-032-17083-5_26

is the bread and butter of most economies and supports the majority of the world's population. With the increasing demand for accurate and sustainable farming methods, the incorporation- The share of intelligent technologies like the Internet of Things (IoT), robots, and cloud computing in agriculture has gained greater significance. This paper intro- The system presents an intelligent farm rover that can navigate autonomously through farmland. In collecting environmental information, like temperature and humidity, and Soil moisture. The system is a two-block modular platform. The first module drives the rover through GPS coordinates, as well as by a magnetometer heading orientation sensor, and Bluetooth-enabled waypoint input. The second module, which is an ESP32 microcontroller, collects environmental information from the field and uploads it to the cloud in real-time. In this project, with the utilization of recent IoT technologies, information will be preserved in the cloud, that efficiency in agriculture [1] Agriculture is the underpinning support for many economies and a source of food for a significant percentage of the world's population. With the increasing demand for accurate and sustainable agriculture, the incorporation of new technologies like the Internet of Things (IoT), robotics, and cloud computing in the agricultural industry has become more and more necessary. This paper introduces a smart farm rover system that can navigate autonomously across farmland while collecting environmental information such as temperature, humidity, and moisture content of the soil.

The system is implemented as a two-block modular platform. The first module manages the motion of the rover through GPS coordinates, a magnetometer sensor for heading orientation, and Bluetooth-based waypoint entry. The second module, based on an ESP32 microcontroller, gathers environmental data from the field and transfers it to the cloud in real-time. In this project, with the utilization of recent IoT technologies, data will be stored in the cloud and later utilized for analysis in order to furnish correct information for forecast, apex harvesting and hence high standard of farming will be made. Some Scientists suggest to contemplate bioengineered foods, i.e., foods those are genetically altered to satisfy the food need of the world. However, many studies highlight their potential health risks, including infertility, insulin disruption, and other physiological issues [2].

Therefore, the optimal solution is to enhance the agriculture system. In this regard, great research projects and experiments have been conducted for decades in this domain and sensors and IoT-based technologies are helping to enhance traditional farming procedures to increase yield output [3].

Conventional field monitoring relies on human labor, resulting in inefficient data collection and slow reaction to environmental conditions. Existing systems are often non-autonomous and do not integrate real-time data. AgriRover overcomes these limitations by combining GPS-based autonomous navigation with real-time environmental monitoring and cloud connectivity. The key contributions of this work are:

- A modular design separating navigation and sensing, improving fault isolation and scalability.

- Integration of GPS, magnetometer, and Bluetooth for autonomous waypoint traversal.
- Real-time environmental data acquisition and cloud-based visualization for precision farming applications.

2 System Design and Architecture

The intended smart agriculture rover system consists of the following major hardware units:

- Arduino Mega for navigation and motor control of the rover
- L293D Motor Driver for driving the rover's DC motors
- u-blox NEO-6M GPS Module for position tracking
- HMC5883L magnetometer for heading and orientation
- HC-06 Bluetooth Module for Mobile App Communication
- ESP32 Microcontroller for sensing data collection and cloud upload
- DHT22 temperature-humidity sensor and Soil Moisture Sensor for environmental sensing
- HW-130 Motor Control Board employed in the robot body

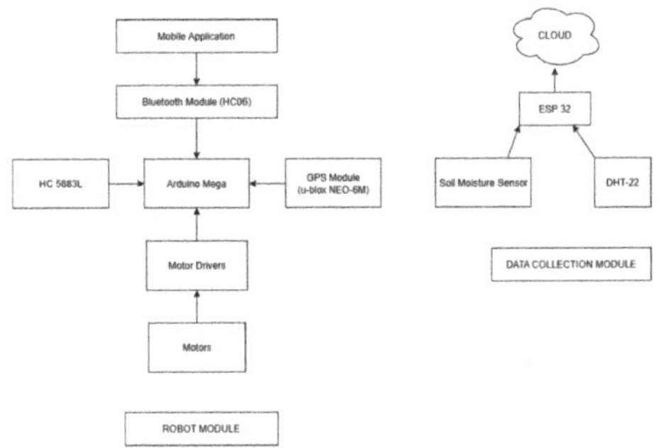

Fig. 1. Block Diagram of the Rover Navigation and Cloud Monitoring Units

The rover gets GPS waypoints via the Bluetooth module from the Arduino Mega. It traverses autonomously based on GPS information and orients itself with the aid of the magnetometer. The ESP32 also accumulates sensor data and streams it to a cloud server (e.g., Blynk IoT) via Wi-Fi at the same time [4] (Fig. 1).

3 Implementation Details

In the case of environmental issues, smart farming practices The use of IoT has several benefits, such as the effective utilization of natural resources, such as soil and water. IoT assists agriculture in precision farming with the use of data on the Internet [5]. The rover is programmed to follow a set of waypoints provided via a mobile app. The ESP32 at each waypoint takes temperature, humidity, and soil moisture readings. The readings are dated prior to cloud upload. Scientists, researchers, and engineers all over the world are working on the creation and invention of various tools, means to attain smart agriculture, and an automatic system in agricultural practices today. Many articles and Publications have emphasized the roles of Robotics and the Internet of issues related to the agricultural industry; however, virtually all of They were focusing only on the applications without offering any insight or outcome. They have concentrated on different designs based on the Internet of Things, better models, and more sophisticated equipment [6]. Motor operation is provided by using the L293D driver, which allows for bidirectional movement and rotation. The interfacing of the motor signal is provided by the HW-130 board. The ESP32 is independent and interfaced with the DHT22 temperature-humidity sensor and capacitive soil moisture sensors.

4 Experimental Setup and Results

4.1 Experimental Setup

In the present time, agriculture has been progressing unceasingly to precision agriculture [7] The prototype rover was tested in an open field of about 30 m × 30 m. The mobile app was employed to create several GPS waypoints around the field. The rover operated entirely on a 14.8 V/3000 mAh Li-ion battery, which ensured adequate runtime for 2–3 h of operation. Major details of the setup are as follows:

- **Testing Site:** Open grassy land with soft ground and few growths.
- **Connectivity:** A local Wi-Fi hotspot formed via a smartphone for providing ESP32 internet access.
- **GPS Accuracy:** Positional accuracy averaged to within ±2.5 m during clear sky conditions.

4.2 Sample Data Collected

Table 1. Environmental Data over Time

Time (HH:MM)	Latitude	Longitude	Temp (°C)	Humidity (%)	Soil Moisture (%)
2.20	15.9240905	80.1863809	39.2	46	60.2
2.45	15.9240907	80.1863810	39.0	46	60.0
2.50	15.9240908	80.1863811	39.1	47	60.1

Fig. 2. (i) Mobile Application Interface, (ii) AgriRover Prototype, (iii) Real-time Data on Blynk IoT Platform

4.3 Validation and Limitations

The system was tested on a 30 m × 30 m field with data collected at 10-min intervals. While results confirm basic functionality and cloud integration, the validation was limited in scope. No comparative metrics such as coverage efficiency, energy consumption, or navigation accuracy were evaluated. Future work will incorporate larger test plots, statistical performance metrics, and comparisons with manual methods or existing autonomous systems to assess improvements quantitatively (Fig. 2 and Table 1).

4.4 Observations

- **Navigation:** The rover followed the waypoints with acceptable deviation, which was corrected using heading information from the HMC5883L.
- **Data Consistency:** The ESP32 reliably collected and uploaded sensor values every 60 s.
- **Modularity:** The separation of the control and sensing modules enabled better debugging and system reliability (Table 2).

4.5 Related Work and Comparative Analysis

Table 2. Comparison of Existing Systems

Ref	Navigation	Sensing	Cloud Upload	Modularity	Platform
[X]	No	Yes	No	No	Arduino
[Y]	Semi-auto	Yes	Yes	No	Raspberry Pi
Ours	Yes	Yes	Yes	Yes	ESP32 + Arduino Mega

5 Conclusion and Future Work

This paper introduced a modular smart agricultural rover system that is autonomous in navigation and environmental sensing. The system combines GPS-waypoint navigation, magnetometer-based heading correction, and real-time data logging based on an ESP32 and sensors for environmental measurements. Dynamic movement control of the rover is supported by the mobile application interface, while cloud connectivity provides easy data collection and visualization for remote users.

The field testing results show that the rover efficiently navigates the designated route and correctly sends environmental data like temperature, humidity, and soil moisture to the cloud. This device can benefit farmers with precision agriculture by targeting farmland conditions, particularly in remote or expansive agricultural areas.

5.1 Limitations and Future Work

The current prototype was tested in a small field, limiting generalizability. No direct comparisons with existing rover systems, drones, or manual data collection methods were performed. In future, experiments across diverse terrains, larger fields, and benchmark comparisons will be conducted. Additionally, integrating AI for crop health assessment and autonomous obstacle avoidance is planned.

5.2 Future Enhancements

- **Weed Detection and Removal:** A computer vision module based on convolutional neural networks (CNN) is planned to detect and classify weeds in real time using onboard camera input. The system could activate a mechanical arm or targeted spraying unit to eliminate detected weeds, extending the rover's capabilities beyond sensing to active field intervention [9].
- **AI-Driven Decision Making:** Future versions will integrate machine learning algorithms that analyze multi-modal sensor inputs (e.g., soil moisture, temperature, weed presence) to make real-time decisions. These may include prioritizing zones for irrigation or fertilization, adjusting navigation paths, or triggering alerts for human intervention.

- **Edge Computing and Sensor Fusion:** Implementing lightweight AI models on the ESP32 or other microcontrollers for local processing will reduce cloud dependency and improve responsiveness. Sensor fusion techniques will combine GPS, IMU, and magnetometer data to improve navigation accuracy in noisy environments.
- **Large-Scale Communication:** Integration of long-range wireless technologies such as LoRa or NB-IoT will be explored to support real-time data transmission across larger fields without relying on local Wi-Fi infrastructure.

The following bibliography gives an example reference list with entries for a blog post on IoT applications in agriculture [1], a journal paper on GM food/feed impacts [2], a review paper on smart farming and crop data management [3], a conference paper on autonomous rover systems based on GPS and IoT [4], an IEEE Access survey on the deployment of smart farming [5], a conference paper on IoT-based solutions for improved crop yield [6], an IEEE Access paper on IoT-based smart agriculture [7], a conference paper on autonomous robot navigation [8], and a conference paper on robotic weed removal based on deep learning [9]. Several citations are clustered [1–4,6–9].

Acknowledgments. We would like to thank our institution and mentors for their valuable support and guidance throughout this project.

Disclosure of Interests. The authors have no competing interests to declare that are relevant to the content of this article.

References

1. Chalimov, A.: IoT in agriculture: 8 technology use cases for smart farming (and challenges to consider), 7 July 2020
2. Sanchez, M.A., Parrott, W.A.: Characterization of scientific studies usually cited as evidence of adverse effects of GM food/feed. Plant Biotech. J. **15** (2017)
3. Saiz-Rubio, V., Rovira-Más, F.: Smart farming towards agriculture 5.0: a review on crop data management. Agronomy (2020)
4. Islam, M.S., et al.: Smart autonomous rover using GPS and IoT for precision agriculture. In: 2021 International Conference on Robotics, Electrical and Signal Processing Techniques (ICREST), pp. 431–436. IEEE (2021). https://doi.org/10.1109/ICREST51555.2021.9331091
5. Farooq, M.S., Riaz, S., Abid, A., Abid, K., Naeem, M.A.: A survey on the role of IoT in agriculture for the implementation of smart farming. IEEE Access **7** (2019)
6. Gayathri, M.K., Jayasakthi, J.: Providing smart agriculture solutions to farmers for better yielding using IoT. In: IEEE International Conference on Technological Innovations in ICT for Agriculture and Rural Development, July 2015
7. Ayaz, M., Uddin, M.A., Sharif, Z., Mansour, A., Aggoune, E.-H.M.: Internet of Things (IoT) based smart agriculture: toward making the fields talk. IEEE Access **7** (2019)
8. Alittappe, R.J., Mahmoudi, N., Jafari, M.R., Foladi, A.: Autonomous robot navigation: deep learning approaches for line following and obstacle avoidance. IEEE (2024)

9. Saini, P., Nagesh, D.S.: Robotic weed removal using deep learning for precision farming. In: 2024 2nd International Conference on Advancements and Key Challenges in Green Energy and Computing. IEEE (2024). https://ieeexplore.ieee.org/document/10869035
10. Miller, R.: How to build a GPS guided robot. Instructables (2020). https://www.instructables.com/How-to-Build-a-GPS-Guided-Robot/
11. ThingSpeak Cloud Platform. https://thingspeak.com. Accessed April 2025
12. u-blox NEO-6M GPS Module Datasheet. Accessed April 2025

Optimized Conv-LSTM Model for Analyzing Negative Affective State Vocalization in Dairy Cattle for an Edge Device

Hitesh Arjunbhai Ramrakhiyani[1(✉)], Himashri Deka[1], Sandeep Kumar Pandey[2], N. S. Sreenivasalu[4], Hanumant Singh Shekhawat[1], and Ravi Jasuja[3]

[1] Indian Institute of Technology Guwahati, Guwahati, India
{ahitesh,himashri.deka,h.s.shekhawat}@iitg.ac.in
[2] Acuity Knowledge Partners, Bengaluru, India
sandeeppandey456@gmail.com
[3] Brigham and Women's Hospital, Harvard Medical School, Boston, MA, USA
rjasuja@bw.harvard.edu
[4] XYonetx Therapeutics, Canton, USA

Abstract. Edge devices play a crucial role in applications where computing can be done locally. As edge devices have limited computing power and memory due to which there is a necessity to design computationally small models that can be deployed on edge devices like Raspberry Pi, Jetson, Intel Edison, etc. In dairy farming livestock's affective states can be identified with their vocalizations. Cows have separate vocalizations for close-range and long-distance communication. Additionally, each individual's vocalizations are unique across circumstances. Distinguishably Understanding dairy cows' vocalizations under anguish, pain, or fear is crucial. This study is conducted to optimize the known CNN-LSTM deep learning model on cow vocalization dataset with model optimization techniques to deploy it on edge devices for local computing on farms to monitor the emotional affective states of livestock animals.

Keywords: Animal Affective States · Model Optimization · Pruning

1 Introduction

In our daily lives, we consume many products produced by the dairy industry [1]. These products are an integral part of our lives, as we need milk to make those products. The raw milk is produced by dairy cattle living in the farms maintained and owned by farmers. Nowadays, the dairy industry has a huge demand for milk products from consumers, which has caused farm sizes to increase, and it isn't easy to maintain such a big farm with proper attentive care for the individual animals. Animals feel various emotions [2] from time to time based on the surrounding environment.

H. S. Shekhawat et al. (Eds.): ICA 2025, CCIS 2795, pp. 314–322, 2026.
https://doi.org/10.1007/978-3-032-17083-5_27

It is observed that it is difficult for big-sized farms to provide individual care with a limited workforce. As technology is getting more advanced, many types of sensors are available, like Wearable sensors for activity monitoring, Microphones for vocalization [3], milk sensors, and imaging sensors like cameras & thermal imaging [4] to sense the behavior of dairy cattle in daily life as these sensors collect data which can be helpful to us to be aware of the present state of animals for providing the good care early. The data we collect from the various sensors will have complex patterns [4]. Machine Learning algorithms excel at finding hidden patterns in vast amounts of data. It can also assist the farmers in decision-making and analysis of the mental state of farm animals [5]. Utilising machine learning in the field of dairy farming can help by enhancing animal welfare through early disease detection, optimized nutrition and feeding, stress monitoring and mitigation, and improved reproduction care.

Cattle vocalizations are a primary form of communication among these animals, conveying important information about their physiological and emotional states. Research has consistently shown that the acoustic characteristics of these sounds change significantly in response to both positive and negative experiences. This direct connection between vocal expression and emotional state makes vocalizations a valuable resource for understanding the internal experiences of dairy cattle [6].

The vocalization of farm animals has been recognized as a valuable indicator of their state of health [7]. Analysis of vocalizations of farm animals, such as pigs and horses, has shown good results regarding distress calls and stress levels [8]. In the context of precision dairy, a fog-enabled wireless sensor network system has been proposed for animal behavior analysis, focusing on mobility and behavior [9]. Research on decoding animal vocalizations using deep learning natural language processing and transformer models has highlighted the importance of understanding the semantics of these calls for interpreting their functional vocabulary and social interactions [10].

Yoshihara and Oya [11] studied cow vocal reactions in four states: feed anticipation, estrus, communication, and parturition. In [12], They gave a survey on video recordings, measured duration, intensity, pitch, and formant, and measured salivary cortisol concentrations. Dinu Gavojdian et al. [6] collected a cattle vocalization dataset, extracted spectrograms, and trained it using explainable machine learning and GRU deep learning models.

Long-term negative affective states in dairy cattle have been shown to have a substantial impact on milk output and increase susceptibility to disease, perhaps resulting in a shorter lifespan [13]. To integrate domestic animals into future on-farm assessment processes, a non-invasive technique to identify their emotions must be developed. For the Automated monitoring of farm animals and predict the animal's affective state [14], we need small hardware to deploy a trained machine learning algorithm that takes the real-time data from sensors [15] attached to it. It can minimize data transfer to the servers by making computations at the site. Many small hardware boards, like NVIDIA Jetson Nano, Raspberry Pi, Intel Neural Stick, etc., are available. In our previous study, we experimented with the NVIDIA Jetson Nano board by deploying [16] a CNN-LSTM deep learning model with many parameters. This study aims to optimize the model that can further be implemented on the ESP32 board, and we are searching for smaller, lighter algorithms with low latency to make our solution more cost-effective for the farmers.

As these small hardware are resource-constrained [13] and have limited computation power with less energy requirement, we are optimizing our developed CNN-LSTM algorithm for BovineTalk Cow Affective State Vocalization by applying the pruning method, and further details of the conducted study are given in the subsequent sections.

2 Methodology

This section discusses the experimental design and our approach to optimize the CNN-LSTM model by reducing its parameters while maintaining accuracy. We also discuss other classification metrics to make the model tiny enough to be deployed on edge devices for real-time applications to identify animals' affective states. The experimental framework is given in Fig. 1 below. We trained two models: (1) Gaussian Naive-Bayes and (2) CNN-LSTM model, which classifies the low-frequency and high-frequency vocalization with 3D Mel frequency cepstral coefficient features as input to the model.

The BovineTalk cow vocalization dataset is divided into two divisions: Train set (80%) and Test set (20%). As far as we know, this is the first tiny optimized CNN-LSTM model to classify cow vocalizations for edge devices, which can help develop real-time monitoring solutions in the future.

2.1 BovineTalk Dataset

This study used the negative affective state cow vocalization dataset called BovineTalk [6] to train our deep learning model, which contains two classes of vocal audio recordings of 192 calls with low-frequencies and 952 calls with high-frequencies. During collection, the cows were isolated and monitored for 240 min post-milking period. The audio recording is at sampling rate of 44.1 kHz with a 16-bit amplitude resolution.

2.2 3D MFCC Feature Extraction

In our study, we used cow vocalization recordings, and each recording was of a different duration. To make it symmetric, we did zero padding, so each recording is of 3 s. To provide input to our CNN-LSTM model, we extracted Mel Frequency Cepstral Coefficients (MFCC), MFCC-Δ, and MFCC-$\Delta\Delta$ using Librosa library [17]. We extracted 32-dimension features by taking 1024 sample windows at a sampling frequency of 44.1 kHz. The MFCC-Δ and MFCC-Δ Δ can be calculated using the first and second differentiation of the MFCC coefficients with respect to time.

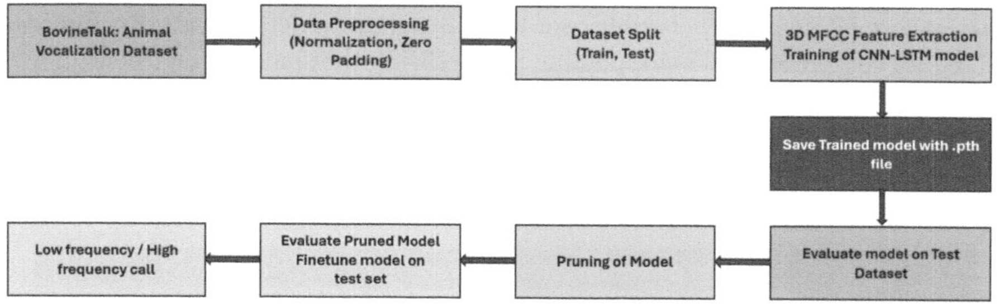

Fig. 1. Experimental Framework of this study

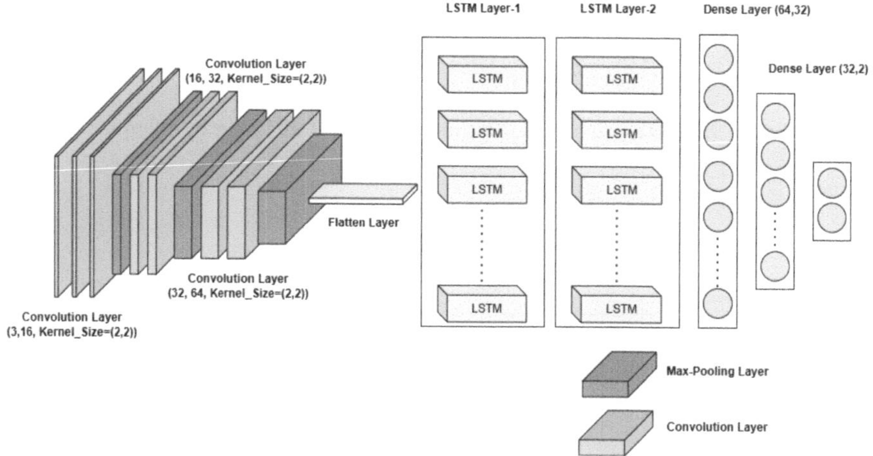

Fig. 2. Model Architecture of CNN-LSTM network

2.3 Classification Model

The prime objective of the study is to develop an optimized small-sized model that can be deployed later on edge devices for non-invasive monitoring of animal emotions. We trained two models: (1)Gaussian Naive Bayes Classifier and (2)CNN-LSTM Model with Model Optimization techniques on 3D MFCC features extracted from cattle vocalization recordings.

For optimization of the CNN-LSTM model, we used the pruning approach to reduce its parameters, which has reduced the computing and size of the model while keeping nearly the same accuracy. We used the PyTorch library with Adam optimizer to train our CNN-LSTM model and Scikit-Learn to train the Gaussian Naive Bayes model.

In the Gaussian Naive Bayes model, we first reshaped the 3D MFCC features into a 26,496-D feature array and gave them as input to the model for training. This model has 1,05,986 parameters, and the model size is 848.6 Kb.

The model architecture of the CNN-LSTM network implemented is shown in Fig. 2. Given Model has Convolutional layers responsible for local feature extraction, where filters or kernels slide over the input data to detect patterns. In order to reduce computational load, pooling layers shrink the spatial dimensions of feature maps. Fully connected layers then combine the features that convolutional layers have learned to generate predictions. In all three convolutional layers, we use ReLU as an activation function.

Model	Number of Model Parameters			Model Size		
	Before Pruning	After Pruning	Reduction	Before Pruning	After Pruning	Reduction
Gaussian Naïve Bayes	1,05,986	-	-	0.85 MB	-	-
CNN-LSTM	79,474	40,434	49.122%	0.31 MB	0.16 MB	48.38%

Fig. 3. Reduction in Model Parameters and Model Size by Pruning

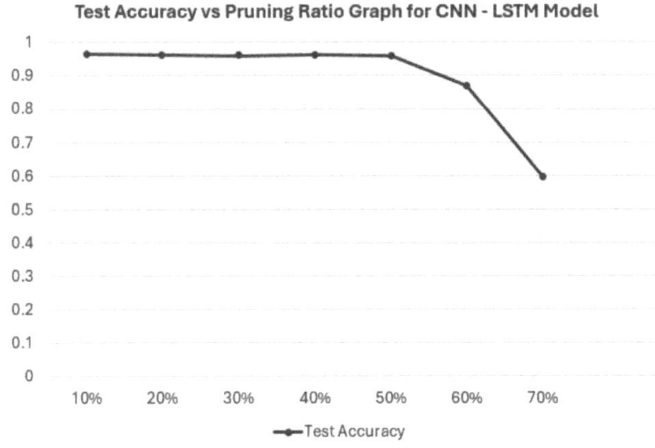

Fig. 4. Test Accuracy vs. Pruning Ratio Graph for CNN-LSTM Model

2.4 Pruning and FineTuning

As per the literature survey, many pruning methods are available [18,19], from which we used unstructured magnitude-based pruning, which prunes the lowest magnitude weights and can remove individual weights rather than removing the entire channel or filter as in structured pruning.

We first converted all our model weights to absolute values in the unstructured magnitude pruning. The weight parameter's absolute value will decide the weight's importance in the model to get the output. Smaller absolute weights typically have the least impact on network output. The threshold is calculated based on the input "percentage of pruning" in which all other weights are zero under the threshold absolute weight value. In Fig. 3, we compared the number of parameters between the Gaussian Naive Bayes model and the CNN-LSTM model.

In this, we did iterative experiments by taking different "percentage of pruning" input as shown in Fig. 4, from which on 50% pruning (50% weights are zero), we observed the best results with 96% test accuracy and then after we fine-tuned our pruned model with 0.0001 learning rate, stochastic gradient descent optimizer and cosine learning rate scheduler for two epochs.

By applying pruning, we zeroed out the weights that are less than the threshold value that do not significantly affect output, due to which model size will be reduced. We can achieve faster inferencing to help optimize resource utilization while deploying the model on edge devices.

3 Results and Discussion

The CNN-LSTM network was trained for 20 epochs at a learning rate of 0.0015 and a batch size of 16. When 3D Mel frequency cepstral coefficients (MFCC) features are used as input, the results demonstrate improved train and test accuracy.

Subsequently, we pruned the model parameters, evaluated the performance with the test set, and got little change in test accuracy with 50% of the original model parameters, significantly reducing the computation and model size. We fine-tuned the pruned model and tested again with an observed accuracy of 97.73%, as shown in Fig. 5.

Model	Features	Train Accuracy	Test Accuracy
Explainable Model [6]	23 Acoustic Features	89.9 %	87.2 %
Deep GRU [6]	Spectrogram	91.5 %	89.4 %
Gaussian Naïve Bayes	3D MFCC	87.46 %	86.93 %
CNN-LSTM	3D MFCC	97.44 %	96.59 %
CNN-LSTM (Pruned-50%)	3D MFCC	-----------	96 %
CNN-LSTM (Finetuned)	3D MFCC	-----------	97.73 %

Fig. 5. Performance Results of Proposed CNN-LSTM model

TARGET / OUTPUT	LFC	HFC
LFC	19 10.80%	1 0.57%
HFC	5 2.84%	151 85.80%

(a) Model without Pruning

TARGET / OUTPUT	LFC	HFC
LFC	19 10.80%	1 0.57%
HFC	6 3.41%	150 85.23%

(b) Pruned Model

TARGET / OUTPUT	LFC	HFC
LFC	18 10.23%	2 1.14%
HFC	2 1.14%	154 87.50%

(c) Finetuned Model

TARGET / OUTPUT	Class0	Class1
Class0	15 8.52%	5 2.84%
Class1	18 10.23%	138 78.41%

(d) Gaussian Naive Bayes

Fig. 6. Test Data Confusion Matrix on (a,b,c) CNN-LSTM Model, (d)Gaussian Naive Bayes

Due to the imbalanced dataset, we analyzed and reported precision: 0.95, Recall: 0.7917, and F1 Score: 0.8636 on the test data from the confusion matrices shown in Fig. 6 for different configurations of CNN-LSTM and Gaussian Naive Bayes model. These findings demonstrate that a system of edge devices to detect animals' emotional states can be created.

According to the findings, our optimized CNN-LSTM model performs well compared to state-of-the-art methodologies with reduced parameters and computation. This accomplishment facilitates future research in this area and enables farmers to promptly and precisely detect negative affective states in animals.

4 Future Scope

This investigation aims to move forward to the design and deployment of a tiny deep learning model with good performance metrics on edge devices to utilize for real-time applications. However, this study does not cover the deployment of edge devices like Raspberry Pi and microcontrollers. However, optimizing the known CNN-LSTM model for cattle vocalization classification study sets the platform for moving forward. In the future, we will deploy it on edge devices and check its performance, along with developing a complete non-invasive solution with animal vocalization and animal movement in farms, which will help identify the illnesses of animals at an initial stage. As we observed that the BovineTalk dataset is insufficient to identify all emotional characteristics of animals at the initial stage, we need to collect more data to make our model robust to identify the affective emotions of animals in real-time.

References

1. Ventura, B.A., von Keyserlingk, M.A., Weary, D.M.: Animal welfare concerns and values of stakeholders within the dairy industry. J. Agric. Environ. Ethics **28**(1), 109–126 (2015)
2. Neethirajan, S., Reimert, I., Kemp, B.: Measuring farm animal emotions-sensor-based approaches. Sensors **21**(2), 553 (2021)
3. Berckmans, D.: Precision livestock farming technologies for welfare management in intensive livestock systems. Rev. Sci. Tech. **33**(1), 189–196 (2014)
4. Eckelkamp, E.A.: Invited review: current state of wearable precision dairy technologies in disease detection. Appl. Anim. Sci. **35**(2), 209–220 (2019)
5. Neethirajan, S.: The role of sensors, big data and machine learning in modern animal farming. Sens. Bio-Sens. Res. **29**, 100367 (2020)
6. Gavojdian, D., Mincu, M., Lazebnik, T., Oren, A., Nicolae, I., Zamansky, A.: Bovinetalk: machine learning for vocalization analysis of dairy cattle under the negative affective state of isolation. Front. Vet. Sci. **11**, 1357109 (2024)
7. Manteuffel, G., Puppe, B., Schön, P.C.: Vocalization of farm animals as a measure of welfare. Appl. Anim. Behav. Sci. **88**(1-2), 163–182 (2004)
8. Browning, D.G., Scheifele, P.M.: Artiodactyl and perissodactyl acoustics: identifying distress calls by farm animals. J. Acoust. Soc. Am. **115**(5_Supplement), 2485–2485 (2004)
9. Bhargava, K., Ivanov, S., Kulatunga, C., Donnelly, W.: Fog-enabled wsn system for animal behavior analysis in precision dairy. In: 2017 International Conference on Computing, Networking and Communications (ICNC), pp. 504–510 (2017)
10. Manikandan, V., Neethirajan, S.: Decoding poultry vocalizations-natural language processing and transformer models for semantic and emotional analysis. *bioRxiv*, pp. 2024–12 (2024)
11. Yoshihara, Yu., Oya, K.: Characterization and assessment of vocalization responses of cows to different physiological states. J. Appl. Anim. Res. **49**(1), 347–351 (2021)
12. García, R., Aguilar, J., Toro, M., Pinto, A., Rodríguez, P.: A systematic literature review on the use of machine learning in precision livestock farming. Comput. Electron. Agric. **179**, 105826 (2020)
13. Ben Sassi, N., Averós, X., Estevez, I.: Technology and poultry welfare. Animals **6**(10), (2016)
14. Sellier, N., Guettier, E., Staub, C.: A review of methods to measure animal body temperature in precision farming. Am. J. Agric. Sci. Technol. **2**(2), 74–99 (2014)
15. Kaur, U., et al.: Invited review: integration of technologies and systems for precision animal agriculture–a case study on precision dairy farming. J. Anim. Sci. **101**, skad206 (2023)
16. Ramrakhiyani, H.A., Pandey, S.K., Sreenivasalu, N.S., Shekhawat, H.S., Jasuja, R.: Negative affective state vocalization analysis of dairy cattle using 3d mfcc features with cnn-lstm model on an edge device. In: International Conference on Agriculture-Centric Computation, pp. 34–42. Springer (2024)
17. McFee, B., Raffel, C., Liang, D., Ellis, D.P., McVicar, M., Battenberg, E., Nieto, O.: librosa: audio and music signal analysis in python. In: Proceedings of the 14th Python in Science Conference, vol. 8 (2015)

18. Han, S., Pool, J., Tran, J., Dally, W.: Learning both weights and connections for efficient neural networks. In: Proceedings of the 29th International Conference on Neural Information Processing Systems - Volume 1, NIPS 2015, pp. 1135–1143, Cambridge, MA, USA, MIT Press (2015)
19. Cheng, H., Zhang, M., Shi, J.Q.: A survey on deep neural network pruning: taxonomy, comparison, analysis, and recommendations. IEEE Trans. Pattern Anal. Mach. Intell. (2024)

Drone RCS Statistical Behaviour and Its Implications for Agriculture Drones Air Traffic Management

Jan Pidanic[1](✉), Pavel Sedivy[1], Gaurav Trivedi[2], and Stephen Paine[3]

[1] University of Pardubice, Studentska 95, 53210 Pardubice, Czech Republic
{jan.pidanic,pavel.sedivy}@upce.cz

[2] Indian Institute of Technology Guwahati, Guwahati, Assam 781039, India
trivedi@iitg.ac.in

[3] University of Cape Town, Rondebosch, Cape Town 7700, South Africa
stephen.paine@uct.ac.za

Abstract. This paper explores UAS (copter-type drone) radar cross-section behaviour impacting drone traffic control support by independent surveillance. The proliferation of drone traffic worldwide includes agriculture and state challenges in airspace safety. Dependent surveillance based on data communication is susceptive to cyber-attacks and should be augmented by more reliable methods. Primary radars are capable of providing relevant data for safety enforcement. Radar observation is challenged by small RCSs, influenced by copter flight behaviour. The paper further presents the measurement and analysis results of several UAV Radar Cross Sections (RCS), focusing on different flight parameters, such as pitch and roll. Understanding RCS statistical behaviour is crucial for primary radar drone detection. Insights into RCS fluctuations can enhance drone surveillance and drone traffic control for agricultural applications.

Keywords: drone · RCS measurement · RCS statistical behaviour · detection · pulse radar · UAV/UAS

1 Introduction

The integration of drones in agriculture has transformed modern farming practices, offering significant improvements in monitoring and thus providing precision and efficiency. Drones provide real-time data collection and are used for tasks such as crop monitoring, soil analysis, pest detection, wildlife tracking, and irrigation management [1].

The growing scale of drone operation poses challenges to airspace traffic management. Compliance enforcement of these flight rules is performed mainly by dependent surveillance monitoring drone emissions [2] or evaluation of drone position reports [3]. These methods are susceptive to cyber-attacks and should be augmented by independent surveillance. The most proven long-range method is primary radar surveillance. Even civilian law enforcement radar systems must be nowadays designed to be highly

H. S. Shekhawat et al. (Eds.): ICA 2025, CCIS 2795, pp. 323–330, 2026.
https://doi.org/10.1007/978-3-032-17083-5_28

resilient for operation in dense radio emission environments which face jamming [4]/ interference challenges [5].

The detection of small reflections, slow speed, and low flight altitude are the main challenges of drone primary radar surveillance on the background of the reflection from the surface (ground clutter). RCS varies with drone flight, especially for copter-type drones, which generate thrust by alternating drone body pitch and roll. Pitch and roll are affected by wind compensation as well.

Any sensor supporting law enforcement and safety, including primary radar, must focus on output information quality. Optimizing radar performance to achieve reasonable relevance requires understanding the impact of the environment and drone behaviour on radar systems. This impact is, by nature, random and must be described statistically. The key part of the detection performance is caused by the drone's radar cross section (RCS) and statistical model of it. The model is crucial for predicting and optimizing sensor accuracy [6].

The RCS of a drone is affected by several factors, including the drone's size, shape, material, number of rotors, radar carrier frequency, and polarization [7].

Another key factor influencing RCS is drone orientation relative to radar, pitch, and roll orientation.

The balancing power of propeller pairs controls pitch and roll. Opposite propellers are spinning in contra-rotating directions, compensating reactive torque. This principle is essential for the drone's stability and control.

A typical flight of a drone, especially a quadcopter, involves different phases and movements that can be analyzed based on the pilot's trajectory, speed, altitude, and manoeuvres. The movements of drones based on trajectory movement and statistical tilt can be divided into several phases:

1. Vertical takeoff/landing – all propellers generate the same lift, and the statistical tilt is close to 0° due to vertical upwards/downwards.
2. Climb (vertical movement) – The drone ascends by increasing the speed of the propellers, generating more lift. Statistical tilt remains relatively small (max. 2°–5°) due to wind during moving straight up.
3. Horizontal flight (trajectory movement) – The horizontal movement involves a combination of speed changes and tilt adjustments to make the drone move forward, backwards, or sideways. During horizontal flight, the tilt changes depending on the desired trajectory and speed. The drone typically tilts forward by about 5°–10° (depending on speed) for a smooth forward flight. At higher speeds, where more aerodynamic force is needed, the tilt can increase to 15°–20°. Statistical analysis shows normal distribution, with most flights showing a small tilt (under 10°) in calm conditions. More aggressive manoeuvres or fast direction changes can cause a greater tilt (over 20°), but the tilt would be lower on average.
4. Deceleration and Direction Change (Turns, Maneuvering) – during the drone direction changes (turning the drone sideways or performing a roll), the drone tilts its body in the direction of the turn. This movement is accomplished by adjusting the speeds of the opposing propellers. During sharp turns, the tilt can range from 30°-40°, which is typical for quick changes in direction.

Drone tilt (both pitch and roll) can be biased by the tilt necessary for wind compensation. From the radar point of view, even short-range drone targets are typically observed at very low elevations (target altitudes are much lower than range, "front observation"). As a result, drone flight behaviour influences drone RCS, such as tilting for movement control [8].

The RCS signature is typically and inexactly characterized in primary radar systems as a single number (average RCS) that does not consider RCS variation by target state fluctuation. The drone RCS depends on wavelength, polarization, material, shape, number of rotors, pitch, yaw, roll, etc. The RCS's description by the single "average" figure is far from the truth. Primary radar sensor performance estimation with reasonable precision requires a more accurate RCS description, where a statistical one makes the most sense.

The paper presents a real drone's flight control parameters and RCS statistical behaviour in the X-band within real flight pitch/roll extent.

2 Drone Flight Behaviour

Real drone behaviour may vary according to the pilot's mode selection from very agile to steady. The drone brain–autopilot interprets input control into drone acceleration according to the selected mode. Even in the case of steady modes, abrupt and substantial changes compensating impact of wind gusts may occur.

Drone flight behaviour is presented by analysis of a DJI mini SE telemetry log. The example of the trajectory is shown in Fig. 1.

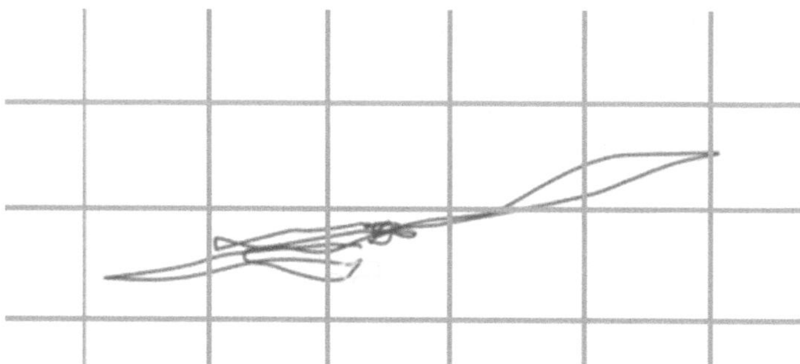

Fig. 1. DJI mini SE flight trajectory (grid 100 m), normal mode.

A DJI Mini SE telemetry log was deeply analyzed, and the probability and cumulative density function of drone pitch and roll were evaluated. Figure 2 illustrates pitch changes over time in agile mode, and Fig. 3 for steady mode. The extent of pitch/roll is reduced in steady mode (most values are up to $\pm 10°$), and high angles are rare. Pitch and roll up to 30° occur and should be considered for RCS measurement.

More analytic insight into flight data brings probability density (PDF) and cumulative density functions (CDF) estimates. The PDF and CDF results for agile flight mode are presented in Fig. 4a (pitch) and 4b (roll).

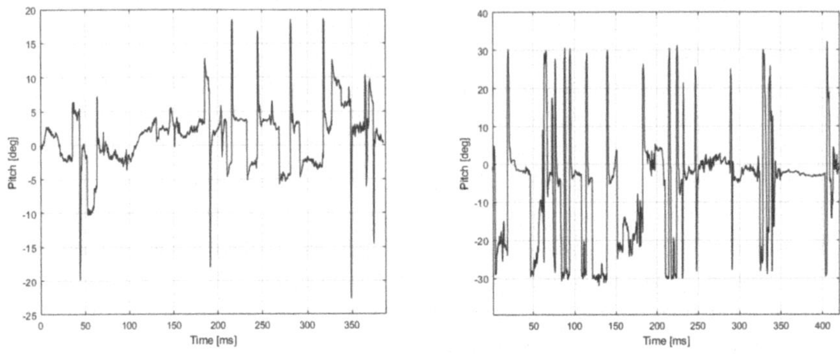

Fig. 2. Pitch changes over time: a) flight in steady (left), b) flight in agile (right)

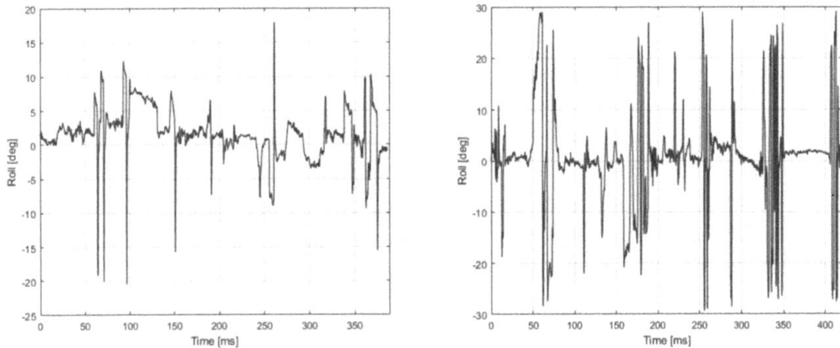

Fig. 3. Roll changes over time: a) flight in steady (left), flight in agile (right)

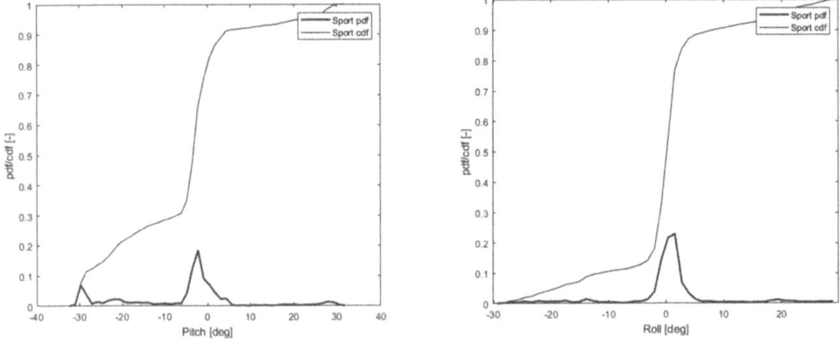

Fig. 4. PDF and CDF for distinct agile (SPORT) flight mode: a) pitch (left), b) roll (right)

In a similar way, the results for steady flight mode are shown in Fig. 5a (pitch) and 5b (roll).

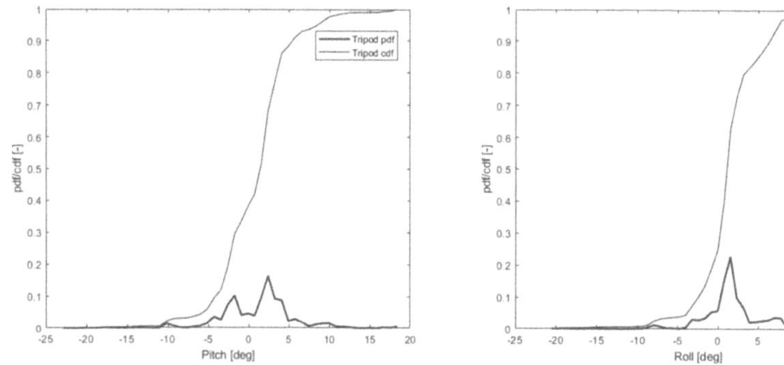

Fig. 5. Probability density for distinct steady (Tripod) flight mode: a) pitch (left), b) flight in agile (right)

3 RCS Measurement and Data Processing

The Radar Cross Section (RCS) was measured in a far-field setup (spherical wavefront sufficiently close to planar wave). The measurement was performed using separate transmit and receive antennas. A Vector Network Analyzer was connected to these antennas and configured to evaluate the transfer function. The measurement setup is shown in Fig. 6. The method presented in [8] is based on an isotropic reflector (stainless steel ball) calibration. Antenna cross-talk and most of the environment reflection are suppressed by time-domain reflectometry.

Fig. 6. RCS drone measurement setup

The drones were installed on a low RCS jig connected to a turntable hidden behind pyramidal absorbers. The turntable covers target yaw. Antennas and drone altitude offset set pitch and roll. A vector network analyzer generates a wideband sweep and transmits an antenna that radiates it toward the measured drone. The receiving antenna feeds the vector network analyzer with a drone echo. These data sets are recorded for every drone azimuth (yaw) position. Frequency response (echo power) is converted to the time domain, and all range components of the measured drone are suppressed. The result is converted back into the frequency domain, and received power for distinct frequencies (considering normalization by isotropic reference target) is directly proportional to RCS.

The measurement of the RCS has been a key topic since the first military applications of radar. A typical radar observation scenario is the detection of far targets; thus, the measurements are virtually always performed in a horizontal plane over all azimuths. Results are often further simplified to a single (average) figure. This approach supports reasonable estimates of radar performance for conventional aircraft and helicopters, while for drones, thrust is generated by roll and yaw. In any case, X-band RCS values of small to medium size drones are extremely low, e.g. for the DJI Phantom family, as they are in a magnitude of hundredths, about 0.02 sqm [9, 10].

Our research brings RCS measurement results for multiple pitches and rolls that are much better for representing a copter-type drone's real radar observation scenarios. Measurement was performed on hobby and DIY small drones. Drones used for agriculture support activities would often be bigger but share the same flight principles and behaviour. The article presents results for DJI Phantom [11], Robodrone Hornet [12], and EACHINE ex5 [13]).

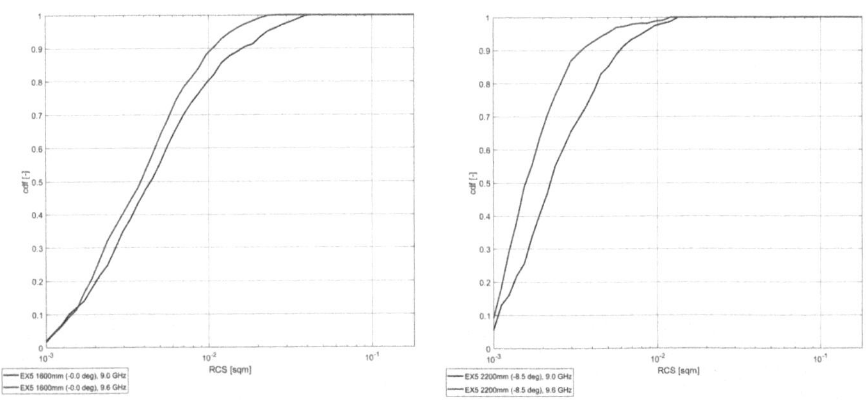

Fig. 7. Eachine EX5 RCS Cumulative density function over azimuth for: (a) 0 degrees elevation (left), (b) −8.5 degrees elevation (right)

A planar carbon fibre frame influences hornet drone RCS elevation dependence. Illumination of drones from low elevation angles results in forward scatter and reduced back-scatter (RCS). Examples of RCS PDF are on Fig. 7 and RCS over azimuth are shown in Fig. 8 for various types of common frequencies.

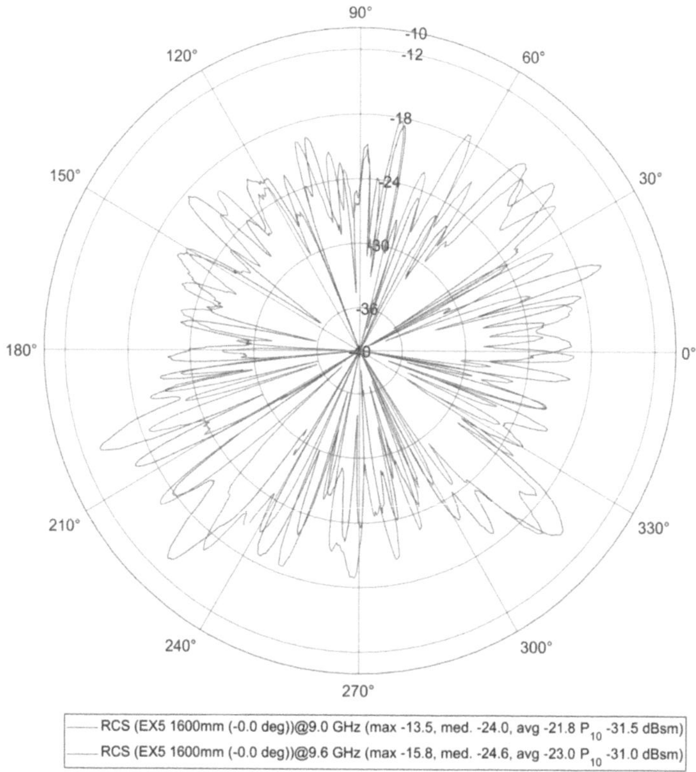

Fig. 8. RCS over azimuth (log scale)

4 Conclusion

With the proliferation of different types of drones and their use in a wide range of applications, detecting them using independent surveillance resistant to cyber attacks is becoming increasingly important. The only long-range, all-weather option is primary radar. The drones pose new challenges to the radar systems. The key aspect of drone detection is the small radar cross-section.

The paper presented measurement scenarios relevant to copter-type drone flight behaviour. Measured radar cross-section over azimuth for zero and a few negative elevation values were subsequently subject to statistical analysis. These results enable primary radar systems detection performance analysis and optimization. Research extended a common, inaccurate understanding of RCS as a single figure into a proper description of random variables. This enables research and design of detection methods that are more appropriate for drones.

Acknowledgments. The work was supported from ERDF "Multi-sector and Interdisciplinary Cooperation in Research and Development of Communication, Information and Detection Technologies for Control and Signalling Systems (CIDET)" (No. CZ.02.01.01/00/23_021/0008402).

Disclosure of Interests. The authors have no competing interests to declare that are relevant to the content of this article.

References

1. Zhang, C., Kovacs, J.M.: The application of small UAVs for precision agriculture: a review. Precision Agric. **13**(6), 693–712 (2012). https://doi.org/10.1007/s11119-012-9274-5
2. Coherent Direction Finder Model With Quadrature Signal Processing in the HF Band. Miroslav Pacek, Jozef Perďoch, Zdeněk Matoušek, Communication and Information Technologies 2023: Conference Proceedings ISBN: 979-8-3503-3838-6, ISBN 979-8-3503-3839-3, pp. 97–102
3. Dronetag. Online. 2025. https://dronetag.com/. [cit. 2025-03-31]
4. Radar Waveforms Based on Costas and Walsh-Hadamard Codes in Active Noise Jamming. J. Perďoch, Z. Matoušek and M. Pacek, 2024 New Trends in Signal Processing (NTSP), Demanovska Dolina, Slovakia (2024). https://doi.org/10.23919/NTSP61680.2024.10726302, Electronic ISBN:978-80-8040-637-0, USB ISBN:978-80-8040-636-3, pp. 1–7. Registrované v IEEE, WoS, Scopus
5. Chernyshov, P., Hessner, K., Zavadsky, Andrey a Toledo, Y.: On the Effect of Interferences on X-Band Radar Wave Measurements. Online. Sensors (2022), roč. 22, č. 10. ISSN 1424-8220. Dostupné z: https://doi.org/10.3390/s22103818. [cit. 2025-03-31]
6. Chen, Y., Wang, L.: Radar cross-section measurement and analysis of UAV in agriculture applications. Remote Sens. Let. **11**(7), 669–679 (2020). https://doi.org/10.1080/2150704X.2020.1797919
7. Sgorbissa, A., Franco, A., Miele, A.: Radar cross-section modeling for UAVs: performance analysis in agricultural applications. J. Field Robot. **32**(10), 1431–1448 (2015). https://doi.org/10.1002/rob.21876
8. SEDIVY, Pavel a NEMEC, Ondrej. Drone RCS statistical behaviour. In: MSG-SET-183 Specialists' Meeting on Drone Detectability: Modelling the Relevant Signatureability: Modelling the Relevant Signature. STO - Science & Technology Organization, 2021, s. 9. ISBN 978-92-837-2357-8
9. Li, Ch.J., Ling, H.: Radar signatures of small consumer drones, AP-S/USNC-URSI IEEE, Purto Rico (2016)
10. de Quevedo, Á.D., Urzaiz, F.I., Menoyo, J.G., López, A.A.: Drone Detection and RCS measurements with ubiquitous radar, International Conference on Radarm Brisbane (2018)
11. DJI. (2025). https://www.dji.com/. [cit. 2025-03-28]
12. Robodrone. (2025https://www.robodrone.com/. [cit. 2025-03-28]
13. Eachine. (2025). https://www.eachine.com/. [cit. 2025-03-31]

Author Index

H. S. Shekhawat et al. (Eds.): ICA 2025, CCIS 2795, pp. 331–332, 2026.
https://doi.org/10.1007/978-3-032-17083-5

The manufacturer's authorised representative in the EU is Springer Nature Customer Service Centre GmbH, Europaplatz 3, 69115 Heidelberg, Germany. If you have any concerns regarding our products, please contact ProductSafety@springernature.com

Printed and bound by CPI Group (UK) Ltd, Croydon, CR0 4YY
07/07/2026
02160906-0008